The Chemical Incident Management Handbook

Chemical Incident Response Service
Medical Toxicology Unit
Guy's & St Thomas' Hospital Trust

Chemical Incident Management Series

Chemical Incident Response Service
Medical Toxicology Unit,
Guy's & St Thomas' Hospital Trust

The Chemical Incident Management Handbook

Editors

Catherine Farrow
Henrietta Wheeler
Nicola Bates
Dr Virginia Murray

London: The Stationery Office

First published 2000

ISBN 0 11 322252 1

On-line Access

The full text of this publication has also been made available on the Internet. You can find this at:

https://www.the-stationery-office.co.uk/document/chemical/chem.htm

The data is held on a secure site that is password protected. The following information will be required to access the site:

USER NAME: chemical
PASSWORD: c4bb4ge

Please note that both fields are case sensitive and contain no spaces.

Contents

Editors

Catherine Farrow gained a Joint Honours degree in Anatomy and Physiology and worked as a study director for a large, multinational contract research organisation before joining the National Poisons Information Service, London and the Medical Toxicology Unit in 1995. She worked initially as a Poisons Information officer before joining the Chemical Incident Response Service as information specialist in 1998. Her work involves advising on chemical toxicology and co-ordinating and preparing chemical information sheets for rapid response to incident enquiries. She is taking a key role in the maintenance of the chemical information on the Chemical Incident Response Service website.

Henrietta Wheeler trained as a Registered General Nurse and gained a degree in Environmental Chemistry before joining the National Poisons Information Service, London and the Medical Toxicology Unit in 1994. During her time at the Unit she has also gained an MSc in Environmental Risk Assessment. She has taken a key role in assisting with setting up the Chemical Incident Response Service and has gained experience in advising on response to chemical incidents.

Nicola Bates gained a degree in Applied Biology before joining the National Poisons Information Service, London and the Medical Toxicology Unit in 1989. During her time at the Unit she has also gained a degree in Earth Science and has recently completed an MSc in Human Evolution and Behaviour. She has been involved in co-ordinating many projects for the Unit and was one of the editors of *Paediatric Toxicology*.

Dr Virginia Murray FFOM FRCP FRCPath trained in occupational medicine before joining the National Poisons Information Service, London, and the Medical Toxicology Unit in 1980. Initially she was involved in projects organised by the Unit and the International Programme on Chemical Safety (WHO/ILO/UNEP). In 1989, she started the Chemical Incident Research Programme at the Unit and is currently the Director of the Chemical Incident Response Service. As a result she has had considerable experience in advising on response to chemical incidents.

Contributors

The chemical summaries have been produced by the National Poisons Information Service (London) and staff of the Chemical Incident Response Service. Those who have contributed are:

Peter Barber B Med Sci

Nicola Bates BSc (Brunel) BSc (Open) MSc

Jennifer Butler BA DBS Dip Med Tox

Alexander Campbell BSc

Grainne Cullen BSc Dip Med Tox

Catherine Farrow BSc

Robie Kamanyire BSc

Sarah McCrea MSc Dip Med Tox

Frances Northall BSc RGN PG Dip

Marie Pickford BSc Dip Med Tox

Elizabeth Schofield BSc

Nicola Scott BSc

Simon Ward MA MSc MPhil MRCP FRCS

Henrietta Wheeler MSc RGN

Foreword

Since 1985, the International Programme on Chemical Safety (WHO/ILO/UNEP) has encouraged Poisons Information Services to be proactive in developing resources to assist in the investigation and management of chemical incidents. The National Poisons Information Service, London (NPIS [London]), part of Guy's and St Thomas' Hospital Trust, has been proud to have been one of the Poisons Information Services to respond to this challenge. As a result, since 1988, we have worked towards the development of the Chemical Incident Response Service (CIRS). This service specifically provides information and support to health care professionals and emergency services during chemical incidents. During this period we have been involved in a learning curve about incidents and their management. We have been active to this end both nationally and internationally. The Chemical Incident Management series has been prepared as a result of our experience.

The chemical summaries have been found to be of considerable value to health care professionals and emergency services during chemical incidents. It is at their request that these are now published for wider access. Our only proviso to their use is that each incident is different and each will require individual risk assessment and management. Therefore, do not hesitate to contact one of the Chemical Incident Provider Units or Poisons Information Services for assistance.

Virginia Murray and Henrietta Wheeler

Acknowledgements

Helaina Checketts, Librarian, Medical Toxicology Unit

Brian Widdop, Laboratory Director, Medical Toxicology Unit

Alison Dines, Information Scientist, National Poisons Information Service (London), Medical Toxicology Unit

Nick Edwards, Manager, National Poisons Information Service (London), Medical Toxicology Unit

Faith Goodfellow, Research Engineer, Water. Chemical Incident Response Service, University of Surrey

Fiona Welch, Research Engineer, Air. Chemical Incident Response Service, University of Surrey

Emma Woodey, Research Engineer, Land. Chemical Incident Response Service, University of Surrey

Joan Bennett, Secretary to Dr Murray

The many other friends and colleagues at the National Poisons Information Service (London) who have read and commented on this book

Designed by TSO Graphic Design

Abbreviations

Abbreviation	Definition	Abbreviation	Definition
%	percent	l	litre
®	registered trademark	LFT	liver function test
≤	less than or equal to	M	molar
µg	microgram	m	metre
µm	micrometer	max	maximum
<	less than	mEq	milliequivalent
>	greater than	mg	milligram
°C	Celsius, degrees centigrade	min	minute
ARDS	adult respiratory distress syndrome	ml	millilitre
AST	aspartate aminotransferase	mmol	millimole
ATM	atmosphere	mol	mole
AV	atrioventricular	mOsm	milliosmole
bpm	beats per minute/breaths per minute	N	normal (equivalents per litre, as applied to concentration)
CAS	Chemical Abstracts Services	NADPH	nicotinamide adenine dinucleotide phosphate
cm	centimetre	NB	*nota bene* (note well)
CNS	central nervous system	ng	nanogram
COHb	carboxyhaemoglobin	NIOSH	National Institute for Occupational Safety and Health
COSHH	Control of Substances Hazardous to Health	OH	oxygen–hydrogen
CPAP	continuous positive airway pressure	oz	ounce
CT	computed tomography	PEEP	positive expiratory end pressure
d	day	pH	negative logarithm of the hydrogen ion concentration
DIC	disseminated intravascular coagulation	pO_2	oxygen partial pressure
dl	decilitre	ppb	parts per billion
DMPS	sodium-(2,3)-dimercaptopropane-(1)-sulphonate	ppm	parts per million
e.g.	*exempli gratia* (for example)	PT	prothrombin time
ECG	electrocardiogram	RTECS	Registry of the Toxic Effects of Chemical Substances: produced by NIOSH
EEG	electroencephalogram	SC	subcutaneous
EMD	electromechanical dissociation	sec	second
ET	endotracheal	sol.	solution
g	gram	UK	United Kingdom
h	hour	UN	United Nations
Hb	haemoglobin	USA	United States of America
IARC	International Agency for Research on Cancer	UV	ultraviolet
i.e.	*id est* (that is)	WBC	white blood cell count
INR	International Normalised Ratio	WHO	World Health Organisation
ITU	Intensive Therapy Unit	WW1	World War 1
IV	intravenous		
kg	kilogram		

Introduction

The Chemical Incident Management Handbook has been compiled to support the other volumes in this series. The other volumes are designed to be used by specific medical professionals, such as Accident and Emergency[1] or Public Health doctors,[2] for preparation and management of chemical incidents. The chemical information in the Chemical Incident Management Handbook has been created for use in conjunction with the appropriate book.

Information about individual chemicals has been written in an easily accessible format to assist in chemical incident management. The data can be used as a quick reference guide during an incident. Naturally, there is more detailed information available about chemicals which may be of use for chronic chemical exposure or for report writing, but during an acute chemical incident it is essential to have the facts laid out clearly and precisely. Further information is available for medical professionals and emergency services about individual chemicals either from a Chemical Incident Provider Unit or a Poisons Information Service.

Chemical Incident Provider Units

Chemical Incident Provider Units are available to assist with identification of chemical hazards, to determine toxic risks and to provide relevant medical information.

The main factors in chemical incident management are to be prepared, and to ensure that communication channels are open and that information is available. Chemical Incident Provider Units, who generally have close connections with local poisons information services, should be informed and available for help during an incident. The contact telephone numbers for these Units is shown in Table 1 and emergency numbers of the NPIS in the UK and Ireland is shown in Table 2.

The Chemical Incident Response Service (CIRS) is available to make rapid risk assessments during a chemical incident, and to assist with identification and the toxic hazard of chemicals. CIRS is able to provide specialised toxicological advice on environmental factors, decontamination, treatment, incident investigation and documentation, as well as information concerning laboratory sampling, analysis, and follow-up. CIRS is available to support and assist with site visits and field service, epidemiology and surveillance. CIRS places a large emphasis on training, with many training days arranged, exercise participation, research with publications, and quarterly chemical incident reports.

Chemicals in this Series

The chemicals chosen for this book are those about which the CIRS has been contacted most frequently, either for chemical incident management, or in preparation for a major incident where the potential for this has been identified.

Table 1: Chemical incident provider units

London, South East, Eastern, North West, Trent, South West. Health authorities in these regions are contracted with the Chemical Incident Response Service at the Medical Toxicology Unit, Guy's and St Thomas' Hospital Trust, Avonley Road, London SE14 5ER, Tel . 020 7771 5383. Fax. 020 7771 5363. 24-hour emergency via NPIS (London): Tel. 020 7635 9191. Fax. 020 7771 5309

West Midlands. Health authorities in this region are contracted with the Chemical Hazard Management and Research Centre at the Institute of Public and Environmental Health, University of Birmingham, Birmingham B15 2TT. Tel. 0121 414 3985/6547. Fax. 0121 414 3827/3630. 24-hour emergency: Tel. 0121 394 5112.

Northern and Yorkshire Region. Health authorities in this region are contracted with the Chemical Incident Service at The Department of Environmental and Occupational Medicine, The Medical School, University of Newcastle, Newcastle Upon Tyne NE2 4HH. Tel. 0191 222 7195. Fax. 0191 222 6442.

Scotland. Scottish Centre for Infection and Environmental Health at Clifton House, Clifton Place, Glasgow G3 7LN. Tel. 0141 300 1100. Fax.0141 300 1170. Covers all health authorities in Scotland.

Wales and Northern Ireland. Health authorities in these areas are contracted with the Chemical Incident Management Support Unit at University of Wales College of Medicine, Therapeutics and Toxicology Centre, Llandough Hospital, Cardiff CF64 2XX. Tel. 029 20 709901. 24-hour emergency: Tel. 029 20 715278.

Table 2:
Emergency 24-hour telephone numbers:
National Poisons Information Service

Belfast	028 9024 0503
Birmingham	0121 507 5588
Cardiff	029 2070 9901
Dublin	(+353) 1 837 9964 or (+353) 1 837 9966
Edinburgh	0131 536 2300
London	020 7635 9191
Newcastle upon Tyne	0191 232 5131
National NPIS number	0870 600 6266

On-line Access

The full text of this publication has also been made available to you on the Internet. You can find this at: https://www.the-stationery-office.com This website also contains links to many of the other documents and organisations referenced in the publication, making it easy for you to cross-refer between texts and make contact with associated bodies. We hope that you will find this feature useful.

The data is held on a secure site that is password protected. The following information will be required to access the site:

USER NAME: chemical
PASSWORD: c4bb4ge

Please note that both fields are case sensitive and contain no spaces.

REFERENCES

1. Fisher J, Morgan-Jones D, Murray V, Davies G. *Chemical Incident Management for Accident and Emergency Clinicians*. London: The Stationery Office, 1999

2. Irwin DJ, Cromie DT, Murray V. *Chemical Incident Management for Public Health Physicians.* London: The Stationery Office, 1999

How to use this book

As far as possible, each chemical summary is laid out in the same order to ensure ease of use. The summaries have been used regularly by CIRS for chemical incident management and advice since 1994, and have evolved through feedback from CIRS users. Through experience at CIRS they are generally thought to be easy to use during an emergency.

The various sections are specifically designed for the different health care professionals that may be involved in a chemical incident. For example, the Key Point and First Aid sections can be used at the site of a chemical incident and for triage purposes in Accident and Emergency departments, whereas the more detailed acute clinical effects and treatment are more appropriate for health care professionals in hospital.

Each sheet is laid out using the following headings:

Key Points

This section gives a brief summary of the nature of the chemical, the major clinical effects and any specific hazards.

First Aid

Although a standardised first aid format has been used for all the chemicals there are changes made for those chemicals that require specific first aid. For example where chemicals potentially cause frostbite injuries, or where specific antidotes are required and may need to be made available at the site of an incident.

Detailed Information: Gives a brief outline of the common synonyms and identification numbers. Occasionally during an incident the only form of chemical identification is a number, for example CAS, UN and NIOSH/RTECS numbers. Thus, these have been identified where available.

Additional information in this section includes common interactions with other chemicals or substances, products of combustion and any other chemical information that may be of use during an emergency.

Summary of Human Toxicity: A brief insight to routes of absorption, mechanism and target organs has been indicated. This section has been left as brief and as basic as possible to give an overview in an emergency. There is more detailed information available but this may not be necessary immediately to assist in a chemical incident. Where available, exposure data has been given, indicating clinical effects that may be expected.

Occupation data is also shown in this section

Occupational Exposure Limits: The standards used in this book are produced by the Health and Safety Executive. All the standards quoted are from EH40/98.[1] EH40 is produced annually and contains up-to-date lists of occupational exposure limits for use with the Control of Substances Hazardous to Health Regulations 1994 (COSHH).[2] The list contains two different types of 'occupational exposure limits' (OELs) for substances hazardous to health in the air in workplaces. The list of OELs have legal status under the COSHH Regulations 1994.

Under COSHH there are two types of occupation exposure limit for hazardous substances: occupation exposure standards (OESs) and maximum exposure limits (MELs). OESs and MELs are set to help protect the health of workers. Both types of limits are concentrations of hazardous substances in the air, averaged over a specified period of time and referred to as a time weighted average (TWA). Two time periods are used: long-term (8 hours) and short-term (15 minutes). Short-term exposure limits (STELs) are set to help prevent effects such as eye irritation, which may occur following exposure for a few minutes.

OESs are set at levels based on current scientific literature, that will not damage the health of workers exposed to it by inhalation on a longer term basis.

MELs are set only for substances that may cause more serious health effects, such as cancer and occupational asthma, and for which a 'safe' level of exposure can not be determined or for substances for which safe levels may exist but control of those levels is not reasonably practicable. Where it is not possible to identify with confidence a level of exposure to such a chemical which is judged to be both safe and realistically achievable, the MEL is listed as not yet set.

Acute Clinical Effects: All the summaries are provided in the same order of inhalation, dermal, eye and oral exposure to assist quick referencing. Bullet points have been used to ensure a clear lay out. The routes of exposure have been clearly separated out to assist quick referencing. Where appropriate, systemic effects have been mentioned in a separate section.

Chronic Clinical Effects: This is a brief section as the summaries are mainly directed towards acute effects. However, this section indicates some of the major chronic and long-term health effects.

Management: Where appropriate the need for monitoring concentrations has been mentioned. Urgent analysis may not be necessary to influence management of the acute phase. It may not be possible to get urgent results. However, **there is no reason to delay taking samples**. It is better to take and store samples which may be discarded later than to miss the opportunity to take early samples which may provide evidence of the highest concentrations of a toxic substance.

This section is laid out in the same order as the clinical effects and should assist ease of access to management

Decontamination: Decontamination of casualties should be undertaken at the incident site or outside of the Accident and Emergency (A&E) department, ideally near to the entrance to the department.[3] Unfortunately, few A&E departments have facilities for decontaminating chemically contaminated patients, such as specific rooms with stretcher access and self contained ventilation and drainage systems. Makeshift arrangements usually have to be utilised.[4] Detailed information about the need for clean and dirty areas is laid out in the A&E volume.

Where appropriate, specific decontamination needs have been indicated, but generally it is essential to ensure that all contaminated clothing is removed and placed into double, sealed, clear plastic bags, labelled and stored in a secure area away from patients and staff, followed by thorough washing of the contaminated areas with tepid water and a mild detergent. Special attention must be paid to eyes, skin and hair.

Management is generally symptomatic and supportive and there is rarely specific management for poisoning. However, where appropriate, antidotes have been mentioned and adult dosages given. Contact a poisons information service for paediatric dosages.

Antidotes: There are few specific antidotes available for poisoned individuals. Information about the availability and location of these can be provided by a chemical incident provider unit or a poisons information service.

Summary of Environmental Hazards: A brief summary of the fate in air, water and soil has been given. Again only a brief outline for the environmental hazards has been given indicating, where appropriate, half-lives in the various media.

Drinking Water Standards: Two drinking water standards have been quoted; those from the United Kingdom (UK) and from the World Health Organisation (WHO). The UK drinking water standards are those specified in the Water Supply (water quality) Regulations 1989 (SL 1989/1147).[5] Those quoted for WHO are from the guidelines for drinking water quality.[6]

Soil Guideline: The soil guidelines used in this book are the Dutch Criteria for Soil.[5] Where available two concentrations have been quoted, the target and the intervention levels.

Air Quality Standards: Two air quality standards have been quoted; those from the United Kingdom (UK) and from the World Health Organisation (WHO). The UK air quality standards are those specified in the United Kingdom National Air Quality Strategy, 1997. The WHO guidelines are quoted from Air Quality Guidelines for Europe.[7]

References: At the end of each chemical summary the main references that were used have been given; these will, hopefully, be of use for further research or if more information is required.

REFERENCES

1. Health and Safety Executive. EH40/98 *Occupational Exposure Limits.* HMSO, London, 1998

2. Control of Substances Hazardous to Health Regulations 1994 (SI 1994/3246) as amended by the Control of Substances Hazardous to Health Regulations (Amendment) 1996 (SI 1996/3138) and the Control of Substances Hazardous to Health (Amendment) Regulations 1997 (SI 1997/11). HMSO

3. Schonfield S, Cummins A & Murray VSG. *Toolkit for handling chemical incidents for trainees in public health medicine.* Medical Toxicology Unit, Guy's & St Thomas' Hospital NHS Trust, London

4. Murray VSG & Volans GN, 1991. Management of injuries due to chemical weapons. *Brit Med J* 302: 129–30

5. Taylor D (ed.) *Croner's Substances Hazardous to the Environment.* Croner Publication Ltd, London 1998

6. WHO. *Guidelines for drinking-water quality,* 2nd edn. Vol 2: *Health criteria and other supporting information.* World Health Organisation, Geneva, 1993

7. Air Quality Guidelines for Europe. European series No. 23. World Health Organisation Regional Publications. Copenhagen, 1987

Chemical list

ACETIC ACID

Key Points

- Acetic acid is a flammable, colourless liquid with a pungent, acrid (vinegar-like) odour
- The vapours and liquid are irritant or corrosive to the eyes, skin, mucous membranes and respiratory tract
- Acetic acid can be fatal on ingestion, inhalation and from dermal exposure
- *In the event of a large spill, stay upwind and out of low areas. Ventilate closed spaces. Protective clothing, eye protection and breathing apparatus should be worn*

FIRST AID

- Terminate exposure and support vital functions
- The casualty should be moved to an uncontaminated area
- Rescuers should, ideally, be trained personnel and must be careful *not to put themselves at risk* and *so wear appropriate protective clothing and, if available, breathing apparatus*
- If the casualty is unconscious a clear airway should be established and maintained; give 100% oxygen if available
- **Inhalation Exposure:** If the patient stops breathing, expired air resuscitation should be started immediately using a pocket mask with a one way valve, if available. It is important where the face is contaminated that expired air resuscitation is NOT attempted unless an airway with rescuer protection is used
- **Dermal Exposure:** Remove contaminated clothing, if possible under a shower, and place in double, sealed, clear bags and label; store the bags in a secure area away from patients and staff
- Wash the skin thoroughly with copious amounts of water
- **Eye Exposure:** Irrigate thoroughly with water or saline for 15 minutes
- **Oral Exposure:** Encourage small quantities of oral fluids (no more than 50–100ml in total) unless perforation is suspected

Detailed Information

- Acetic acid is a flammable, colourless liquid with a pungent, acrid (vinegar-like) odour

- *Common synonyms* ethanoic acid, glacial acetic acid (100%), vinegar (4–6% solution in water)

- CAS 64-19-7

- UN 2789 (>80%)

- UN 2790 (10–80%)

- NIOSH/RTECS AF 1225000

- Molecular formula $C_2H_4O_2$

- Molecular weight 60.05

- Acetic acid is a medium strong organic acid used in the production of cellulose and vinyl acetate; dyeing, pharmaceuticals, and food processing

- Vapours mix readily with air; harmful atmospheric concentrations can build up fairly rapidly on evaporation at 20°C

- Miscible with water, alcohol, glycerol and ether; insoluble in carbon disulphide

- Acetic acid reacts violently with bases; attacks many metals giving off flammable hydrogen gas

- It forms explosive air-vapour mixtures above 50°C; it reacts violently or explosively with strong oxidants and exothermically with bases

Summary of Human Toxicity

- Acetic acid is a low molecular weight organic acid but represents a hazard similar to the strong acids, e.g. sulphuric acid[1]

- It is a severe irritant to the gastric mucosa, respiratory tract, eyes and skin

- Ingestion of as little as 1 ml of glacial acetic acid has resulted in oesophageal perforation and later oesophageal and pyloric stricture formation[1,3,4]

- The lowest lethal dose published (exposure route unreported) is 308 mg.kg[-1] [2]

- Conjunctivitis has been reported following exposure to vapour concentrations less than 10 ppm

- Unacclimatised humans experience extreme eye and nasal irritation at concentrations above 25 ppm

- Most persons find concentrations greater than 50 ppm intolerable[3]

- Immediate danger to life and health follows exposure to 1000 ppm[5,6]

Dermal Toxicity of Acetic Acid[3]

Concentration	Clinical effects
<10%	No injury likely
<50%	Relatively mild injury
50–80%	Moderate to severe burns
>80%	Severe burns

- Occupational Exposure Standards:
  Long-term exposure limit: 10 ppm (25 mg.m^{-3})
  Short-term exposure limit: 15 ppm (37 mg.m^{-3})

Acute Clinical Effects

Inhalation Effects

- Acetic acid vapours are severely irritant or corrosive to the respiratory tract[5,7]

- Initially there may be irritation to the mucous membranes with sore throat, coughing, wheezing and shortness of breath

- Headache, salivation, nausea, ulceration of mucous membranes and muscle weakness may occur

- Severe breathing difficulties, pharyngeal oedema, chronic bronchitis, and pulmonary oedema may develop in serious cases[3]

Dermal Effects

- Acetic acid is irritating or corrosive to skin (depending on the concentration)[3]

- Irritation with redness, tingling or burning sensation, pain and swelling may occur

- Burns may become white and extremely painful; blisters may develop[3]

Eye Effects

- Acetic acid is severely irritant or corrosive (depending on the concentration) to the eyes[3]

- Immediate pain, intense lacrimation and irritation, with inflammation, hyperaemia, conjunctivitis, and impaired vision, with the risk of corneal epithelial injury[8]

- The severity of injury may not be so evident immediately following exposure as a day or two later[8]

- Glacial acetic acid will cause severe injury with permanent corneal opacification

Oral Effects

- Ingestion of small quantities or low concentrations may lead to local mucosal irritation, burning sensation of the lips, mouth and throat; epigastric pain, abdominal pain, nausea and vomiting, which may be mucoid and coffee ground in nature[9]

- Salivation, ulceration of mucous membranes and colour change of the tongue may occur

- Ingestion of a larger quantity may lead to haematemesis, perioral and pharyngeal burns, respiratory distress, oesophageal corrosion, stricture, necrosis and perforation of the gastrointestinal tract, typically more severe in the stomach and intestinal tract rather than the oesophagus

- Severe metabolic acidosis and shock may occur; haemolysis with haemoglobinuria which may lead to acute renal failure and disseminated intravascular coagulation[4]

- Pyloric stenosis may follow several weeks to years later in patients who survive an acute episode of ingestion

Chronic Clinical Effects

Occupational exposures for 7–12 years to concentrations of 80–200 ppm at peaks caused blackening and hyperkeratosis of the skin and hands, conjunctivitis (but no corneal damage), bronchitis and pharyngitis and erosion of the exposed teeth (incisors and canines). Digestive disorders with heartburn and constipation have been reported at unspecified prolonged exposures.[3]

Management

Inhalation Management

- Maintain a clear airway, give humidified oxygen and ventilate if necessary

- If respiratory irritation occurs, assess respiratory function and if necessary perform chest X-rays to check for chemical pneumonitis

- Consider the use of steroids to reduce the inflammatory response

- Treat pulmonary oedema with PEEP or CPAP ventilation

- Symptomatic and supportive care

Dermal Management

- Remove any remaining contaminated clothing, place in double, sealed, clear bags and label; store in a secure area away from patients and staff

- Irrigate with copious amounts of water

- Treat burns symptomatically

Eye Management

- Irrigate thoroughly with running water or saline for 15 minutes

- Stain with fluorescein and refer to an ophthalmologist if there is any uptake of the stain

Oral Management

- NO GASTRIC LAVAGE OR EMETIC

- Encourage oral fluids, unless perforation is suspected

- Give plasma expanders/blood or IV fluids for shock and analgesics for pain

- Consider the use of steroids to reduce the inflammatory response

- Take abdominal X-ray to check for perforation

- Symptomatic and supportive care

- If facilities are available, early gastro–oesophagoscopy should be undertaken within 12 hours of the event to assess the extent and severity of the injury

Summary of Environmental Hazards

- If released to the atmosphere acetic acid is degraded in the vapour-phase by reaction with photochemically produced hydroxyl radicals (estimated typical half-life of 22 days)[2]

- It occurs in atmospheric particulate matter in acetate form and physical removal from air can occur via wet and dry deposition

- Natural waters will neutralise dilute solutions of acetic acid

- Spills of acetic acid on soil will readily biodegrade

- It is not expected to bioconcentrate in the aquatic system

- Low concentrations of acetic acid are harmful to fish

- Drinking Water Standards: no data available

- Soil Guidelines: no data available

- Air Quality Standards: no data available

REFERENCES

1. Gosselin RE, Smith RP & Hodge HC. *Clinical Toxicology of Commercial Products*, 5th edn. Williams & Wilkins, Baltimore, 1984

2. Hall AH & Rumack BH (eds.) *TOMES System®*, Micromedex, Englewood, Colorado, CD ROM. vol 41 (expires 31 July 1999)

3. Hathaway GJ, Proctor NH & Hughes JP. *Proctor and Hughes' Chemical Hazards of the Workplace*, 4th edn. Van Nostrand Reinhold, New York, 1996

4. Greif F & Kaplan O, 1986. Acid ingestion: Another cause of disseminated intravascular coagulopathy. *Crit Care Med*; 14 (11): 990–1

5. Ellenhorn MJ, Schonwalds S, Ordog G & Wasserberger J. *Ellenhorn's Medical Toxicology – Diagnosis and Treatment of Human Poisoning*, 2nd edn. Williams & Wilkins, London, 1997

6. Clayton GD & Clayton FE (eds.) *Patty's Industrial hygiene and toxicology*, 4th edn. John Wiley & Sons, Inc. New York, 1994

7. Sax NI. *Dangerous Properties of Industrial Materials*, 6th edn. Van Nostrand Reinhold, New York, 1984

8. Grant MW & Schuman JS. *Toxicology of the Eye*, 4th edn. Charles C Thomas, Springfield, 1993

9. Luxon SG (ed.) *Hazards in the Chemical Laboratory*, 5th edn. Royal Society of Chemistry, 1992

Catherine Farrow

ACETONE

Key Points

- Acetone is a highly volatile, flammable, colourless liquid with a distinctive sweet, pungent odour
- It is an irritant of the eyes and mucous membranes; at very high concentrations it is a CNS depressant and may cause convulsions and coma
- Systemic effects may occur from inhalation and ingestion
- *In the event of a large spill, stay upwind and out of low areas. Ventilate closed spaces. Protective clothing, eye protection and breathing apparatus should be worn*

FIRST AID

- Terminate exposure and support vital functions
- The casualty should be moved to an uncontaminated area
- Rescuers should, ideally, be trained personnel and must be careful *not to put themselves at risk* and *so wear appropriate protective clothing and, if available, breathing apparatus*
- If the casualty is unconscious a clear airway should be established and maintained; give 100% oxygen if available
- **Inhalation Exposure:** If the patient stops breathing, expired air resuscitation should be started immediately using a pocket mask with a one way valve, if available. It is important where the face is contaminated that expired air resuscitation is NOT attempted unless an airway with rescuer protection is used
- **Dermal Exposure:** Remove contaminated clothing, if possible under a shower, and place in double, sealed, clear bags and label; store the bags in a secure area away from patients and staff
- Wash the skin thoroughly with copious amounts of water
- **Eye Exposure:** Irrigate thoroughly with water or saline for 15 minutes
- **Oral Exposure:** Encourage small quantities of oral fluids (no more than 50–100 ml in total)

Detailed Information

- Acetone is a colourless, highly volatile liquid with a distinctive sweet, pungent odour
- *Common synonyms* dimethyl formaldehyde, dimethyl ketone, ketone propane, methyl ketone, 2-propanone
- CAS 67-64-1
- UN 1090
- NIOSH/RTECS AL 3150000
- Molecular formula C_3H_6O
- Molecular weight 58.08
- Acetone is an organic solvent, found in nail polish removers, varnishes, glues and rubber cement
- It is highly flammable and reacts vigorously with oxidising materials and acids
- It forms potentially explosive reactions with nitric acid and sulphuric acid, bromine trifluoride, nitrosyl chloride and platinum, nitrosyl perchlorate, chromyl chloride, thiotrithiazyl perchlorate, 2,4,6-trichloro-1,3,5-triazine and water, and hexachloromelamine
- Acetone reacts to form explosive peroxide products with 2-methyl-1,3-butadiene, hydrogen peroxide and peroxomonosulphuric acid

Summary of Human Toxicity

- Acetone is absorbed readily following inhalation and ingestion but is poorly absorbed through the skin although there have been reports of systemic poisoning after using synthetic plaster casts[1]
- Inhaled acetone is largely excreted unchanged via the lungs and in urine
- The concentration of the acetone in expired air and blood correlate positively with the degree of exposure[2]
- The lowest reported lethal dose in man (route unreported) is 1159 mg.kg^{-1} [2]
- Ingestion of 200 ml (2–3 mg.kg^{-1}) caused coma, hyperglycaemia and acetonuria in an adult
- Lethal concentration in human blood: 550.0 µg.ml^{-1} (55 mg %)[2]

Toxicity of Acetone[2]

Concentration	Clinical effects
500 ppm	Odour threshold, no effects
1000 ppm	Mild, transient irritation of the nose, throat and eyes
>1,000 ppm	Headache, light-headedness, nose, throat and eye irritation
>10,000 ppm	CNS depression, dizziness, weakness, and loss of consciousness

- Occupational Exposure Standards:
  Long-term exposure limit: 750 ppm (1810 mg.m^{-3})

  Short-term exposure limit: 1500 ppm (3620 mg.m^{-3})

Acute Clinical Effects

Inhalation Effects

- Irritation of the respiratory tract with chest tightness and coughing
- High vapour concentrations can produce systemic effects; see below
- Chemical pneumonitis or pulmonary oedema are theoretical possibilities in severe exposures

Dermal Effects

- Acetone is a mild skin irritant and may cause erythema and dermatitis
- Prolonged or repeated contact can cause defatting and drying of the skin

Eye Effects

- Acetone vapours may cause irritation, lacrimation and a burning sensation but injury is unlikely from vapour concentrations tolerable to humans

- Splashes in the eye will cause immediate stinging with a transient foreign body sensation and mild epithelial injury (presence of microscopic grey dots)

- Prolonged contact may cause deep damage to the cornea and temporary or permanent corneal opacity[3]

Oral Effects

- Erythema and irritation of the pharynx, vomiting or haematemesis[2]

- See systemic effects, below

Systemic Effects

- CNS depression, drowsiness, light-headedness, incoherent speech, ataxia and stupor

- Coma, respiratory depression and, rarely, convulsions may occur following large ingestion

- There may be hypotension, tachycardia, metabolic acidosis, hyperglycaemia and ketosis

- Rarely convulsions and tubular necrosis may be observed

Chronic Clinical Effects

Some accumulation was reported following chronic occupational exposure to 2100 ppm for $8\,h.d^{-1}$ for three days. A steady state blood concentration of $182\,mg.l^{-1}$, (with trough level of $91\,mg.l^{-1}$ sixteen hours post exposure) was achieved after 3 days. Intoxication is not observed with a blood acetone concentration of $330\,mg.l^{-1}$ [4]

Management

Measurement of serum and urine acetone concentrations may be useful to monitor the severity of ingestion or inhalation.

Inhalation Management

- Maintain a clear airway, give humidified oxygen and ventilate if necessary

- If respiratory irritation occurs assess respiratory function and if necessary perform chest X-rays to check for chemical pneumonitis

- Consider the use of steroids to reduce the inflammatory response

- Treat pulmonary oedema with PEEP or CPAP ventilation

- For systemic management, see below

Dermal Management

- Remove any remaining contaminated clothing, place in double, sealed, clear bags; label and store in a secure area away from patients and staff

- Irrigate with copious amounts of water

- An emollient may be required

- For prolonged or large exposure see systemic management below

Eye Management

- Irrigate thoroughly with running water or saline for 15 minutes

- Stain with fluorescein and refer to an ophthalmologist if there is any uptake of the stain

Oral Management

- NO GASTRIC LAVAGE OR EMETIC

- Encourage oral fluids

- For systemic management, see below

Systemic Management

- Monitor blood glucose and arterial pH

- Ventilate if respiratory depression occurs

- If patient unconscious, monitor renal function

- Symptomatic and supportive care

Summary of Environmental Hazards

- In air acetone is lost by photolysis and reaction with photochemically produced hydroxyl radicals; the estimated half-life of these combined processes is about 22 days

- This relatively long half-life allows acetone to be transported long distances from its emission source

- Acetone is highly soluble and slightly persistent in water, with a half-life of 20 hours and minimally toxic to aquatic life

- If released into the soil acetone will volatilise although some may leach into the ground and rapidly biodegrade; acetone biodegrades fairly rapidly in soil

- Acetone does not concentrate in the food chain

- Drinking Water Standards: no data available

- Soil Guidelines: no data available

- Air Quality Standards: no data available

REFERENCES

1. Hift W & Patel PL, 1961 Acute Acetone Poisoning due to synthetic plaster cast – with a discussion of acetone metabolism. *S Afr Med J*; 35(12): 246–50

2. Hall AH & Rumack BH (eds.) *TOMES System* ® Micromedex, Englewood, Colorado. CD ROM. vol 41 (exp. 31 July 1999)

3. Grant MW & Schuman JS. *Toxicology of the Eye*, 4th edn. Charles C Thomas, Springfield, 1993

4. Baselt RC & Cravey RH. *Disposition of Toxic Drugs and Chemicals in Man*, 4th edn. Chemical Toxicology Institute, 1995

Catherine Farrow

ACETONITRILE

Key Points

- Acetonitrile is a clear, colourless, volatile, flammable liquid with a sweet, ether-like odour
- It is absorbed following inhalation, ingestion and dermal exposure
- Acetonitrile is not intrinsically toxic, however it is slowly metabolised to release cyanide
- Clinical effects may be delayed in onset due to the conversion of acetonitrile to its toxic metabolite cyanide
- *In the event of a large spill, stay upwind and out of low areas. Ventilate closed spaces. Protective clothing, eye protection and breathing apparatus should be worn*

FIRST AID

- Terminate exposure and support vital functions
- Move the casualty to an uncontaminated area
- Rescuers should, ideally, be trained personnel and must be careful *not to put themselves at risk* and *so wear appropriate protective clothing and, if available, breathing apparatus*
- If the casualty is unconscious a clear airway should be established and maintained; give 100% oxygen if available
- **Inhalation Exposure:** If the patient stops breathing, expired air resuscitation should be started immediately using a pocket mask with a one way valve, if available. It is important where the face is contaminated that expired air resuscitation is NOT attempted unless an airway with rescuer protection is used
- If a physician is present at the scene then blood should be taken for measurement of cyanide concentration, then 50 ml of 25% sodium thiosulphate solution (12.5 g) should be given IV over 10 minutes
- Details of antidotal therapy administered should accompany the patient to hospital
- **Dermal Exposure:** Remove contaminated clothing, if possible under a shower, and place in double, sealed, clear bags and label; store the bags in a secure area away from patients and staff
- Wash the skin thoroughly with copious amounts of water
- **Eye exposure:** Irrigate the eyes thoroughly with water or saline for 15 minutes
- **Oral exposure:** Encourage small quantities of oral fluids (no more than 50–100 ml in total)

Detailed Information

- Acetonitrile is a clear, colourless, volatile, flammable liquid with a sweet, ether-like odour
- *Common synonyms* methyl cyanide, cyanomethane, ethanenitrile, ethyl nitrile, methanecarbonitrile
- CAS 75-05-8
- UN 1648
- NIOSH/RTECS AL 7700000

- Molecular formula CH_3CN
- Molecular weight 41.05
- Acetonitrile is a highly polar solvent used as a chemical intermediate in various industrial extraction processes, and as a false nail glue remover; it is a by-product in the manufacture of acrylonitrile
- It is soluble only in some organic salts (e.g. silver nitrate, lithium nitrate, magnesium bromide)
- It is miscible with water and immiscible with many saturated hydrocarbons (petroleum fractions)
- Acetonitrile burns with a luminous flame

Summary of Human Toxicity

- Acetonitrile is absorbed by inhalation, ingestion and percutaneously[1]
- Effects may occur within 20 minutes or be delayed for 8 to 14 hours post ingestion,[1] but typically 3 to 12 hours [2]
- The onset of effects is delayed due to the conversion to cyanide via hepatic cytochrome P-450–dependent metabolism and then until endogenous thiosulphate has been depleted[3]
- Cyanide acts by inhibiting cytochrome oxidase which impairs cell respiration[4]
- Concomitant exposure to acetone and acetonitrile delays severe toxic effects but increases toxicity; acetone initially inhibits then induces the metabolism of acetonitrile to cyanide[5]
- Toxic effects may be prolonged; further deterioration after initial treatment has been reported up to 3 days post ingestion[1]
- Acetonitrile has a low volume of distribution and low molecular weight, with a half-life of 32 hours for acetonitrile and 15 hours for cyanide[1]
- Cyanide antidotes are considered to be less effective in the treatment of acetonitrile toxicity than with inorganic cyanide poisoning[6]

Toxicity of Acetonitrile[7]

Concentration	Clinical Effects
40 ppm	Odour threshold
500 ppm	Irritation of the mucous membranes
>500 ppm	Weakness, nausea, convulsions, death

- Occupational Exposure Standards:
    Long-term exposure limit: 40 ppm (68 mg.m^{-3})
    Short-term exposure limit: 60 ppm (102 mg.m^{-3})

Acute Clinical Effects

Inhalation Effects

- Asphyxia, chest pain and tightness
- Systemic effects are likely to develop; see below

Dermal Effects

- Irritation, causing transient, faint erythema may occur

Eye Effects

- Irritation resulting in superficial reversible injury may occur

Oral Effects

- Acetonitrile causes nausea and vomiting within several hours of ingestion which, in most cases, precedes serious toxicity

- Haematemesis may occur

- Systemic effects may develop; see below

Systemic Effects

- Systemic effects may develop following inhalation, ingestion or dermal exposure

- Effects may be prolonged for several days

- Exposure to low concentrations can cause headache and lassitude, weakness and stupor

- Tachycardia, tachypnoea, chest tightness and dizziness may occur

- More severely, extreme weakness or lassitude, respiratory depression, anion gap metabolic acidosis, hypotension, shock, coma, convulsions and death

- A lack of normal venous blood haemoglobin desaturation (i.e. venous oxygen tension is similar to arterial oxygen tension) may be noted (because of reduced tissue oxygen extraction)

Chronic Clinical Effects

Anorexia, and macular papular vesicular dermatitis have been reported following chronic exposure.[1]

Management

Blood cyanide concentrations are a useful indicator of toxicity.[7] Plasma cyanide concentration correlates better with toxic effects than whole blood assays.[4] Blood concentrations of <2 mg.l^{-1} indicate mild toxicity, 2 to 3 mg.l^{-1} moderate toxicity and 3 to 4 mg.l^{-1} severe toxicity.

Inhalation Management

- Maintain a clear airway, give humidified oxygen and ventilate if necessary

- If respiratory irritation occurs assess respiratory function and if necessary perform chest X-rays to check for chemical pneumonitis

- Consider the use of steroids to reduce the inflammatory response

- Treat pulmonary oedema with PEEP or CPAP ventilation

- For antidote therapy see systemic management and antidote therapy below

Dermal Management

- Remove any remaining contaminated clothing, place in double, sealed, clear bags and label; store in a secure area away from patients and staff

- Irrigate with copious amounts of water

- Skin burns should be treated as a thermal injury

- For antidote therapy see systemic management and antidote therapy, below

Eye Management

- Irrigate thoroughly with running water or saline for 15 minutes

- Stain with fluorescein and refer to an ophthalmologist if there is any uptake of the stain

Oral Management

- Consider gastric lavage followed by 50g activated charcoal (using a cuffed ET tube to protect the airway) within one hour of ingestion, or if the patient is unconscious

- Contact a poisons information service for further guidance on gut decontamination

- For antidote therapy see systemic management and antidote therapy below

Systemic Management

- All cases should be observed for a minimm of 24 hours post exposure[4]

- Maintain a clear airway, give humidified 100% oxygen using a mask and bag with a non-return valve to prevent inspiration of exhaled gases

- Oxygen therapy should be maintained while the patient is receiving antidotal therapy

- Monitor blood cyanide concentration and arterial blood gases

- Early use of PEEP or CPAP ventilation may be required

- Acidosis should be corrected with sodium bicarbonate

- Low sodium content IV solutions should be used because of the risk of hypernatraemia from antidotal sodium thiosulphate

- Monitor electrolytes, especially sodium

- Haemodialysis or haemoperfusion should be considered to enhance elimination of acetonitrile, thiocyanate and cyanide in the severely poisoned patient or to correct hypernatraemia

- Symptomatic and supportive care

Antidote Therapy

- Contact a poisons information service for further guidance and paediatric doses

- Sodium thiosulphate should be given by injection (50 ml of 25% sodium thiosulphate solution (12.5 g) IV over 10 minutes) repeated at regular intervals or by continuous infusion because of the prolonged duration of effects[8]

- The use of sodium thiosulphate is limited by its sodium content; repeated doses may result in hypernatraemia

- Nitrite antidotes are of ambiguous benefit; sodium nitrite may exacerbate hypotension and cause bradycardia[2,4]

Summary of Environmental Hazards

- Acetonitrile is unlikely to undergo direct photolysis in air and the half-lives for its reaction with hydroxyl radicals and ozone have been estimated to be 535 days and 860 days, respectively

- It will persist in the troposphere for a long time and may be transported a long distance from its source of emission

- Wet deposition may remove some of the atmospheric acetonitrile

- Biodegradation is expected to be a major loss process in water

- Acetonitrile released to soil is likely to undergo aerobic biodegradation; it is expected to be mobile in soil and may evaporate from soil surfaces[1]

- Drinking Water Standards: no data available

- Soil Guidelines: no data available

- Air Quality Standards: no data available

REFERENCES

1. Hall AH & Rumack BH (eds.) *TOMES system ® Micromedex*, Englewood, Colorado. CD ROM. Vol 41 (exp. 31 July 1999)

2. Turchen SG, Manoguerra AS & Whitney C, 1991. Severe cyanide poisoning from the ingestion of an acetonitrile-containing cosmetic. *Am J Emerg Med*; 9(3): 264–7

3. Geller RJ, Ekins BR & Iknoian RC, 1991. Cyanide toxicity from acetonitrile-containing false nail remover. *Am J Emerg Med*; 9(3): 268–70

4. Ellenhorn MJ, Schonwalds S, Ordog G & Wasserberger J. *Ellenhorn's Medical Toxicology – Diagnosis and Treatment of Human Poisoning*, 2nd edn. Williams & Wilkins, London, 1997

5. Clayton GD & Clayton FE (eds.) *Patty's Industrial hygiene and toxicology*, 4th edn. John Wiley & Sons, Inc., New York, 1994

6. Losek JD, Rock AL & Boldt RR, 1991. Cyanide poisoning from a cosmetic nail remover. *Pediatrics*; 88(2): 337–40

7. Baselt RC & Cravey RH. *Disposition of Toxic Drugs and Chemicals in Man*, 4th edn. Chemical Toxicology Institute, 1995

8. Michaelis HC, Clemens C, Kijewski H, Neurath H & Eggert A, 1991. Acetonitrile serum concentrations and cyanide blood levels in a case of suicidal oral acetonitrile ingestion. *Clin Toxicol*; 29(4): 447–58

Catherine Farrow

ACRYLONITRILE

Key Points

- Acrylonitrile is a colourless, explosive, flammable, volatile liquid with a faintly pungent odour; a yellow colour may develop slowly on standing particularly after excessive exposure to light
- It is readily absorbed following inhalation, ingestion and dermal exposure
- Acrylonitrile is a respiratory tract and eye irritant; it causes corrosive damage to the skin which may be delayed in onset
- Acrylonitrile is partly metabolised to release cyanide, however, acute toxicity is caused by the whole molecule not as a result of cyanide alone
- Clinical effects are non-specific but may include central nervous, hepatic, renal, cardiovascular, respiratory and gastrointestinal systems
- *In the event of a large spill, stay upwind and out of low areas. Ventilate closed spaces. Protective clothing, eye protection and breathing apparatus should be worn*

First Aid

- Terminate exposure and support vital functions
- The casualty should be moved to an uncontaminated area
- Rescuers should, ideally, be trained personnel and must be careful *not to put themselves at risk* and *so wear appropriate protective clothing and, if available, breathing apparatus*
- If the casualty is unconscious a clear airway should be established and maintained; give 100% oxygen if available
- **Inhalation Exposure:** If the patient stops breathing, expired air resuscitation should be started immediately using a pocket mask with a one way valve, if available. It is important where the face is contaminated that expired air resuscitation is NOT attempted unless an airway with rescuer protection is used
- If a physician is present at the scene, then 50 ml of 25% sodium thiosulphate solution (12.5 g) should be given IV over 10 minutes
- Details of antidotal therapy should accompany the patient to hospital
- **Dermal Exposure:** Remove contaminated clothing, if possible under a shower, and place in double, sealed, clear bags and label; store the bags in a secure area away from patients and staff
- Wash the skin thoroughly with copious amounts of water
- **Eye exposure:** Irrigate the eyes thoroughly with water or saline for 15 minutes
- **Oral exposure:** Encourage small quantities of oral fluids (no more than 50–100 ml in total)

Detailed Information

- Acrylonitrile is a colourless, explosive, flammable, volatile liquid with a faintly pungent odour; a yellow colour may develop slowly on standing particularly after excessive exposure to light

- *Common synonyms* acritet, acrylon, carbacryl, cyanoethylene, fumigrain, 2-propenenitrile, ventox, vinyl cyanide
- CAS 107-13-1
- UN 1093
- NIOSH/RTECS AT 5250000
- Molecular formula CH_2CHCN
- Molecular weight 53.06
- Acrylonitrile is used in the production of acrylic and modacrylic fibres, nitrile rubbers and various plastics, and as a grain store fumigant
- Acrylonitrile mixes readily with air forming explosive mixtures; it is heavier than air and will spread along ground level and collect in low or confined areas
- Acrylonitrile is water soluble; it is miscible with most organic solvents
- It polymerises spontaneously, particularly in the absence of oxygen or if exposed to light, readily when moderately heated and violently in the presence of bases or peroxides
- It reacts violently with strong oxidising agents
- Acrylonitrile attacks copper and its alloys
- Acrylonitrile is extremely flammable; the product of combustion is cyanide

Summary of Human Toxicity

- Acrylonitrile is readily absorbed by inhalation, ingestion and percutaneously[1]
- Systemic effects are non-specific but may include pathological changes in central nervous, hepatic, renal, cardiovascular and gastrointestinal systems and skin[2]
- Acrylonitrile is partly converted *in vivo* to cyanide by hepatic microsomal reactions[2]
- Cyanide acts by inhibiting cytochrome oxidase which impairs cell respiration[3]
- Although cyanide is released, acute toxicity is not due to cyanide alone and is thought to result from the whole acrylonitrile molecule[1]
- Acrylonitrile conjugates with glutathione, proteins and nucleic acids[1]
- Detoxification of radical species, reactive vinyl groups and epoxide intermediaries can deplete glutathione stores which may lead to liver toxicity[1]
- Cyanide antidotes are considered to be less effective in the treatment of acrylonitrile toxicity than with inorganic cyanide poisoning[1]
- Lowest dose causing toxic effects in man by inhalation is 16 ppm for 20 minutes[4]
- Inhalation, ingestion and dermal exposure to acrylonitrile may cause death but the fatal dose is not known[2]

- Maximum Exposure Limits:
    Long-term exposure limit: 2 ppm (4.4 mg.m^{-3})
    Short-term exposure limit: no data available

Acute Clinical Effects

Inhalation Effects

- Vapour initially causes sneezing, sore throat, inflammation of mucous membranes, severe breathing difficulties

- Respiratory distress, asphyxia and pulmonary oedema may develop in severe cases

- Systemic effects may develop; see below

Dermal Effects

- Burning sensation, redness, erythema, blistering, vesiculation, dermatitis and chemical burns which may develop after a latent period of several hours[2, 5]

- Allergic contact dermatitis may develop several days after exposure[5]

- Systemic effects may develop; see below

Eye Effects

- Immediate but transient irritation, redness and pain may occur, followed by lacrimation[2, 4]

Oral Effects

- Initially sore throat and severe breathing difficulties

- Systemic effects may develop; see below

Systemic Effects

- Initially headache and dizziness, then nausea, vomiting, diarrhoea, weakness and tachycardia

- Lactic acidosis may develop due to an increased rate of glycolysis[4]

- Liver dysfunction, characterised by jaundice and liver tenderness, anorexia and leukocytosis has been reported[5]

- Low-grade anaemia and kidney dysfunction may occur[5]

- In severe cases, coma, convulsions, respiratory arrest, cardiovascular collapse and death occur

- Convulsions and cardiac arrest may occur without warning[4]

Chronic Clinical Effects

Repeated exposure is not considered likely to result in cumulative toxicity. Acrylonitrile has been shown to be carcinogenic in animals and has been associated with an increase in risk of lung cancer and other malignant tumours in humans following chronic exposure.[5]

Acrylonitrile is a skin and eye sensitiser.[1] Repeated dermal exposure may cause scaling dermatitis due to its solvent effect. Skin and eye irritation, nausea, vomiting, weakness, fatigue, jaundice, anaemia, leucocytosis, bilirubinuria, increased serum thiocyanate concentration, hepatic and renal irritation have all been described following chronic occupational exposure.[1, 2]

Management

Monitoring cyanide concentrations would not be useful as acrylonitrile is only partly metabolised to cyanide.

Inhalation Management

- Maintain a clear airway, give humidified 100% oxygen and ventilate if necessary

- If respiratory irritation occurs assess respiratory function and if necessary perform chest X-rays to check for chemical pneumonitis

- Consider the use of steroids to reduce the inflammatory response

- Treat pulmonary oedema with PEEP or CPAP ventilation

- For antidote therapy see systemic management and antidote therapy below

Dermal Management

- Leather and rubber absorb acrylonitrile

- Remove any remaining contaminated clothing, place in double, sealed, clear bags and label; store in a secure area away from patients and staff

- Irrigate with copious amounts of water

- Skin burns should be treated as thermal injury

Eye Management

- Irrigate thoroughly with running water or saline for 15 minutes

- Stain with fluorescein and refer to an ophthalmologist if there is any uptake of the stain

Oral Management

- Consider gastric lavage followed by 50g activated charcoal (using a cuffed ET tube to protect the airway) within one hour of ingestion, or if the patient is unconscious

- Contact a poisons information service for further guidance on gut decontamination

- For antidote therapy see systemic management and antidote therapy below

Systemic Management

- All cases should be observed for a minimum of 24 hours post exposure

- Maintain a clear airway, give humidified 100% oxygen using a mask and bag with a non-return valve to prevent inspiration of exhaled gases

- Oxygen therapy should be maintained while the patient is receiving antidotal therapy

- Monitor blood cyanide concentration and arterial blood gases

- Acidosis should be corrected with sodium bicarbonate

- Low sodium content IV solutions should be used because of the risk of hypernatraemia from antidotal sodium thiosulphate

- Monitor electrolytes, especially sodium

- Symptomatic and supportive care

Antidote Therapy

• Contact a poisons information service for further guidance and paediatric doses

• Cyanide antidotes are considered to be less effective in the treatment of acrylonitrile toxicity; they have been used in rat studies with varying degrees of success but there are no reports of efficacy in cases of human poisoning

• Sodium thiosulphate should be given by injection (50 ml of 25% sodium thiosulphate solution (12.5 g) IV over 10 minutes) repeated at regular intervals or by continuous infusion because of prolonged duration of effects

Summary of Environmental Hazards

• In the atmosphere acrylonitrile degradation is primarily by reacting with photochemically produced hydroxyl radicals

• It has a half-life of 3.5 sunlit days under relatively clean atmospheric conditions; therefore there would be an opportunity for dispersal from the source

• Acrylonitrile in wastewater evaporates slowly with a half-life of 1–6 days; biodegradation would be complete in approximately one week in receiving water with micro-organisms present

• On soil it would rapidly volatilise although some may leach into ground water where its fate is unknown

• Drinking Water Standards:
    cyanide: 70 μg.l^{-1} (WHO guideline)

• Soil Guidelines: Dutch Criteria:
    free cyanide: 1 mg.kg^{-1} (target)
                20 mg.kg^{-1} (intervention)

    complex cyanide (pH<5): 5 mg.kg^{-1} (target)
                650 mg.kg^{-1} (intervention)

    complex cyanide (pH>5): 5 mg.kg^{-1} (target)
                50 mg.kg^{-1} (intervention)

• Air Quality Standards:
    no safe level recommended due to carcinogenic properties (WHO guideline)

REFERENCES

1. GDCh Advisory Committee on Existing Chemicals of Environmental Relevance (BUA) (eds.) *BUA report 142: Acrylonitrile* (August 1993). S. Hirzel Wissen Schaftliche Verlagsgesellschaft, 1995

2. Willhite CC, 1982. Toxicology updates: acrylonitrile. *J Appl Toxicol*; 2(1): 54–6

3. Hathaway GJ, Proctor NH & Hughes JP. *Proctor and Hughes' Chemical Hazards of the Workplace*, 4th edn. Van Nostrand Reinhold, New York, 1996

4. Hall AH & Rumack BH (eds.) *TOMES System®*, Micromedex, Englewood, Colorado, CD ROM. vol 41, (exp. 31 July 1999)

5. Buchter A & Peter H, 1984. Clinical toxicology of acrylonitrile. *G Ital Med Lav*; 6: 83–6

Catherine Farrow

ALUMINIUM AND ALUMINIUM COMPOUNDS

Key Points

- Aluminium is a tin–white, malleable, ductile metal with a slightly bluish tint; it is available as bars, leaf, powder, sheets or wire
- Soluble forms of aluminium chloride, aluminium fluoride or aluminium sulphate are potentially toxic; the insoluble forms have no measurable acute response
- Aluminium is thought to be absorbed by inhalation
- Patients with renal failure are prone to aluminium toxicity, either from aluminium in the dialysate or other sources, especially aluminium-containing phosphate binders and antacids
- Aluminium is of low toxicity in eye and dermal exposures
- *In the event of a large spill, stay upwind and out of low areas. Ventilate closed spaces. Protective clothing, eye protection and breathing apparatus should be worn*

FIRST AID

- Terminate exposure and support vital functions
- The casualty should be moved to an uncontaminated area
- Rescuers should, ideally, be trained personnel and must be careful *not to put themselves at risk* and *so wear appropriate protective clothing and, if available, breathing apparatus*
- **Inhalation Exposure:** If the casualty is unconscious a clear airway should be established and maintained; give 100% oxygen if available
- **Dermal Exposure:** Remove contaminated clothing
- **Eye Exposure:** Irrigate thoroughly with water or saline for 15 minutes
- **Oral Exposure:** Encourage small quantities of oral fluids (no more than 50–100 ml in total)

Detailed Information

- Aluminium is a tin–white, malleable, ductile metal with a slightly bluish tint; it is available as bars, leaf, powder, sheets or wire

- *Common synonyms* aluminium metal, aluminium powder, elemental aluminium

- CAS 7429-90-5

- UN 1309 (powder, coated)

  UN 1396 (powder, uncoated)

  UN 9260 (molten)

- NIOSH/RTECS BD 0330000

- Atomic symbol Al

- Atomic weight 26.98

Synonyms of Common Aluminium Compounds

Compound	Synonyms
Aluminium chloride	aluminium trichloride
Aluminium chlorohydrate	aluminium chlorohydroxide, basic aluminium chloride, chlorhydrol, polyaluminium chloride, aluminium hydroxychloride
Aluminium fluoride	aluminium trifluoride
Aluminium hydride	aluminium trihydride
Aluminium hydroxide	aluminium hydrate, aluminium trihydrate
Aluminium nitrate	aluminium trinitrate
Aluminium oxide	aluminium trioxide
Aluminium phosphate	aluminium monophosphate
Aluminium sulphate	alum, aluminium trisulphate, cake alum
Sodium aluminate	–

Identification Numbers of Common Aluminium Compounds

Compound	CAS	UN	NIOSH/RTECS
Aluminium chloride	7446-70-0	1726 (anhydrous) 2581 (solution)	BD 0525000
Aluminium chlorohydrate	1327-41-9	n.a.	BD 0549500
Aluminium fluoride	7784-18-1	n.a.	n.a.
Aluminium hydride	7784-21-6	246 3	BD 0930000
Aluminium hydroxide	21645-51-2	n.a.	n.a.
Aluminium nitrate	13473-90-0	1438	BD 1040000
Aluminium oxide	1302-74-5	n.a.	BD 1200000
Aluminium phosphate	7784-30-7	1760	TB 6450000
Aluminium sulphate	10043-01-3	1760	BD 1700000
Sodium aluminate	1302-42-7	n.a.	n.a.

n.a. = number not available

Molecular Formula and Weight of Common Aluminium Compounds

Compound	Molecular formula	Molecular weight
Aluminium chloride	$AlCl_3$	113.34
Aluminium chlorohydrate	$AlClH_5O_5$	174.46
Aluminium fluoride	AlF_3	83.98
Aluminium hydride	AlH_3	30.01
Aluminium hydroxide	AlH_3O_3	78.00
Aluminium nitrate	$Al(NO_3)_3$	213.01
Aluminium oxide	Al_2O_3	101.96
Aluminium phosphate	AlO_4P	121.95
Aluminium sulphate	$Al_2(SO_4)_3$	342.14
Sodium aluminate	$AlNaO_2$	81.97

- Aluminium is used for building and construction; in corrosion-resistant chemical equipment, auto parts, the electrical industry; photoengraving plates, permanent magnets, powder paints; as pyro powders for fireworks; food packaging, food additives, cooking utensils, cans and dental materials

- Aluminium hydroxide is used in the treatment of drinking water; other compounds have a wide variety of uses including antacids, and in cosmetics, anti-perspirants, hardening agents and varnishes

- Aluminium occurs as organic and inorganic compounds; inorganic compounds have the valence Al^{3+}

- When exposed to air an aluminium surface becomes oxidised to form a thin coating of aluminium oxide, which protects against corrosion

- Pure aluminium oxide is soluble in water; aluminium hydroxide is amphoteric, i.e. binds with hydrogen ions when dissolved in an acid and releases hydrogen ions in alkali[1]

- At neutral pH, aluminium hydroxide has limited solubility but solubility increases markedly with increasing or decreasing pH

- Aluminium is insoluble in water; soluble in hydrochloric acid, sulphuric acid and alkalies

- Aluminium sulphate and aluminium chloride dissolve in water and hydrolyse water to acid

- If phosphate is present in water, aluminium phosphate which has low solubility in water will be precipitated; aluminium fluoride is slightly more soluble in water

- Aluminium halogenides, hydrides and lower aluminium alkyls react violently with water

- Powder and flake aluminium are flammable and can form explosive mixtures in air, especially when treated to reduce surface oxidation (pyro powders)

- Aluminium powder is a strong reducing agent which reacts violently with oxidants; it reacts violently with strong bases, acids and some halogenated hydrocarbons, giving off hydrogen and corrosive fumes

- Aluminium powder reacts violently with nitrates, sulphates and halogens, metal oxides and mercury and its compounds

Summary of Human Toxicity

- Aluminium is the most abundant metal in the earth's crust and is found in all human tissues, but its biological role is unknown

- Aluminium daily intake from food and beverages ranges between 2.5–13 mg; drinking water may contribute to approximately $0.4\ mg.d^{-1}$ at the present international guidelines, although this is more likely to be about $0.2\ mg.d^{-1}$

- Aluminium and its compounds appear to be poorly absorbed through the gastrointestinal tract although the rate and extent of absorption have not been adequately studied [2]

- Normal daily intake of aluminium is about 3 to 5 $mg.d^{-1}$; approximately 15 µg is absorbed in the gastrointestinal tract [3]

- Patients on antacid or phosphate binding therapy ingest up to $5\ g.dl^{-1}.d^{-1}$

- Only about 2–6% of aluminium in antacid preparations is absorbed from the gastrointestinal tract; citric acid has been associated with enhanced absorption of aluminium from aluminium-containing antacids

- Aluminium concentrations in foods are low, usually less than $5\ mg.kg^{-1}$; boiling water in aluminium pans causes a marked increase in aluminium intake

- Individuals drinking 1.5 litres per day of water with $100\ µg.l^{-1}$ of aluminium, absorbed approximately 3% of their total intake of aluminium from the water[2]

- The total body burden is maintained at 30 mg; the kidneys can excrete up to 0.5 mg in 24 hours [3]

- Acute aluminium toxicity is unlikely; there are two main categories of aluminium toxicity:

  1. patients with chronic renal failure

  2. occupational exposure to aluminium

- Patients with renal failure are prone to aluminium toxicity, either from aluminium in the dialysate or other sources, especially aluminium-containing phosphate binders and antacids or sucralfate[4]

- In the 1970s the high aluminium content in dialysate water was recognised as the cause of increased morbidity and mortality in chronic renal failure patients; clinical effects such as dialysis encephalopathy syndrome (dialysis dementia); osteomalacia with fracturing osteodystrophy and microcytic anaemia were noted

- Studies suggesting a link between aluminium in drinking water and the occurrence of Alzheimer's disease have been conducted with varying results; based on current knowledge of the pathogenesis of Alzheimer's disease and evidence from these epidemiology studies, it has been concluded that the evidence does not support a causal association[2]

- Aluminium is thought to be absorbed by inhalation[5]

- The actual absorption rate of aluminium is unknown; elimination takes place mainly via urine; urinary elimination half-lives vary from days to years[5]

- Inhalation of aluminium powder with particle size of 1.2 µm (96%), over 10–20 minutes several times a week, resulted in no adverse health effects over several years[6]

- A baseline >200 µg.l⁻¹ was associated with aluminium bone disease (specifically 93%), however lower values did not preclude the diagnosis (sensitivity 43%)[7]

Toxicity and Serum Aluminium Concentrations

Concentration	Effect
>60 µg.l⁻¹	Excess aluminium absorption
>100 µg.l⁻¹	Potential clinical concern; associated with signs of toxicity
>200 µg.l⁻¹	Associated with clinical effects

- Occupational Exposure Standards:
  Long-term exposure limit:

  Aluminium metal: 10 mg.l⁻¹ (total inhalable dust)
  4 mg.l⁻¹ (respirable dust)

  Aluminium oxides: 10 mg.l⁻¹ (total inhalable dust)
  4 mg.l⁻¹ (respirable dust)

  Aluminium salts, soluble: 2 mg.l⁻¹

  Short-term exposure limit: no data available

Acute Clinical Effects

Inhalation Effects

- Welding aluminium may lead to coughing, increased phlegm production and some irritation; these effects may be due to increased ozone concentrations formed by the welding of aluminium

- Inhalation of very fine aluminium powder in large concentrations occasionally leads to pneumonitis

Dermal Effects

- Aluminium is not considered to be a skin irritant

- Excess aluminium sulphate in drinking water may lead to skin rashes and ulcers

Eye Effects

- Aluminium dust in the eye may cause irritation

Oral Effects

- Gastrointestinal upset is the most likely consequence of ingestion of aluminium

- Excess aluminium sulphate in drinking water may lead to nausea, vomiting, diarrhoea and mouth ulcers

Chronic Clinical Effects

Occupational exposure may result in asthma, chronic obstructive lung disease and pulmonary fibrosis. Long-term overexposure may cause dyspnoea, cough, pneumothorax, variable sputum production and nodular interstitial fibrosis; death has been reported.[8] Chronic interstitial pneumonia with severe cavitations in the right upper lung and small cavities in the rest of the lungs has been reported.

Shaver's Disease is caused by occupational exposure to fumes or dust leading to respiratory distress and fibrosis with large blebs; spontaneous pneumothorax is common. Silicon is often inhaled with aluminium, so the role of each element is unclear. Chronic

exposure may result in asthma which is probably related to the inhalation of fumes and particulate matter.

Aluminium ingestion has been associated with Alzheimer's disease; it is not known whether aluminium is a causal agent or if the neurodegenerative disease allows more aluminium to accumulate in the brain.[6] Neurological syndromes including impairment of cognitive function, motor dysfunction and peripheral neuropathy have been reported in limited studies of workers exposed to aluminium fumes.[2]

There is insufficient information to allow for classification of cancer risk from human exposure to aluminium and its compounds. Animal studies do not indicate that aluminium or aluminium compounds are carcinogenic.

Dialysis Exposure

Aluminium toxicity may be characterised by hypercalcaemia, a reversible microcytic anaemia and vitamin D refractory osteodystrophy. A progressive encephalopathy may be seen, characterised by speech disturbance, asterixis, tremors, myoclonus, dementia, focal convulsions, bone pain, proximal myopathy, decline in visual memory, attention and concentration and arthropathy. Symptoms tend to develop over months to years in chronic renal failure patients. Haematopoietic toxicity from aluminium correlates with bone surface staining for aluminium but not with other measures of aluminium overload.[3]

Management

Normal serum concentrations are < 0.4 µmol.l⁻¹ (< 10 µg.l⁻¹). Determination of aluminium concentrations may confirm exposure.

Excessive accumulation: >2.2 µmol.l⁻¹ (>60 µg.l⁻¹)

Clinical concern: >3.7 µmol.l⁻¹ (>100 µg.l⁻¹)

High risk of toxicity: >7.4 µmol.l⁻¹ (>200 µg.l⁻¹)

Inhalation Management

- If respiratory irritation occurs assess respiratory function

- Symptomatic and supportive care

Dermal Management

- Remove any remaining contaminated clothing, place in double, sealed, clear bags and label; store in a secure area away from patients and staff

- Symptomatic and supportive care

Eye Management

- Irrigate thoroughly with running water or saline for 15 minutes

- Stain with fluorescein and refer to an ophthalmologist if there is any uptake of the stain

Oral Management

- Encourage oral fluids

- Symptomatic and supportive care

- Symptoms tend to be short-lived with no long lasting effects

Summary of Environmental Hazards

- Aluminium is ubiquitous in the environment in the form of silicates, oxides and hydroxides, combined with other elements such as sodium, fluorine and arsenic complexes with organic matter[2]

- Upon acidification of soils aluminium can be released into transportable solution

- Mobilisation of aluminium by acid rain results in more aluminium being available for plant uptake

- Drinking Water Standards:

  Aluminium : $200\,\mu g.l^{-1}$ (UK max)
  $200\,\mu g.l^{-1}$ (WHO guideline)

  Chloride: $400\,mg.l^{-1}$ (UK max)
  $250\,mg.l^{-1}$ (WHO guideline)

  Fluoride: $1.5\,mg.l^{-1}$ (UK max)
  $1.5\,mg.l^{-1}$ (WHO guideline)

  Nitrate: $50\,mg.l^{-1}$ (UK max)
  $50\,mg.l^{-1}$ (WHO guideline)

  Sulphate: $250\,mg.l^{-1}$ (UK max)

- Soil Guidelines: no data available

- Air Quality Standards: no data available

REFERENCES

1. Friberg L, Nordberg GF & Vouk VB. (eds.) *Handbook on the Toxicology of Metals*. Elsevier, Amsterdam, 1989

2. International Programme on Chemical Safety. *Environmental Health Criteria 194: Aluminium*. WHO, Geneva, 1997

3. Ellenhorn MJ, Schonwalds S, Ordog G & Wasserberger J. *Ellenhorn's Medical Toxicology – Diagnosis and Treatment of Human Poisoning*, 2nd edn. Williams & Wilkins, London, 1997

4. Robertson JA, Salusky IB, Goodman WG, Norris KC & Coburn JW, 1989. Sucralfate, intestinal aluminum absorption, and aluminum toxicity in a patient on dialysis. *Ann Intern Med*; 111:179–81

5. Baselt RC & Cravey RH. *Disposition of Toxic Drugs and Chemicals in Man*, 4th edn. Chemical Toxicology Institute, 1995

6. Hathaway GJ, Proctor NH & Hughes JP. *Proctor and Hughes' Chemical Hazards of the Workplace*, 4th edn. Van Nostrand Reinhold, New York, 1996

7. Hall AH & Rumack BH (eds.) *TOMES System* ® Micromedex, Englewood, Colorado. CD ROM. vol 41 (exp. 31 July 1999)

8. Clayton GD & Clayton FE (eds.) *Patty's Industrial hygiene and toxicology*, 4th edn. John Wiley & Sons, Inc., New York, 1994

Henrietta Wheeler

AMMONIA

Key Points

- Ammonia is a colourless gas or liquids (liquefied compressed gas or aqueous solution) with a characteristic odour of drying urine
- It is extremely alkaline and can cause severe corrosive damage to the eyes and skin
- The extent of dermal injury ranges from mild erythema to full thickness burns
- Ammonia is a respiratory irritant, and may cause local irritation of the mucous membranes, acute pulmonary oedema or chronic obstructive pulmonary disease may occur
- Contact with liquid (compressed gas) may cause frostbite damage to lungs, eyes, skin, mucous membranes, oesophagus or any other tissue with which it is directly in contact
- *In the event of a large spill, stay upwind and out of low areas. Ventilate closed spaces. Protective clothing, eye protection and breathing apparatus should be worn*

First Aid

- Terminate exposure and support vital functions
- The casualty should be moved to an uncontaminated area
- Rescuers should, ideally, be trained personnel and must be careful *not to put themselves at risk* and *so wear appropriate protective clothing and, if available, breathing apparatus*
- If the casualty is unconscious a clear airway should be established and maintained; give 100% oxygen if available
- **Inhalation Exposure:** If the patient stops breathing, expired air resuscitation should be started immediately using a pocket mask with a one way valve, if available. It is important where the face is contaminated the expired air resuscitation is NOT attempted unless an airway with rescuer protection is used
- **Dermal Exposure:** If frostbite occurs, *do not* remove clothing, flush skin with water
- If frostbite is not present, remove contaminated clothing, if possible under a shower and place in double, sealed, clear bags and label; store the bags in a secure area away from patients and staff
- Wash the skin thoroughly with copious amounts of water until the pH is no longer alkaline
- **Eye Exposure:** If the eye tissue is frozen, seek medical advice as soon as possible
- If eye tissue is not frozen, irrigate thoroughly with water or saline until pH is no longer alkaline
- **NB:** Testing the pH of the skin and eyes immediately following irrigation may be misleading. It is recommended that 15 minutes elapse before this is undertaken. Continue the irrigation process if the skin or eyes are still alkaline
- **Oral Exposure:** Encourage small quantities of oral fluids (no more than 50–100 ml in total) unless perforation is suspected

Detailed Information

- Ammonia is a colourless gas or liquids (liquefied compressed gas or aqueous solution); it is a strong alkali with a penetrating, pungent characteristic odour of drying urine
- *Common synonyms*

  Gas: ammonia anhydrous, ammonia gas

  Liquefied compressed gas: liquid ammonia

  Aqueous solution: ammonia solution, ammonia water, ammonium hydroxide, ammonia hydrate, Spirit of Hartshorn
- CAS 7664-41-7 (gas, liquefied compressed gas)
- CAS 8007-57-6 (deleted registry number)
- CAS 1336-21-6 (aqueous solution)
- UN 1005 (liquefied anhydrous ammonia; anhydrous ammonia; ammonia solution > 50% ammonia)
- UN 2073 (ammonia solution > 35% but ≤ 50% ammonia)
- UN 2672 (ammonia solution > 10% but ≤ 35% ammonia)
- UN 3318 (ammonia solution > 50% ammonia)
- NIOSH/RTECS BO 0875000 (gas, liquefied compressed gas)
- NIOSH/RTECS BQ 9625000 (aqueous solution)
- Molecular formula NH_3 (gas, liquefied compressed gas)
- Molecular formula NH_4OH (aqueous solution)
- Molecular weight 17.04 (gas, liquefied compressed gas)
- Molecular weight 35.05 (aqueous solution)
- Ammonia is used as a refrigerant, in explosives, synthetic fibres and fertilisers; in bleaching agents and some household cleaners; ammonia solution is commonly supplied to laboratories as a 35% solution in water
- Ammonia gas is highly water-soluble forming alkaline solutions of ammonium hydroxide
- It is lighter than air and is generally considered non-flammable but mixtures of ammonia and air will explode when ignited under favourable conditions
- When heated to decomposition, ammonia in solution emits toxic fumes of ammonia and oxides of nitrogen
- Liquid ammonia attacks some plastics, rubber and coatings

Summary of Human Toxicity

- Ammonia acts as an alkali
- Anhydrous ammonia reacts with moisture in mucosal surfaces (eyes, skin and respiratory tract) to produce ammonium hydroxide which may cause caustic injury
- Burns tend to feel soapy as a result of saponification of the tissue fat

- Mixing ammonia solution and hypochlorite bleach results in the formation of chloramine, which may cause respiratory irritation

Toxicity of Ammonia gas[1]

Concentrations	Clinical effects
53 ppm	Least detectable odour
408 ppm	Least amount causing immediate irritation to the throat
698 ppm	Least amount causing immediate irritation to the eye
1,720 ppm	Least amount causing coughing
2,500 – 4,500 ppm	Dangerous for even short exposure (30 minutes)
5,000 – 10,000 ppm	Rapidly fatal for short exposure

- Occupational Exposure Standards:
  Long-term exposure limit: 25 ppm (18 mg.m^{-3})
  Short-term exposure limit: 35 ppm (25 mg.m^{-3})

Acute Clinical Effects

Inhalation Effects

- Ammonia gas and mists of solution are severe respiratory tract irritants
- Initially there may be a dry mouth with sore throat and eyes, tight chest, headache, ataxia and confusion
- Following massive exposures, chronic airway hyper-reactivity and asthma, associated with obstructive pulmonary function changes may occur[2]

Dermal Effects

- Exposures to dilute solutions may cause irritation and erythema to the skin
- Concentrated solutions or vapours may cause severe, penetrating burns or necrosis
- Frostbite with severe tissue damage may occur from direct contact with liquid ammonia

Eye Effects

- Ammonia gas and mists of solution may cause irritation, lacrimation and conjunctivitis
- Ammonia has been reported to cause temporary or permanent blindness
- Total corneal epithelial loss may occur, damage may progress deeper into the eye[3]
- Exposure to liquid ammonia may lead to frostbite of the eye tissue

Oral Effects

- Exposures to dilute solutions or brief exposures to the gas lead to local mucosal irritation, epigastric pain, nausea and vomiting, swelling of the lips, mouth and larynx
- Exposures to more concentrated solutions or prolonged exposures to the gas may lead to oral burns or oesophageal corrosion and stricture
- Perforation with subsequent shock may occur

- Pyloric stenosis often occurs after several weeks and cancer may develop years later in patients who survive severe acute oral exposures[4]

Chronic Clinical Effects

Prolonged inhalation of low concentrations of ammonia gas may lead to chronic bronchitis, pulmonary fibrosis and chemical pneumonitis.

Management

Blood ammonia concentrations not a useful indicator of exposure or toxicity.

Inhalation Management

- Maintain a clear airway, give humidified oxygen and ventilate if necessary
- If respiratory irritation occurs assess respiratory function and if necessary perform chest X-rays to check for chemical pneumonitis
- Consider the use of IV steroids to reduce the inflammatory response
- Treat pulmonary oedema with PEEP or CPAP ventilation
- Symptomatic and supportive care

Dermal Management

- *If frostbite has occurred*: remove clothing carefully, these may need to be soaked off with tepid water
- *If frostbite has not occurred*: remove any remaining contaminated clothing
- Place any contaminated clothing in double, sealed, clear bags, and label; store in a secure area away from patients and staff
- Irrigate with copious amounts of water until the skin is no longer alkaline
- Testing the pH of the skin immediately after irrigation may be misleading. It is recommended that 15 minutes elapse before this is undertaken. Continue the irrigation process if the skin is still alkaline
- Treat burns symptomatically

Eye Management

- Irrigate thoroughly with running water or saline for 15 minutes until the eye is no longer alkaline
- Testing the pH of the eyes immediately following irrigation may be misleading. It is recommended that 15 minutes elapse before this is undertaken. Continue the irrigation process if the eye is still alkaline
- Stain with fluorescein and refer to an ophthalmologist if there is any uptake of the stain

Oral Management

- NO GASTRIC LAVAGE OR EMETIC
- Encourage oral fluids, unless perforation is suspected
- Give plasma expanders/blood or IV fluids for shock and analgesics for pain

- Consider the use of IV steroids to reduce the inflammatory response

- Take abdominal X-ray to check for perforation

- If facilities are available, early gastro-oesophagoscopy should be undertaken within 12 hours of the event to assess the extent and severity of the injury

- Symptomatic and supportive care

- Long term follow up is necessary following severe oral exposure due to sequelae

Summary of Environmental Hazards

- In air, ammonia is persistent and in water it biodegrades rapidly to nitrate, causing high oxygen demand; in soil it is strongly adsorbed

- Ammonia gas is highly soluble in water; it is non-persistent with a half-life of 2 days and is moderately toxic to fish under normal temperature and pH

- If ammonia enters the water it is harmful to aquatic life at low concentrations; it does not concentrate in the food chain

- Drinking Water Standards:
  $0.5 \, mg.l^{-1}$ (UK max)
  $1.5 \, mg.l^{-1}$ (WHO level where customers may complain)

- Soil Guidelines: no data available

- Air Quality Standards: no data available

REFERENCES

1. Harbison RD (ed.). *Hamilton and Hardy's Industrial Toxicology*, 5th edn. Mosby-Year Book Inc., St Louis, 1986

2. Hathaway GJ, Proctor NH & Hughes JP. *Proctor and Hughes' Chemical Hazards of the Workplace*, 4th edn. Van Nostrand Reinhold, New York, 1996

3. Grant MW & Schuman JS. *Toxicology of the Eye*, 4th edn. Charles C Thomas, Springfield, 1993

4. Gosselin RE, Smith RP & Hodge HC. *Clinical Toxicology of Commercial Products*, 5th edn. Williams & Wilkins, Baltimore, 1984

Henrietta Wheeler

AMMONIUM NITRATE

Key Points

- Ammonium nitrate is a colourless crystalline solid when pure
- Irritating to eyes, nose, throat and mucous membranes
- Ammonium nitrate is completely absorbed after ingestion, potentially absorbed by inhalation; absorption of nitrates may occur dermally
- Symptoms may be delayed in onset as it is the metabolites of ammonium nitrate which induce methaemoglobinaemia
- *In the event of a large spill, stay upwind and out of low areas. Ventilate closed spaces. Protective clothing, eye protection and breathing apparatus should be worn*

FIRST AID

- Terminate exposure and support vital functions
- The casualty should be moved to an uncontaminated area
- Rescuers should, ideally, be trained personnel and must be careful *not to put themselves at risk* and *so wear appropriate protective clothing and, if available, breathing apparatus*
- If the casualty is unconscious a clear airway should be established and maintained; give 100% oxygen if available
- **Inhalation Exposure:** If the patient stops breathing, expired air resuscitation should be started immediately using a pocket mask with a one way valve, if available. It is important where the face is contaminated that expired air resuscitation is NOT attempted unless an airway with rescuer protection is used
- **Dermal Exposure:** Remove contaminated clothing, if possible under a shower and place in double, sealed, clear bags and label; store the bags in a secure area away from patients and staff
- Wash the skin thoroughly with copious amounts of water
- **Eye Exposure:** Irrigate thoroughly with water or saline for 15 minutes
- **Oral Exposure:** Encourage small quantities of oral fluids (no more than 50–100 ml in total)

Detailed Information

- Ammonium nitrate is a colourless, odourless, hygroscopic, crystalline solid when pure; fertiliser grade may be grey-brown due to impurities
- *Common synonyms* ammonium saltpeter, nitrate of ammonia, Norway saltpeter
- CAS 6484-52-2
- UN 1942 (ammonium nitrate with organic coating)
- UN 1942 (ammonium nitrate with not more than 0.2% combustible substances)
- UN 2426 (ammonium nitrate liquid hot concentrated solution)
- NIOSH/RTECS BR 9050000

- Molecular formula $H_4N_2O_3$
- Molecular weight 80.04
- Ammonium nitrate is used in fertilisers, explosives, pesticide chemical manufacture, solid rocket propellants, nutrient mediums and as a catalyst; available in cold packs, which contain 50 to 234 grams per pack;[1] previously used as cathartic but is no longer utilised
- Soluble in water, alcohol, acetone and alkalis; insoluble in ether
- 0.1M aqueous solutions of ammonium nitrate have a pH of 5.43
- Ammonium nitrate is a strong oxidising agent; it reacts violently with reducing agents, strong acids, powdered metals and organic material with the risk of fire and explosion[2]
- Ammonium nitrate does not readily burn but will accelerate burning of combustible materials
- It emits toxic nitrous oxide fumes when heated to decomposition; ammonia also a combustion by-product[3]

Summary of Human Toxicity

- There is limited information about the toxicity of ammonium nitrate; unless otherwise stated the information is about nitrates in general
- Ammonium nitrate is well absorbed after ingestion, potentially absorbed by inhalation; dermal absorption of nitrates may occur through abraded areas[4]
- The main mode of toxicity of nitrates is the induction of methaemoglobinaemia
- The primary systemic toxicity of nitrates is due to *in vivo* conversion to nitrites
- The minimum lethal human exposure has not been established
- Inhalation exposure to 200 $\mu g \cdot m^{-3}$ for 2 hours caused no adverse health effects[3]
- Five patients who ingested 6 to 234 g (from cold packs) suffered no severe symptoms; three developed mild gastritis and two had mild hypotension[5]
- Occupational Exposure Standards: no data available

Acute Clinical Effects

Inhalation Effects

- Ammonium nitrate is a respiratory irritant causing coughing, irritation to mucous membranes, wheezing and shortness of breath
- There are no reports of significant absorption via inhalation although theoretically it is possible; see systemic effects below

Dermal Effects

- Dermal contact with ammonium nitrate may cause irritation
- In solution ammonium nitrate can cause burns

- Systemic toxicity may occur following contact with damaged skin; see systemic effects below

Eye Effects

- Ammonium nitrate is irritating to the eyes

- Concentrated solutions can cause conjunctivitis, iritis and burns

Oral Effects

- Irritation to mucous membranes, nausea and vomiting may occur

- Diarrhoea and abdominal pain; exposure to a large quantity of ammonium nitrate may lead to gastrointestinal haemorrhage

- Ingestion of ammonium nitrate may lead to systemic effects, see below

Systemic Effects

- Dizziness, weakness, headache and warm flushed sweaty skin; nausea, vomiting, diuresis and fatigue

- Shortness of breath, and drowsiness leading to cyanosis, methaemoglobinaemia and metabolic acidosis

- Both tachycardia and bradycardia have been reported

- Atrial fibrillation, cardiac ischaemia, frequent ventricular premature beats and bigeminy have been reported

- Hypotension, decreased peripheral vascular resistance, cardiovascular collapse, convulsions and coma in severe toxicity[3]

Chronic Clinical Effects

Chronic ammonium nitrate exposure has been reported to cause faintness and hypotension.[6]

Chronic ingestion of 6–12 g.d[-1] may cause gastritis, acidosis, isosmotic diuresis, and nitrite toxicity manifested by methaemoglobinaemia or vasodilatation.[5]

Management

Methaemoglobin concentrations are the best indicator of toxicity; plasma nitrate concentrations are not clinically useful. Urine nitrate concentration depends on the conversion of nitrate to nitrite by bacterial action. There should be a minimum of four hours between exposure and testing.

Inhalation Management

- Maintain a clear airway, give humidified oxygen and ventilate if necessary

- If respiratory irritation occurs assess respiratory function and perform chest X-rays to check for chemical pneumonitis

- Consider the use of steroids to reduce the inflammatory response

- Treat pulmonary oedema with PEEP or CPAP ventilation

- For further treatment, see admission criteria and systemic management below

Dermal Management

- Remove any remaining contaminated clothing, place in double, sealed, clear bags, and label; store in a secure area away from patients and staff

- Irrigate with copious amounts of water

- An emollient may be required

- If exposure is large, prolonged or through abraded skin, see admission criteria and systemic management below

Eye Management

- Irrigate thoroughly with running water or saline for 15 minutes

- Stain with fluorescein and refer to an ophthalmologist if there is any uptake of the stain

Oral Management

- NO GASTRIC LAVAGE OR EMETIC

- Encourage oral fluids

- For further treatment, see admission criteria and systemic management below

Admission criteria:

Patients with no immediate symptoms should be observed for a minimum of 4 hours post exposure.

All patients showing effects should be assessed in hospital where methaemoglobin concentrations should be checked.

Systemic management

- Give 100% oxygen

- If the patient is very drowsy, unconscious or the methaemoglobin level exceeds 30%, methylene blue should be given at the dosage below

- If methaemoglobinaemia is suspected but levels can not be measured, methylene blue (1% solution, i.e. 10 mg.ml[-1]) 1–2 mg.kg[-1] should be given IV over 5 minutes and repeated within 1 hour if there is no response

- Suspect methaemoglobinaemia if the arterial blood is chocolate brown and remains dark on aeration, if cyanosis is unresponsive to oxygen therapy, or pO_2 is normal in the presence of a decreased oxygen saturation

- For methaemoglobinaemia unresponsive to methylene blue, exchange transfusion and/or haemodialysis could be considered

- Give plasma expanders/blood or IV fluids for shock

- Symptomatic and supportive care

Summary of Environmental Hazards

- Ammonium nitrate is soluble in water; nitrate ions are readily leached in water

- Ammonium nitrate is a nutrient in water; spills may cause massive algal eutrophication in static waters and affect the local aquatic population[3]

- Ammonium nitrate will penetrate into soil at a rate dependent on the soil type and its water content

- Ammonium nitrate is not expected to accumulate or concentrate in the food chain

- Drinking water standards:
  ammonia and ammonium ions: 0.5 mg.l^{-1} (UK max)
  ammonia: 1.5 mg.l^{-1} (WHO guideline)
  nitrate: 50 mg.l^{-1} (UK max)
  50 mg.l^{-1} (WHO guideline)

- Soil Guidelines: no data available

- Air Quality Standards: no data available

REFERENCES

1. Challoner KR & McCarron MM, 1988. Ammonium nitrate cold pack ingestion. *J Emerg Med*; 6 (4): 289–93

2. Dutch Institute for the Working Environment and the Dutch Chemical Industry Association (eds.) *Chemical Safety Sheets – Working Safely with Hazardous Chemicals*. Samson Chemical Publishers, The Netherlands, 1991

3. Hall AH & Rumack BH (eds.) *TOMES System®* Micromedex, Englewood, Colorado. CD ROM. vol 41 (exp. 31 July 1999)

4. Mozingo DW, Smith AA, McManus WF, Pruitt BA & Mason AD, 1988. Chemical Burns. *J Trauma*; 28(5): 642–7

5. Ellenhorn MJ, Schonwalds S, Ordog G & Wasserberger J. *Ellenhorn's Medical Toxicology – Diagnosis and Treatment of Human Poisoning*, 2nd edn. Williams & Wilkins, London, 1997

6. Sax NI. *Dangerous Properties of Industrial Materials*, 6th edn. Van Nostrand Reinhold, New York, 1984

Simon Ward

ANILINE

Key Points

- Aniline is a volatile, flammable, colourless, oily liquid that darkens on exposure to air and light. It has a characteristic, peculiar odour and burning taste
- Aniline is rapidly absorbed by all routes of exposure and induces methaemoglobinaemia
- Skin absorption is the mode of greatest toxicological significance in industry as it is frequently the main route of entry
- Symptoms may be delayed in onset as it is the metabolites of aniline which induce methaemoglobinaemia
- *In the event of a large spill, stay upwind and out of low areas. Ventilate closed spaces. Protective clothing, eye protection and breathing apparatus should be worn*

FIRST AID

- Terminate exposure and support vital functions
- The casualty should be moved to an uncontaminated area
- Rescuers should, ideally, be trained personnel and must be careful *not to put themselves at risk* and *so wear appropriate protective clothing and, if available, breathing apparatus*
- If the casualty is unconscious a clear airway should be established and maintained; give 100% oxygen if available
- **Inhalation Exposure:** If the patient stops breathing, expired air resuscitation should be started immediately using a pocket mask with a one way valve, if available. It is important where the face is contaminated that expired air resuscitation is NOT attempted unless an airway with rescuer protection is used
- **Dermal Exposure:** Remove contaminated clothing, if possible under a shower, and place in double, sealed, clear bags and label; store the bags in a secure area away from patients and staff
- Wash the skin thoroughly with copious amounts of water
- **Eye Exposure:** Irrigate thoroughly with water or saline for 15 minutes
- **Oral Exposure:** Encourage small quantities of oral fluids (no more than 50–100 ml in total)

Detailed Information

- Aniline is a volatile, flammable, colourless, oily liquid that darkens on exposure to air and light; it has a characteristic, peculiar odour and burning taste

- *Common synonyms* aminobenzene, aminophen, benzenamine, phenylamine

- CAS 62-53-3

- UN 1547

- NIOSH/RTECS BW 6650000

- Molecular formula $C_6H_5NH_2$

- Molecular weight 93.13

- Aniline is used in the manufacture of dyestuffs, dyestuff intermediates, rubber accelerators and antioxidants; as an intermediate in pharmaceutical manufacturing, photographic developers, plastics, isocyanates, hydroquinones, herbicides, fungicides and ion-exchange resin

- Aniline reacts violently with strong oxidants, alkaline-earth and alkali metals, giving off hydrogen, with a risk of explosion

- When heated, aniline may polymerise forming explosive mixtures with air; nitrous vapours and carbon monoxide are emitted when heated to decomposition

Summary of Human Toxicity

- Aniline is rapidly absorbed by all routes and induces methaemoglobinaemia[1]

- It is transformed *in vivo* producing a metabolite phenylhydroxylamine which is responsible for the production of methaemoglobinaemia and many of the toxic effects of aniline[2]

- Peak methaemoglobin levels may, therefore, occur some hours after aniline exposure while it is converted to its metabolites[1,3]

- The no-effect level in humans following oral administration is 5 to 15 mg (0.07 to 0.21 mg.kg^{-1})[4]

- The minimum toxic dose of aniline in humans has not been defined but 25 to 65 mg orally produced significant elevation of methaemoglobin levels in human volunteers[4]

- The lowest fatal dose reported in humans is 1 g orally, however, much higher doses have been survived[2]

- The mean lethal dose by ingestion in humans is estimated at between 15 and 30g[5]

Toxicity of Aniline[1,2]

Concentration	Clinical effects
0.5 ppm	Odour threshold
7–53 ppm	Slight symptoms
100 ppm	Immediate danger to health
100–160 ppm	Serious effects if inhaled for 1 hour

- Maximum Exposure Limit: not yet set

Acute Clinical Effects

Inhalation Effects

- Inhalation of aniline causes systemic effects which may be delayed some hours after exposure; see systemic effects below

Dermal Effects

- Aniline is rapidly absorbed across intact skin and induces methaemoglobinaemia

- Aniline is irritant to the skin and a mild dermal sensitiser

- Systemic effects may be delayed in onset for 2 to 4 hours after initial skin contact;[2] see below

Eye Effects

- Aniline (liquid or vapours) is a mild irritant to the eyes and may cause corneal damage

- Photophobia, impairment of vision and brown discoloration of the conjunctiva and cornea in the palpebral fissure have been reported

- The vessels of the conjunctiva and retina may appear violet, without affecting vision, due to the discoloration of the blood by the methaemoglobin induced by other routes of exposure[6]

Oral Effects

- Aniline is rapidly absorbed following ingestion and induces methaemoglobinaemia and systemic effects which may be delayed some hours after exposure; see systemic effects below

Systemic Effects

- Methaemoglobinaemia is characterised by navy blue to black lips, tongue and mucous membranes, and slate grey skin

- Tachypnoea, dyspnoea and cyanosis may develop

- Severe headache, nausea and vomiting, paraesthesia, tremor, abdominal pain, drowsiness, and tachycardia may also develop

- Haematuria, haemoglobinuria, methaemoglobinuria, oliguria and mild renal insufficiency, anaemia, cardiac arrhythmias, coma and shock have all been reported

- Convulsions, and a late acute haemolytic crisis with Heinz body formation may occur, but are uncommon[7]

- Death is rare; cardiovascular collapse or asphyxiation are thought to be the main causes

Chronic Clinical Effects

Continuous exposure to small doses of aniline has been suggested to cause anaemia, lethargy, digestive disturbances and headache.[2]

Bladder tumours have been associated with aromatic amines, including aniline, used in the dye industry. Convincing evidence that aniline is a carcinogen could not be established, primarily as a result of exposures to a multiplicity of compounds, including known carcinogens.[2] Aniline is classified as an IARC Group 3 carcinogen, i.e. not classifiable as to its carcinogenicity as the evidence for carcinogenicity is limited in animals and inadequate in humans.[2] NIOSH has determined that there is sufficient evidence to recommend that this substance is a potential occupational carcinogen from animal studies.[8]

In humans aniline crosses the placenta and has been reported to be foetotoxic causing foetal methaemoglobinaemia.[9] Aniline was not found to be teratogenic or foetotoxic in animal studies at maternally toxic oral doses.[8]

Management

Blood methaemoglobin concentrations are a good indicator of aniline toxicity and should be measured 3 to 6 hourly for at least 24 hours post exposure.[1]

Note: Care workers must ensure adequate protection to prevent self-contamination when carrying out decontamination and medical treatment.

Inhalation Management

- Maintain a clear airway and give 100% oxygen

- Determine the blood methaemoglobin concentration

- See admission criteria below

Dermal Management

- Remove any remaining contaminated clothing, place in double, sealed, clear bags and label; store in a secure area away from patients and staff

- Irrigate with copious amounts of water, paying particular attention to hair, scalp, finger and toenails, nostrils and ear canals

- Give 100% oxygen and determine the blood methaemoglobin concentration

- See admission criteria below

Eye Management

- Irrigate thoroughly with running water or saline for 15 minutes

- Stain with fluorescein and refer to an ophthalmologist if there is any uptake of the stain

Oral Management

- Encourage oral fluids

- Consider gastric lavage followed by 50g activated charcoal within 1 hour of a substantial ingestion, ensuring the airway is protected

- Contact a poisons information service for further guidance on gut decontamination

- Give 100% oxygen and determine the blood methaemoglobin concentration

- See admission criteria below

Admission criteria:

Patients with no immediate symptoms should be observed for a minimum of 4 hours post exposure.

All patients showing effects should be assessed in hospital where methaemoglobin levels should be monitored.

Further management

- If the patient is very drowsy, unconscious or the methaemoglobin concentration exceeds 30%, methylene blue should be given at the dosage below

- If methaemoglobinaemia is suspected but levels can not be measured, methylene blue (1% solution, i.e. $10 \, mg \cdot ml^{-1}$) $1-2 \, mg \cdot kg^{-1}$ should be given IV over 5 minutes and repeated within 1 hour if there is no response

- Suspect methaemoglobinaemia if the arterial blood is chocolate brown and remains dark on aeration, if cyanosis is unresponsive to oxygen therapy, or pO_2 is normal in the presence of a decreased oxygen saturation

- For methaemoglobinaemia unresponsive to methylene blue, exchange transfusion and/or haemodialysis could be considered[5,8]

- Check complete blood and reticulocyte counts for evidence of haemolysis in severe cases

- Give plasma expanders/blood or IV fluids for shock

- Symptomatic and supportive care

Summary of Environmental Hazards

- In air aniline reacts with photochemically produced hydroxyl radicals with an estimated half-life of 3.3 hours

- In water and soil aniline undergoes rapid biodegradation

- Aniline is harmful to aquatic life in high concentrations; it is not expected to concentrate in the food chain

- Drinking Water Standards: no data available

- Soil Guidelines: no data available

- Air Quality Standards: no data available

REFERENCES

1. Hall AH & Rumack BH (eds.) *TOMES System®* Micromedex, Englewood, Colorado. CD ROM. vol 41 (exp. 31 July 1999)

2. Hathaway GJ, Proctor NH & Hughes JP. *Proctor and Hughes' Chemical Hazards of the Workplace*, 4th edn. Van Nostrand Reinhold, New York, 1996

3. Baselt RC & Cravey RH. *Disposition of Toxic Drugs and Chemicals in Man*, 4th edn. Chemical Toxicology Institute, 1995

4. Gosselin RE, Smith RP & Hodge HC. *Clinical Toxicology of Commercial Products*, 5th edn. Williams & Wilkins, Baltimore, 1984

5. Mier RJ, 1988. Treatment of aniline poisoning with exchange transfusion. *J Toxicol-Clin Toxicol*; 26(5 & 6): 357–64

6. Grant MW & Schuman JS. *Toxicology of the Eye*, 4th edn. Charles C Thomas, Springfield, 1993

7. Lubash GD, Phillips RE, Shields JD & Bonsnes RW, 1964. Acute aniline poisoning treated by hemodialysis. *Arch Intern Med*; 114: 530–2

8. Clayton GD & Clayton FE (eds.) *Patty's Industrial hygiene and toxicology*, 4th edn. John Wiley & Sons, Inc. New York, 1994

9. Harley JD & Celermajer JM, 1970. Neonatal methaemoglobinaemia and the "red-brown" screening-test. *Lancet*; 2: 1223–5

Catherine Farrow

ARSENIC AND ARSENIC COMPOUNDS

Key Points

- Arsenic occurs in a number of forms: elemental arsenic, and inorganic and organic (trivalent and pentavalent) salts
- Arsenic is a steel-grey brittle metalloid; most arsenic compounds are colourless powders or crystals; arsenic trichloride and arsenic acid are oily liquids
- Arsenic compounds are odourless and tasteless and highly toxic by ingestion, inhalation and percutaneous absorption
- Primary target organs include the gastrointestinal tract, heart, brain and kidneys; secondary targets are the skin, haematopoietic system (bone marrow and spleen), and peripheral nervous system
- Elemental (i.e. metallic) arsenic is less toxic than the arsenic compounds
- *In the event of a large spill, stay upwind and out of low areas. Ventilate closed spaces. Protective clothing, eye protection and breathing apparatus should be worn*

FIRST AID

- Terminate exposure and support vital functions
- The casualty should be moved to an uncontaminated area
- Rescuers should, ideally, be trained personnel and must be careful *not to put themselves at risk* and *so wear appropriate protective clothing and, if available, breathing apparatus*
- If the casualty is unconscious a clear airway should be established and maintained; give 100% oxygen if available
- **Inhalation Exposure:** If the patient stops breathing, expired air resuscitation should be started immediately using a pocket mask with a one way valve, if available. It is important where the face is contaminated that expired air resuscitation is NOT attempted unless an airway with rescuer protection is used
- **Dermal Exposure:** Remove contaminated clothing, if possible under a shower, and place in double, sealed, clear bags and label; store the bags in a secure area away from patients and staff
- Wash the skin thoroughly with copious amounts of water
- **Eye exposure:** Irrigate the eyes thoroughly with water or saline for 15 minutes
- **Oral exposure:** Encourage small quantities of oral fluids (no more than 50–100 ml in total)

Detailed Information

- Elemental arsenic is a steel-grey brittle metalloid; most arsenic compounds are colourless powders or crystals; arsenic trichloride and arsenic acid are oily liquids
- *Common synonyms* arsenic black, arsenic-75, colloidal arsenic, grey arsenic, metallic arsenic
- CAS 7440-38-2
- UN 1558
- NIOSH/RTECS CG 0525000
- Atomic symbol As
- Atomic weight 74.92

Common Synonyms of Arsenic Compounds

Compound	Synonyms
Inorganic arsenic, trivalent	
arsenic (III) oxide	arsenic trioxide, arsenous oxide, white arsenic
arsenic (III) chloride	arsenic trichloride, arsenous trichloride
arsenic (III) sulphide	arsenic trisulphide, orpiment, auripigment
Inorganic arsenic, pentavalent	
arsenic (V) oxide	arsenic pentoxide
arsenic acid	orthoarsenic acid
Organic arsenic	
methylarsonic acid	methanearsonic acid
dimethylarsenic acid	cacodylic acid
arsanilic acid	p-aminobenzenearsonic acid, (4-aminophenyl)–arsonic acid
arsphenamine	4,4-arsenobis(2-amino-phenol)dihydrochloride
carbarsone	–
tryparsamide	–
3-nitro-4-hydroxy-phenylarsonic acid	–
4-nitrophenylarsonic acid	p-nitrophenylarsonic acid

Identification Numbers of Arsenic Compounds

Chemical	CAS number	UN number	NIOSH/RTECS
Inorganic arsenic, trivalent			
arsenic (III) oxide	1327–53–3	1561	CG 3325000
arsenic (III) chloride	7784–34–1	1560	CG 1750000
arsenic (III) sulphide	1303–33–9	1557	CG 2638000
Inorganic arsenic, pentavalent			
arsenic (V) oxide	1303–28–2	1559	CG 2275000
arsenic acid	7778–39–4	1553	CG 0765000
Organic arsenic			
methylarsonic acid	124–58–3	not available	PA 1575000
dimethylarsenic acid	75–60–5	1572	CH 7525000
arsanilic acid	98–50–0	not available	CF 7875000
arsphenamine	139–93–5	not available	SJ 7175000
carbarsone	121–59–5	not available	not available
tryparsamide	554–72–3	not available	not available
3-nitro-4-hydroxy-phenylarsonic acid	121–19–7	not available	CY 5250000
4-nitrophenylarsonic acid	98–72–6	not available	not available

UN 1556 Arsenic compound, liquid, not otherwise stated, inorganic

UN 1556 Arsenic compound, liquid, not otherwise stated

UN 1557 Arsenic compound, solid, not otherwise stated, inorganic

UN1577 Arsenic compound, solid, not otherwise stated

Molecular Formula and Weight of Arsenic Compounds

Compound	Molecular formula	Molecular weight
Inorganic arsenic, trivalent		
arsenic (III) oxide	As_2O_3 (or As_4O_6)	197.84
arsenic (III) chloride	$AsCl_3$	181.28
arsenic (III) sulphide	As_2S_3	246.04
Inorganic arsenic, pentavalent		
arsenic (V) oxide	As_2O_5	229.84
arsenic acid	H_3AsO_4	141.94
Organic arsenic		
methylarsonic acid	$CH_3AsO(OH)_2$	139.97
dimethylarsenic acid	$(CH_3)_2AsO(OH)$	138.00
arsanilic acid	$C_6H_8AsNO_3$	217.06
arsphenamine	$C_{12}H_{14}As_2Cl_2N_2O_2$	439.00
carbarsone	$C_7H_9AsN_2O_4$	260.08
tryparsamide	$C_8H_{10}AsN_2NaO_4$	296.09
3-nitro-4-hydroxy-phenylarsonic acid	$C_6H_6AsNO_6$	263.04
4-nitrophenylarsonic acid	$C_6H_6AsNO_5$	247.04

- Arsenic is used for hardening and increasing heat resistance of copper, lead and alloys, in pigment production, in the manufacture of certain types of glass, in wood preservatives, insecticides, fungicides and rodenticides

- It is a by-product in the smelting of copper, lead and gold ores

- Harmful concentrations of vapour or airborne particles may build up rapidly at 20°C

- Elemental arsenic is insoluble in water; inorganic arsenic is more soluble than organic arsenic

- Arsenic reacts with incandescence or may explode on contact with a wide variety of chemicals

- It reacts violently on heating with powdered aluminium, on ignition it reacts with bromine pentafluoride and at ambient or slightly elevated temperatures with chlorine trifluoride

- Nitrogen trichloride decomposes violently and explosively in the presence of arsenic

- Arsenic trichloride decomposes to chlorine gas and arsenic trioxide when exposed to light

- It reacts violently with bases, strong oxidants and water evolving anhydrous ammonia, chlorine gas, and hydrochloric acid and arsenic acid respectively

- Arsenic trichloride reacts with air and metals producing hydrochloric acid fumes and hydrogen respectively

- Arsenic may burn but does not ignite readily; its dust is slightly explosive when exposed to flame

Summary of Human Toxicity

- Arsenic is absorbed by ingestion, inhalation and across the skin[1]

- Acute toxicity varies with the valency of the element

- Trivalent arsenic is the most toxic because it is more soluble and has a greater affinity for sulphydryl groups; it is lipid soluble and well absorbed across the skin[2,3]

- Pentavalent arsenic is less soluble and much less toxic, however, it may be converted to trivalent arsenic *in vivo*; it is well absorbed by the gut[3,4]

- Inorganic arsenic binds to sulphydryl groups within cells and inhibits enzymes[2]

- Primary target organs include the gastrointestinal tract, heart, brain and kidneys, then skin, haematopoietic system (bone marrow and spleen), and peripheral nervous system

- Organic arsenic inhibits oxidative phosphorylation by substitution of inorganic phosphorus

- Estimated lethal dose for arsenic trioxide ranges from 70 to 300 mg[1,4]

- Arsenic is present in most foodstuffs in concentrations of <1 mg.kg^{-1}; however some marine fish may contain arsenic concentrations of up to 5 mg.kg^{-1}[5]

- Maximum Exposure Limits:
    Long-term exposure limit: 0.1 mg.m^{-3} (as As)
    Short-term exposure limit: no data available

Acute Clinical Effects

Inhalation Effects

- Arsenic compounds are irritant to the mucous membranes of the upper respiratory tract

- Initially causing cough, sore throat, respiratory distress

- Ulceration and perforation of the nasal septum may occur[1]

- Pharyngitis and pulmonary irritation; pulmonary oedema and respiratory failure may develop in severe cases

- In severe cases, systemic effects may also develop; see below

Dermal Effects

- Arsenic compounds are irritant or corrosive to the skin

- Primary irritation and sensitisation dermatitis may result

- Acute dermal exposure may lead later to hyperpigmentation

- Trivalent arsenic compounds are rapidly absorbed across the skin and systemic effects may develop; see below

Eye Effects

- Arsenic compounds are highly irritant or corrosive to the eyes[2]

- Conjunctivitis with itching, burning and watering eyes; photophobia and visual disturbance may develop[6]

- Some organic arsenicals (e.g. arsanilates) have a selective effect on the optic nerve and can cause blindness

Oral Effects

- Oral toxicity depends on the solubility of the arsenical; the more soluble the greater the toxicity

- Onset of effects is usually within 30 minutes; effects may be delayed for several hours with food

- Initially, burning lips, garlic odour on breath, constriction of the throat and dysphagia

- Excruciating abdominal pain, severe nausea, projectile vomiting, profuse bloody or rice-water diarrhoea, and mucosal erosions[1]

- Systemic effects are likely; see below

Systemic Effects

- Dehydration, thirst, fluid-electrolyte disturbances, hypotension and collapse

- Hepatomegaly, insomnia and anorexia may occur

- Fluid or blood loss may lead to shock

- Muscular cramps which may progress to rhabdomyolysis

- Haematuria, acute tubular necrosis and acute haemolysis may occur

- ECG changes may include tachycardia, ventricular fibrillation or tachycardia, QT prolongation or T-wave changes

- In severe cases coma, convulsions, encephalopathy, delirium and death follow within 24 hours

- Possible sequelae include hepatotoxicity, bone marrow depression, central and peripheral neuropathies; gangrene of the extremities may develop[2]

Chronic Clinical Effects

Chronic arsenic ingestion is characterised by weakness, anorexia, gastrointestinal disturbance, impairment of cognitive function, peripheral neuritis and neuropathy, hepatomegaly, jaundice, irritation of nose and throat, perforation of nasal septum, and skin disorders including ulceration, hyperkeratosis of palms and soles, hyperpigmentation, eczemoid and allergic dermatitis.[1,2,6]

Conjunctivitis, keratoconjunctivitis, corneal necrosis and ulceration have been reported. Pigment spots on the corneal and conjunctival epithelia accompany the pigmentation of the skin.[2,6] Brittle nails with Mee's lines (white transverse discoloration) are common. Arsenical dermatoses, epidermal carcinoma and lung, liver, bladder, larynx, lymphoid system, viscera and kidney cancers are associated risks of prolonged or repeated exposure to arsenic compounds.

Myocarditis has been reported following chronic poisoning. Pancytopaenia, aplastic anaemia or leukaemia have been reported following chronic exposure.

Management

Normal whole blood arsenic concentrations are $10 \,\mu g \,.\, l^{-1}$ ($0.13 \mu g \,.\, l^{-1}$). Arsenic is rapidly eliminated from the blood, and urine measurements are the preferred index of exposure. Normal urine arsenic concentrations are $< 10 \mu g$ over 24 hours ($< 0.13 \,\mu mol$ over 24 hours).

N.B. Diets high in seafood can lead to raised analytical results due to the presence of non-toxic organo-arsenic compounds.

Inhalation Management

- Maintain a clear airway, give humidified oxygen and ventilate if necessary

- If respiratory irritation occurs assess respiratory function and if necessary perform chest X-rays to check for chemical pneumonitis

- Consider the use of steroids to reduce the inflammatory response

- Treat pulmonary oedema with PEEP or CPAP ventilation

- For systemic management, see below

Dermal Management

- Remove any remaining contaminated clothing, place in double, sealed, clear bags and label; store in a secure area away from patients and staff

- Irrigate with copious amounts of water

- Skin burns should be treated as a thermal injury

- For systemic management, see below

Eye Management

- Irrigate thoroughly with running water or saline for 15 minutes

- Stain with fluorescein and refer to an ophthalmologist if there is any uptake of the stain

Oral Management

- Consider gastric lavage within 1 hour of ingestion unless profuse vomiting has already occurred

- Contact a poisons information service for further guidance on gut decontamination

- For systemic management, see below

Systemic Management

- Maintain fluid and electrolyte balance

- Monitor ECG and renal and hepatic functions

- Blood and serial urine arsenic concentrations must be measured

- Chelation therapy is required for urine concentrations $> 200 \,\mu g \,.\, l^{-1}$ (see below)

- Periodical follow-up is required following arsenic-induced systemic effects

Chelation Therapy

Contact a poisons information service for further guidance and paediatric doses.

- **DMPS** is the treatment of choice; dosage depends on severity of poisoning:

  - ORAL: Acute – initial dose of 100 to 300 mg then 100 mg every 2 hours for the first day; then 100 mg every 4 to 8 hours

  - Chronic – 100 mg orally three times a day

  - INTRAVENOUS: Acute – $5 \,mg \,.\, kg^{-1}$ 4 hourly for first 24 hours, thereafter intervals may be prolonged to 6, 8 or 12 hours according to the condition of the patient

  - Treatment should be continued until the urine concentration is $< 200 \,\mu g \,.\, l^{-1}$

- **Dimercaprol**

  - $3 \,mg \,.\, kg^{-1}$ intramuscularly (gluteal) every 4 hours day and night for the first 2 days; then every 12 hours for 7 days until recovery is complete or total urine concentration is $< 50 \,\mu g \,.\, l^{-1}$ in 24 hours

• Side effects: urticaria, burning sensation of lips, mouth and throat, fever, headache, conjunctivitis, hypotension, anxiety, muscle cramps; usually resolve with supportive measures or decreased doses

• **Penicillamine**: May also be used, usually in conjunction with dimercaprol because it has fewer side effects.

   • Do not use in penicillin-sensitive patients

   • Dosage: 500 mg orally every 6 hours

   • Penicillamine may be repeated after a 5 day rest period if symptoms recur or arsenic concentrations remain elevated

   • Side effects: fever, rashes, leucopaenia, eosinophilia and thrombocytopenia.

Summary of Environment Hazards

• Airborne concentrations of arsenic in urban areas may range from a few nanograms to a few tenths of a microgram per cubic metre; airborne arsenic is mainly inorganic[5]

• Arsenic is mainly transported into the environment by water

• Methylation of inorganic arsenic to methyl- and dimethylarsenic acids is associated with biological activity in water

• In oxygenated soil, inorganic arsenic is present in the pentavalent form; under reducing conditions it is in the trivalent form

• Drinking Water Standards:
   Arsenic: 50 μg.l^{-1} (UK max)
      0.01 mg.l^{-1} (WHO provisional guideline)

   Chloride: 400 mg.l^{-1} (UK max)
      250 mg.l^{-1} (WHO guideline)

• Soil Guidelines:
   Dutch Criteria: 29 mg.kg^{-1} (target)
      55 mg.kg^{-1} (intervention)

• Air Quality Standards:
   No safe level recommended due to carcinogenic properties (WHO guideline)

REFERENCES

1. Hathaway GJ, Proctor NH & Hughes JP. *Proctor and Hughes' Chemical Hazards of the Workplace*, 4th edn. Van Nostrand Reinhold, New York, 1996

2. Friberg L, Nordberg GF & Vouk VB. *Handbook on the Toxicology of Metals*. Elsevier, Amsterdam, 1989

3. Ellenhorn MJ, Schonwalds S, Ordog G & Wasserberger J. *Ellenhorn's Medical Toxicology – Diagnosis and Treatment of Human Poisoning*, 2nd edn. Williams & Wilkins, London, 1997

4. Hall AH & Rumack BH (eds.) *TOMES System®* Micromedex, Englewood, Colorado. CD ROM. vol 41 (exp. 31 July 1999)

5. International Programme on Chemical Safety. *Environmental Health Criteria 18: Arsenic*. WHO, Geneva 1981

6. Grant MW & Schuman JS. *Toxicology of the Eye*, 4th edn. Charles C Thomas, Springfield, 1993

Catherine Farrow

ARSINE

Key Points

- Arsine is a colourless, non-irritating gas with a mild garlic-like odour but poor olfactory warning properties
- It is absorbed by inhalation and is the most acutely toxic form of arsenic
- Arsine causes severe haemolysis and renal failure
- Symptoms of acute arsenic poisoning may also develop
- *In the event of a large spill, stay upwind and out of low areas. Ventilate closed spaces. Protective clothing, eye protection and breathing apparatus should be worn*

FIRST AID

- Terminate exposure and support vital functions
- The casualty should be moved to an uncontaminated area
- Rescuers should, ideally, be trained personnel and must be careful *not to put themselves at risk* and *so wear appropriate protective clothing and, if available, breathing apparatus*
- If the casualty is unconscious a clear airway should be established and maintained; give 100% oxygen if available
- **Inhalation Exposure:** If the patient stops breathing, expired air resuscitation should be started immediately using a pocket mask with a one way valve, if available. It is important where the face is contaminated that expired air resuscitation is NOT attempted unless an airway with rescuer protection is used
- **Dermal Exposure:** Remove contaminated clothing, if possible under a shower, and place in double, sealed, clear bags and label; store the bags in a secure area away from patients and staff
- Wash the skin thoroughly with copious amounts of water
- **Eye exposure:** Irrigate the eyes thoroughly with water or saline for 15 minutes

Detailed Information

- Arsine is a colourless, non-irritating gas with a mild garlic-like odour but poor olfactory warning properties
- *Common synonyms* arsenic hydride, arsine trihydride, arseniuretted hydrogen, arsenous hydride, hydrogen arsenide
- CAS 7784-42-1
- UN 2188
- NIOSH/RTECS CG 6475000
- Molecular formula AsH_3
- Molecular weight 77.95
- Arsine is used in the electronics industry in the manufacture of solid state electronic components
- It evolves from the addition of an acid to an arsenic compound and is a by-product of metal smelting
- Arsine is heavier than air and spreads at ground level
- It is slightly water soluble; aqueous solutions are neutral

- Arsine reacts violently with chlorine, with a risk of fire and explosion
- It decomposes when heated above 300°C giving off arsenic and arsenic oxides
- Arsine is extremely flammable; it can explode on contact with warm, dry air

Summary of Human Toxicity

- Arsine is absorbed by inhalation and is the most acutely toxic form of arsenic[1]
- It binds with oxidised haemoglobin, causing sudden and profound haemolysis[1,2]
- Renal failure occurs due to precipitation of haemolysis by-products in tubules and/or hypoxic damage caused by the reduced oxygen-carrying capacity of the blood[1]
- Haemolysis also effects myocardial microcirculation, liver, bone marrow, lungs and skeletal muscle[1]

Toxicity of Arsine [1,2,3]

Concentration	Clinical Effects
> 0.004 ppm	Increased arsenic urinary excretion
0.05 ppm	Potential toxicity
0.5 ppm	Odour threshold
3 ppm	Lowest toxic dose reported and immediate danger to life or health
10 to 50 ppm	Anaemia, extensive haemolysis and death after 30 minutes
100 ppm	Extensive haemolysis and death in less than 30 minutes
>150 ppm	Immediate death

- Occupational Exposure Standards:
  Long-term exposure limit: 0.05 ppm (0.16 mg . m^{-3})
  Short-term exposure limit: no data available

Acute Clinical Effects

Inhalation Effects

- Effects may be delayed in onset for 2 to 36 hours post exposure, but may occur 30 minutes post exposure depending on the severity and length of exposure[1,4]
- Initially, thirst, headache, malaise, weakness, dizziness and dyspnoea, then abdominal pain, nausea and vomiting, and possibly a garlic-like odour of the breath[4,5]
- Weakness proceeding to muscle cramps, and hypotension may develop[2]
- Dark red urine, haemoglobinuria, develops 4 to 6 hours post exposure[1,2]
- Progression to brown urine, frequently accompanied by jaundice, 24 to 48 hours post exposure[4,5]
- Tachycardia may occur and T-wave changes (elevation/lowering/inversion) are common[4]

- Persistent haemolysis may result in hyperkalaemia which may cause cardiac arrest

- Reticulocytosis and leukocytosis develop[1]

- Dilatation of the heart due to the effect of haemolysis on cardiac muscle may lead to pulmonary oedema prior to renal failure

- In severe cases, oliguria or anuria may develop within 3 days of the exposure

- Renal failure may result in death

- Some symptoms of arsenic poisoning may also occur; arsenic encephalopathy (agitation, extreme restlessness, memory loss and disorientation), and peripheral neuropathy developing over a few weeks (numbness of hands and feet, severe muscle weakness and photophobia)

- Urine arsenic concentration may remain elevated for several days or until normal renal function is restored

Dermal Effects

- No adverse health effects reported

Eye Effects

- Arsine is not irritant to the eyes

- Red staining of the conjunctiva and sclera has been described due to the blood pigment released by haemolysis as a result of systemic toxicity[6]

Oral Effects

- Not applicable

Chronic Clinical Effects

Chronic exposure to very small concentrations of arsine has cumulative effects. Severe anaemia causing exertional dyspnoea and weakness, in the absence of any other signs, was reported following chronic arsine exposure.[4,5] Marked basophilic stipling and elevated urine arsenic concentration were noted.

Arsine is oxidised *in vivo* to elemental trivalent arsenic and arsenic trioxide after inhalation, chronic (lifetime) exposure to which have been reported to cause cancer.

Management

Blood arsine concentration is not a good indicator of toxicity; arsine binds to oxidised haemoglobin and is oxidised to elemental trivalent arsenic and arsenic trioxide. Plasma and urine haemoglobin, and blood and urine arsenic concentrations are useful indicators of arsine toxicity.

Inhalation Management

- Maintain a clear airway, give humidified oxygen and ventilate if necessary

- Monitor plasma haemoglobin, white cell count, urine haemoglobin, renal function, and ECG

- Blood and urine arsenic concentrations may be checked but are of less value in severe cases

- If serum haemoglobin reaches 1.5 g.dl^{-1} or oliguria develops, exchange transfusion should be performed

- Haemodialysis should be performed if renal failure develops

- Symptomatic and supportive care

- Chelating agents are not effective for arsine exposure as arsine does not cause classic arsenic poisoning and no protection is given against haemolysis

Eye Management

- Irrigate thoroughly with running water or saline for 15 minutes

Oral Management

- Not applicable

Summary of Environmental Hazards

- In water arsine rapidly undergoes hydrolysis to arsenic acids and hydrides

- There is no data available concerning environmental toxicity

- Drinking Water Standards:
    Arsenic: 50 µg.l^{-1} (UK max)
        0.01 mg.l^{-1} (WHO provisional guideline)

- Soil Guidelines:
    Dutch Criteria: Arsenic: 29 mg.kg^{-1} (target)
        55 mg.kg^{-1} (intervention)

- Air Quality Standards: no data available

REFERENCES

1. Hathaway GJ, Proctor NH & Hughes JP. *Proctor and Hughes' Chemical Hazards of the Workplace*, 4th edn. Van Nostrand Reinhold, New York, 1996

2. Hall AH & Rumack BH (eds.) *TOMES System®* Micromedex, Englewood, Colorado. CD ROM. vol 41 (exp. 31 July 1999)

3. Ellenhorn MJ, Schonwalds S, Ordog G & Wasserberger J. *Ellenhorn's Medical Toxicology – Diagnosis and Treatment of Human Poisoning*, 2nd edn. Williams & Wilkins, London, 1997

4. Harbison RD (ed.) *Hamilton and Hardy's Industrial Toxicology*, 5th edn. Mosby-Year Book Inc., St Louis, 1998

5. Friberg L, Nordberg GF & Vouk VB. *Handbook on the Toxicology of Metals*. Elsevier, Amsterdam, 1989

6. Grant MW & Schuman JS. *Toxicology of the Eye*, 4th edn. Charles C Thomas, Springfield, 1993

Catherine Farrow

ASBESTOS

Key Points

- There are different forms of asbestos which differ in their cations; the fibres vary in size, colour and texture but are generally a white, brown or blue colour
- There have been no fatalities reported following acute exposure to asbestos
- The respiratory tract is the main target organ
- Occupational exposure to dust can result in mesothelioma, squamous cell carcinoma and adenocarcinoma of the lung, as well as fibrotic lung disease and asbestosis
- Toxicity usually manifests after a long latent period following chronic exposure

Detailed Information

Types of Asbestos

Chrysotile

- CAS 12001–29–5
- UN 2590
- NIOSH/RTECS CI 6478500
- Molecular formula $Mg_6(Si_4O_{10})(OH)_8$
- Colour: White
- Texture: Usually flexible, silky and tough

Anthophyllite

- CAS 77536–67–5
- UN 2212
- NIOSH/RTECS CI 6478000
- Molecular formula $(Mg,Fe)_7(Si_8O_{22})(OH)_2$
- Colour: White to grey, pale brown
- Texture: Usually brittle

Amosite

- CAS 12172–73–5
- UN 2212
- NIOSH/RTECS BT 6825000
- Molecular formula $Fe_5Mg_2(Si_8O_{22})(OH)_2$
- Colour: Light grey to pale brown
- Texture: Usually brittle

Actinolite

- CAS 77536–66–4
- UN 2590
- NIOSH/RTECS CI 6476000
- Molecular formula $Ca_2(Mg,Fe)_5(Si_8O_{22})(OH)_2$
- Colour: Pale to dark green

Tremolite

- CAS 77536–68–6
- UN 2590
- NIOSH/RTECS CI 656000
- Molecular formula $Ca_2Mg_5(Si_8O_{22})(OH)_2$
- Colour: White to grey
- Texture: Usually brittle

Crocidolite

- CAS 12001–28–4
- UN 2212
- NIOSH/RTECS CI 6479000
- Molecular formula $Na_2Fe_3^{2+}Fe_2^{3+}(Si_8O_{22})(OH)_2$
- Colour: Blue
- Texture: Flexible to brittle and tough

- The term 'asbestos' refers to either the naturally occurring minerals, or the industrial product obtained by mining and processing the minerals[1]
- Different forms of asbestos vary in their content of iron, magnesium, calcium, aluminium, sodium, and trace elements[1]
- *Common synonyms* Asbestos is a generic term applied to a number of hydrated mineral silicates[2,3]
- Asbestos has fibres of various sizes, colours and textures
- It is used for thermal and electrical insulation; fireproofing and cement products
- The use of asbestos in building materials has been discontinued in the USA and the UK since 1975[1]
- Significant environmental exposure can occur from friable asbestos in older buildings which are deteriorated, remodelled, or destroyed[4]
- Public buildings previously used asbestos as a fireproof insulation in ceiling or walls; when the structure is damaged asbestos fibres may be blown into the air space of the building
- Asbestos has high tensile strength and relative resistance to acid and temperature
- It does not evaporate, dissolve, burn, or undergo significant reactions with other chemicals; thus, it is non–biodegradable and is environmentally cumulative
- *Natural sources*: Asbestos is a naturally–occurring mineral and can be released into the air, drinking water, or soil by erosion or mining of deposits; this source usually accounts for a small fraction of total exposure, although some areas have large natural asbestos deposits; living near natural deposits does not appear to confer increased risk for cancer mortality

- *Anthropogenic sources*: Because of its widespread use, asbestos is ubiquitous in the environment; fibre levels > 5 μm in length at remote rural locations are generally below the detection limit (< 1 fibre .l⁻¹), while those in urban air range from < 1 to 10 fibres .l⁻¹ or occasionally more[1,2]

Summary of Human Toxicity

- The major route of exposure is inhalation of airborne particles

- Minor oral exposure occurs from ingestion of contaminated drinking water or after mucociliary clearance from the lungs but the role of ingested asbestos as an aetiological agent is uncertain

- The potential for toxicity is present if individuals are chronically exposed to asbestos fibres

- The greater the exposure the greater the risk of developing a related disease

- The dose–response for developing cancer from exposure to asbestos is not well understood for low exposures; it has been suggested that one large acute exposure may be sufficient to induce cancer which may occur many years later

- The risk of developing lung cancer is increased up to ten-fold if those exposed also smoke cigarettes[5]

- Fibres with a diameter > 3 μm are generally most respirable; particles with a very large or very small airborne diameter are preferentially deposited in the naso-pharynx from which they are cleared within a short time (hours to days)[6]

- Occupational exposure generally involves not only exposure to the highest concentrations of asbestos but also the greatest length of time; therefore most reference levels for exposure are based on the assumption of a normal working week over a period of an average working life, i.e. 20 years or more

- Levels of asbestos in air that lead to adverse health effects depend on the length of the exposure, the length of time since exposure first started, the presence of other risk factors (such as smoking) and the inherent fibre characteristics

- These levels are described as the permissible exposure level below which risk of developing disease is acceptably low

- Occupational Exposure Standards:

For any asbestos consisting of, or containing, one or more of the five amphibole asbestos minerals (amosite, crocidolite, fibrous actinolite, fibrous tremolite and fibrous anthophyllite, or any mixture of these with chrysotile):

Short-term exposure limit: 0.6 fibres .ml⁻¹ averaged over any continuous 10 min

Long-term exposure limit: 0.2 fibres .ml⁻¹ averaged over any 4 h

For chrysotile:

Short-term exposure limit: 1.5 fibres .ml⁻¹ averaged over any continuous 10 min

Long-term exposure limit: 0.5 fibres .ml⁻¹ averaged over any 4 h

Acute Clinical Effects

- Nearly all of the toxicological literature on asbestos is concerned with its activity as a human carcinogen from chronic exposures; asbestos is not acutely toxic, except for possible irritation of exposed areas due to mechanical action of the fibres

Chronic Clinical Effects

- Asbestos causes chronic lung disease (asbestosis), inflammation of the pleura, mesothelioma, squamous cell carcinoma and adenocarcinoma of the lung

- Asbestosis is a disorder characterised by a diffuse interstitial pulmonary fibrosis, at times including pleural changes of fibrosis and calcification

- In a study of insulation workers, 7% developed asbestosis caused by inhaling fibres less than about 3 μm in diameter and 200 μm in length; in these same workers 8% developed mesothelioma of the pleura and peritoneum, and 21% developed carcinoma of the lung[2]

- Mesothelial tumours may have a long latency period of up to 50 years, and may occur with exposures of less than one year[7]

- Besides increasing the risk of these cancers, other kinds of biological effects have been reported to be associated with occupational exposure to asbestos; rheumatoid factor and anti-nuclear antibodies (found in some autoimmune diseases) have also been found with asbestosis

- Diagnosis should be based on a history of known exposure and presence of diffuse interstitial pulmonary fibrosis, but when the exposure is difficult to ascertain diagnosis must be based on pathological findings

Public Health Risk

- Population exposure can result from a wide variety of activities or situations; damage to buildings containing asbestos is the most common source

- Earthquakes, fires, demolition or renovation work may all cause large quantities of asbestos fibres to be liberated in the air; this type of exposure may be intense and short-lived

- More prolonged lower grade exposure is possible when the source of contamination is slowly eroded, as in asbestos cement roofing or pipes, or worn asbestos insulated panelling

- Assessment of risk should be carried out for each individual incident

- The following questions are important in determining the extent of the public health risk:

  1. The condition of the asbestos containing materials, i.e. is it friable?

  2. Where is it located and what is the potential for human contact?

  3. How much is incorporated into the structure and in what form?

  4. What is the function of the building? (a risk evaluation for a school building would differ from that of a storage warehouse)

5. What provisions are there for asbestos monitoring and control?

Summary of Environmental Hazards

- Asbestos is considered to be non-biodegradable by aquatic organisms

- There is no evidence to show that asbestos bioaccumulates in aquatic organisms

- In soil it does not have an adsorptive affinity for the solids normally found in natural water systems

- Drinking Water Standards:
    No recommended health-based guideline value as not hazardous to human health at concentrations normally found in drinking water (WHO)

- Soil Guidelines: no data available

- Air Quality Standards:
    No safe level recommended because of carcinogenic properties (WHO guideline)

REFERENCES

1. Hall AH & Rumack BH (eds.) *TOMES System®* Micromedex, Englewood, Colorado. CD ROM. vol 41 (exp. 31 July 1999)

2. International Programme on Chemical Safety. *Environmental Health Criteria 53: Asbestos and Other Natural Mineral Fibres*. WHO, Geneva 1986

3. Sax NI. *Dangerous Properties of Industrial Materials*, 6th edn. Van Nostrand Reinhold, New York, 1984

4. Gaensler EA, 1992. Asbestos exposure in buildings. *Clin Chest Med*; 13(2): 231–42

5. Montizaan GK, Knaap AG & Van der Herijden CA, 1989. Asbestos: Toxicity and risk assessment for the general population in The Netherlands. *Food Chem Toxicol*; 27(1): 53–63

6. European series No 23. *Asbestos. Air quality guidelines for Europe*. WHO Regional Publications, 1987

7. Doll R & Peto J. *Asbestos: Effects on health of exposure to asbestos*. HSE, 1996

Henrietta Wheeler

BENZENE

Key Points

- Benzene is a clear, colourless, highly volatile, flammable liquid
- It is absorbed rapidly by ingestion and inhalation
- Exposure to high concentrations of benzene can cause death by respiratory depression or cardiac arrhythmias
- At concentrations of 20,000 ppm and upwards death can occur within 5–10 minutes
- Chronic exposure can result in bone marrow depression
- Benzene is a carcinogen in man
- *In the event of a large spill, stay upwind and out of low areas. Ventilate closed spaces. Protective clothing, eye protection and breathing apparatus should be worn*

FIRST AID

- Terminate exposure and support vital functions
- The casualty should be moved to an uncontaminated area
- Rescuers should, ideally, be trained personnel and must be careful *not to put themselves at risk* and *so wear appropriate protective clothing and, if available, breathing apparatus*
- If the casualty is unconscious a clear airway should be established and maintained; give 100% oxygen if available
- **Inhalation Exposure:** If the patient stops breathing, expired air resuscitation should be started immediately using a pocket mask with a one way valve, if available. It is important where the face is contaminated that expired air resuscitation is NOT attempted unless an airway with rescuer protection is used
- **Dermal Exposure:** Remove contaminated clothing, if possible under a shower and place in double, sealed, clear bags and label; store the bags in a secure area away from patients and staff
- Wash the skin thoroughly with copious amounts of water
- **Eye Exposure:** Irrigate thoroughly with water or saline for 15 minutes
- **Oral Exposure:** Encourage small quantities of oral fluids (no more than 50–100 ml in total)

Detailed Information

- Benzene is a naturally occurring, clear, colourless, highly volatile liquid which at low concentrations possesses a characteristic sweet odour but at high concentrations the odour is disagreeable and irritant[1]
- *Common synonyms* benzol, coal naphtha, cyclohexatriene, phenyl hydride, pyrobenzole
- CAS 71-43-2
- UN 1114
- NIOSH/RTECS CY 1400000
- Molecular formula C_6H_6
- Molecular weight 78.11

- Benzene is used as an intermediary in the production of styrene, phenol, cyclohexane, and other organic chemicals; in the manufacture of detergents, pesticides, solvents and paint removers, it is a component of petrol[2]
- Benzene is a naturally occurring chemical found in crude petroleum at levels of up to $4 \, g.l^{-1}$ and it is found in cigarette smoke[3]
- Commercial benzene has three standard grades usually containing various concentrations of toluene, xylene and phenol
- Benzene is slightly soluble in water and is incompatible with strong oxidisers, many fluorides, perchlorates and nitric acid
- Containers may explode when heated and form highly flammable and explosive mixtures with air

Summary of Human Toxicity

- Benzene is absorbed rapidly by ingestion and inhalation; only minimal absorption is expected from dermal exposure
- On inhalation, 12% of a dose is exhaled unchanged and 0.1% is excreted unchanged in the urine
- Due to its high lipid solubility, absorbed benzene is found in high concentrations in the body fat and brain; it is eventually converted by the liver into water soluble metabolites and excreted in the urine
- 51 to 87% is excreted as phenol in the urine, 6% as catechol and 2% as hydroquinone[4,5]
- In animals approximately 100% of benzene is absorbed after oral exposure
- Exposure to benzene causes CNS depression, coma and death
- Death is either due to the anaesthetic properties of benzene or through prolongation of fatal arrhythmias
- Oral exposures of 9–30 g have been fatal[3]
- Chronic exposure can cause depression of the haematopoietic system causing leukaemia

Toxicity of Benzene[1,3]

Concentration	Clinical effects
1–5 ppm	Odour threshold in air
25 ppm	No clinical effects after 8 h
50–150 ppm	Headache, lethargy and weakness after 5 h
500–1,500 ppm	Serious symptoms after 60 min
7,500 ppm	Dangerous symptoms after 30 min
>20,000 ppm	Fatal after 5–10 min

- Maximum Exposure Limits:
  Long-term exposure limit: 5 ppm ($16 \, mg.m^{-3}$)
  Short-term exposure limit: no data available

Acute Clinical Effects

Inhalation Effects

- The main signs of toxicity are CNS effects: dizziness, ataxia, drowsiness, headache, delirium, convulsions and coma

- Bronchial irritation, hoarseness, coughing and pulmonary oedema

- It has an anaesthetic effect, consisting of a preliminary stage of excitation followed by depression and, if exposure is continued, death through respiratory failure occurs, often associated with muscular twitching and convulsions

- Death may occur from ventricular arrhythmias as at high concentrations the solvent sensitises the myocardium to adrenaline and other catecholamines

- Recovery from the narcotic effects of benzene is usually rapid following cessation of the exposure

Dermal Exposure

- Benzene has a defatting action on the skin

- Direct contact with the liquid may cause erythema, blistering, and dermatitis

- Prolonged or repeated contact has been associated with the development of a dry scaly dermatitis or with secondary infections

- Absorption through the skin is slow and is unlikely to cause systemic toxicity

Eye Exposure

- Splash contact causes a moderate burning sensation with slight transient epithelial cell injury but recovery is rapid[6]

- Exposure to benzene vapour may cause a smarting sensation at high concentrations

Oral Exposure

- Burning sensation of the oral mucous membranes, oesophagus and stomach

- Nausea, abdominal pain and vomiting with a risk of aspiration

- At moderate concentrations symptoms are dizziness, excitation, and pallor, followed by flushing, weakness, headache, breathlessness, constriction of the chest and fear of impending death; visual disturbances and convulsions are common

- At higher concentrations, the clinical signs are excitement, euphoria, and tachycardia followed by drowsiness, coma and death

- Death may occur from ventricular arrhythmias as at high concentrations the solvent sensitises the myocardium to adrenaline and other catecholamines

- Recovery from the narcotic effects of benzene is usually rapid following cessation of the exposure

Chronic Clinical Effects

Benzene causes progressive degeneration of the bone marrow, aplastic anaemia, leukaemia and dysfunction of the immune system. In chronic exposure, workers exhibit signs of CNS lesions and impairment of hearing.

Benzene is a well established human carcinogen; occupational exposures have shown a relationship between exposure to benzene and production of myelogenous leukaemia. There may also be a relationship between benzene exposure and the production of lymphoma and multiple myeloma.[3]

Prolonged dermal exposure may produce lesions resembling first or second degree burns.

Management

Benzene concentrations can be measured in blood and exhaled air; metabolites can be measured in urine.

Inhalation Management

- Maintain a clear airway, give humidified oxygen and ventilate if necessary

- Control convulsions with diazepam

- If respiratory irritation occurs assess respiratory function and if necessary perform chest X-rays to check for chemical pneumonitis

- Patients should be kept at rest

- With severe exposure the patient should be kept on a cardiac monitor and all stimulants should be avoided except for resuscitation

- Treat pulmonary oedema with PEEP or CPAP ventilation

- Symptomatic and supportive care

Dermal Management

- Remove any remaining contaminated clothing, place in double, sealed, clear bags and label; store in a secure area away from patients and staff

- Irrigate the skin with water

- Emollients may ease irritation

- Symptomatic and supportive care

Eye Management

- Irrigate thoroughly with running water or saline for 15 minutes

- Stain with fluorescein and refer to an ophthalmologist if there is any uptake of the stain

Oral Management

- Encourage oral fluids

- DO NOT INDUCE EMESIS due to risk of aspiration

- Consider gastric lavage within 1 hour of a substantial ingestion, ensuring the airway is protected

- Contact a poisons information service for further guidance on gut decontamination

- If symptomatic institute cardiac monitoring. Do not give stimulants except for resuscitation

- In severe exposures perform chest X-rays to check for chemical pneumonitis

- Treat pulmonary oedema with PEEP or CPAP ventilation

- Symptomatic and supportive care

Summary of Environmental Hazards

- If benzene is released to the atmosphere, it will exist predominantly in the vapour phase

- Vapour-phase benzene will not be subject to direct photolysis but it will react with photochemically produced hydroxyl radicals with an approximate half-life of 13.4 days[5]

- The reaction time in polluted atmospheres which contain nitrogen oxides or sulphur dioxide is accelerated with the half-life being reported as 4–6 hours; products of photo-oxidation include phenol, nitrophenols, nitrobenzene, formic acid and peroxyacetyl nitrate

- In water benzene rapidly volatilises with an estimated half-life of 2.7 hours

- In soil benzene undergoes rapid volatilisation; it is not expected to be adsorbed by sediments

- Benzene does not bioconcentrate in the food chain

- Drinking Water Standards:
  hydrocarbon total: $10\,\mu g.l^{-1}$ (UK max)

  benzene: $10\,\mu g.l^{-1}$ (WHO guideline)

- Soil Guidelines: Dutch Criteria:

  $0.05\,mg.kg^{-1}$ (detection limit) (target)
  $1\,mg.kg^{-1}$ (intervention)

- Air Quality Standards:
  1 ppb averaging time 1 year (UK)
  No safe level recommended due to carcinogenic properties (WHO guideline)

REFERENCES

1. Clayton GD & Clayton FE (eds.) *Patty's Industrial hygiene and toxicology*, 4th edn. John Wiley & Sons, Inc., New York, 1994

2. Hathaway GJ, Proctor NH & Hughes JP. *Proctor and Hughes' Chemical Hazards of the Workplace*, 4th edn. Van Nostrand Reinhold, New York, 1996

3. International Programme on Chemical Safety. *Environmental Health Criteria 150: Benzene*. WHO, Geneva, 1993

4. Baselt RC & Cravey RH. *Disposition of Toxic Drugs and Chemicals in Man*, 4th edn. Chemical Toxicology Institute, 1995

5. Hall AH & Rumack BH (eds.) *TOMES System®* Micromedex, Englewood, Colorado. CD ROM. vol 41 (exp. 31 July 1999)

6. Grant MW & Schuman JS. *Toxicology of the Eye*, 4th edn. Charles C Thomas, Springfield, 1993

Nicola Scott

BROMINE

Key Points

- Bromine is a dark reddish-brown, volatile liquid with a suffocating odour which, at room temperature, gives off brown vapours; may also present as rhomboid-shaped crystals
- Liquid or vapours cause severe irritation and corrosive damage to the eyes, skin, mucous membranes and lungs
- It is similar to chlorine in biological effect but is reported to be a more potent respiratory irritant
- *In the event of a large spill, stay upwind and out of low areas. Ventilate closed spaces. Protective clothing, eye protection and breathing apparatus should be worn*

FIRST AID

- Terminate exposure and support vital functions
- The casualty should be moved to an uncontaminated area
- Rescuers should, ideally, be trained personnel and must be careful *not to put themselves at risk* and *so wear appropriate protective clothing and, if available, breathing apparatus*
- If the casualty is unconscious, a clear airway should be established and maintained; give 100% oxygen if available
- **Inhalation exposure:** If the patient stops breathing, expired air resuscitation should be started immediately using a pocket mask with a one way valve, if available. It is important where the face is contaminated that expired air resuscitation is NOT attempted unless an airway with rescuer protection is used
- **Dermal exposure:** Remove contaminated clothing, if possible under a shower, and place in double, sealed, clear bags and label; store the bags in a secure area away from patients and staff
- Wash the skin thoroughly with copious amounts of water
- **Eye exposure:** Irrigate thoroughly with water or saline for 15 minutes
- **Oral exposure:** Encourage small quantities of oral fluids (no more than 50–100 ml in total) unless perforation is suspected

Detailed Information

- Bromine is a dark reddish-brown, volatile liquid with a suffocating odour which, at room temperature, gives off brown vapours; may also present as rhomboid-shaped crystals
- CAS 7726-95-6
- UN 1744
- NIOSH/RTECS EF 9100000
- Molecular formula Br_2
- Molecular weight 159.81
- Bromine is used in water disinfection; anti-knock compounds for gasoline, fire retardants, bleaching of fibres and silk; in the manufacture of medicinal bromine compounds and dyestuffs
- The vapour is heavier than air

- Bromine is slightly soluble in water forming hydrobromic acid and oxygen; it is freely soluble in alcohol, ether, trichloromethane, carbon tetrachloride, carbon disulphide, concentrated hydrochloric acid and aqueous solutions of bromides
- It is incompatible with alkali hydroxides, arsenites, ferrous and mercurous salts, hypophosphites, other reducing agents and combustible organic materials
- Bromine is a halogen and a strong oxidising agent, less reactive than chlorine
- Contact with metals may evolve hydrogen gas with a risk of explosion
- Bromine is not combustible but may decompose on heating to produce toxic fumes (and corrosive fumes in the presence of water/steam); spontaneous reaction with reducing agents may cause fire or explosion

Summary of Human Toxicity

- Bromine liquid or vapours cause severe irritation and corrosive damage to the eyes, skin, mucous membranes and lungs
- Its biological effect is similar to chlorine but it is reported to be more toxic and a more potent respiratory irritant[1]
- The lowest published fatal doses in humans by oral and inhalation exposures are 14 mg.kg[-1] and 1000 ppm respectively[2]

Toxicity of Bromine[3,4]

Concentration	Clinical Effects
0.2 – 0.5 ppm	Eye irritation and lacrimation
10 ppm	Intolerable, severe irritation of the upper respiratory tract
40 – 60 ppm	Brief exposures are dangerous to life
1000 ppm	Choking, glottal and pulmonary oedema, rapid death

- Occupational Exposure Standards:
  Long-term exposure limit: 0.1 ppm (0.66 mg.m^{-3})
  Short-term exposure limit: 0.3 ppm (2.0 mg.m^{-3})

Acute Clinical Effects

Inhalation Exposure

- Dizziness, headache, epistaxis, coughing, bronchospasm, glottal and upper airway oedema may occur
- Abdominal pain and diarrhoea may develop several hours later
- Chemical pneumonitis and delayed pulmonary oedema and pneumonia may develop in severe cases
- Death may be delayed and has been associated with peribronchiolar abscesses due to deep tissue penetration and damage[3]

Dermal Exposure

- Measle-like eruptions and severe burns may develop following exposure to the vapour[5]

- Liquid causes initial cooling then burning sensation

- Brown discoloration, vesicle and pustule formation, leading to severe, deep surface burns with deep, slow-healing ulcers[3,6,7]

- Burns may be delayed up to 24 hours post exposure

Eye Exposure

- The vapour causes irritation, lacrimation and inflammation

- Higher concentrations may cause blepharospasm and photophobia

- Severe, painful burns may develop from contact with the liquid or vapour

Oral Exposure

- Brown discoloration of the lips, tongue and mucous membranes may occur

- Even dilute solutions may cause severe corrosive injury to the gastrointestinal mucosa, abdominal pain, haemorrhagic gastroenteritis leading to hypotension, shock and circulatory collapse and death[4]

- Haemorrhagic nephritis with oliguria or anuria may develop 1 to 2 days post ingestion secondary to shock or haemolysis

- If survived, oesophageal and pyloric stenosis may be sequelae

Chronic Clinical Effects

Chronic exposure to bromine has been associated with headache, joint pain, chest pain, loss of appetite, increasing irritability, loss of corneal reflexes, pharyngitis, vegetative disorders, hypertension and myocardial degeneration. Gastro-intestinal secretory disorders, leukopoiesis and leukocytosis inhibition and thyroid dysfunction have also been reported.[4]

Management

Blood bromine concentrations are not a good indicator of toxicity and should not be measured.

Inhalation Management

- Maintain a clear airway, give humidified 100% oxygen and ventilate if necessary

- If respiratory irritation occurs assess respiratory function and if necessary perform chest X-rays to check for chemical pneumonitis

- Consider the use of steroids to reduce the inflammatory response

- Treat pulmonary oedema with PEEP or CPAP ventilation

Dermal Management

- Remove any remaining contaminated clothing, place in double, sealed, clear bags and label; store in a secure area away from patients and staff

- Irrigate with copious amounts of water

- Skin burns should be treated symptomatically

Eye Management

- Irrigate thoroughly with running water or saline for 15 minutes

- Stain with fluorescein and refer to an ophthalmologist if there is any uptake of the stain

Oral Management

- NO GASTRIC LAVAGE OR EMETIC

- Encourage oral fluids, unless perforation is suspected

- Consider nasogastric aspiration of the stomach contents within 1 hour of a substantial ingestion

- Give plasma expanders/blood or IV fluids for shock and analgesics for pain

- Consider the use of steroids to reduce the inflammatory response

- Take abdominal X-ray to check for perforation

- If facilities are available, early gastro-oesophagoscopy should be undertaken within 12 to 24 hours of the event to assess the extent and severity of the injury

Summary of Environmental Hazards

- In the atmosphere bromine destroys ozone catalytically; the bromine cycle is believed to be more efficient at destroying ozone than the chlorine cycle

- In water it is slowly reduced to bromide by natural oxidisable materials

- Bromine is harmful to aquatic life in low concentrations

- Drinking Water Standards: no data available

- Soil Guidelines: no data available

- Air Quality Standards: no data available

REFERENCES

1. Harbison RD (ed.) *Hamilton and Hardy's Industrial Toxicology*, 5th edn. Mosby-Year Book Inc., St Louis, 1998

2. Sax NI. *Dangerous Properties of Industrial Materials*, 6th edn. Van Nostrand Reinhold, New York, 1984

3. Hathaway GJ, Proctor NH & Hughes JP. *Proctor and Hughes' Chemical Hazards of the Workplace*, 4th edn. Van Nostrand Reinhold, New York, 1996

4. Hall AH & Rumack BH (eds.) *TOMES System®* Micromedex, Englewood, Colorado. CD ROM. vol 41 (exp. 31 July 1999)

5. Budavari S, O'Neil MJ, Smith A, Heckelman PE & Kinneary JF (eds.) *The Merck Index*, 12th edn. Merck & Co., Inc., Whitehouse Station, 1996

6. Gosselin RE, Smith RP & Hodge HC. *Clinical Toxicology of Commercial Products*, 5th edn. Williams & Wilkins, Baltimore, 1984

7. Reynolds JEF (ed.) *Martindale The Extra Pharmacopoeia*, 31st edn. Royal Pharmaceutical Society, London, 1996

Catherine Farrow

BUTANE

FIRST AID

- Terminate exposure and support vital functions
- The casualty should be moved to an uncontaminated area
- Rescuers should, ideally, be trained personnel and must be careful *not to put themselves at risk* and *so wear appropriate protective clothing and, if available, breathing apparatus*
- If the casualty is unconscious a clear airway should be established and maintained; give 100% oxygen if available
- **Inhalation Exposure:** If the patient stops breathing, expired air resuscitation should be started immediately using a pocket mask with a one way valve, if available. It is important where the face is contaminated that expired air resuscitation is NOT attempted unless an airway with rescuer protection is used
- **Dermal Exposure:** If frostbite occurs *do not* remove clothing, flush skin with water
- *If frostbite is not present* remove contaminated clothing, if possible under a shower and place in double, sealed, clear bags and label; store the bags in a secure area away from patients and staff
- Wash the skin thoroughly with copious amounts of water
- **Eye Exposure:** If the eye tissue is frozen seek medical advice as soon as possible
- If eye tissue is not frozen, irrigate thoroughly with water or saline for 15 minutes
- **Oral Exposure:** Encourage small quantities of oral fluids (no more than 50–100 ml in total)

Detailed Information

- Butane is a colourless gas with a natural gas or gasoline-like odour
- *Common synonyms* butyl hydride, methylethylmethane, *n*-butane
- CAS 106-97-8
- UN 1011
- NIOSH/RTECS EJ 4200000
- Molecular formula C_4H_{10}
- Molecular weight 58.12

- Butane is mainly used in cigarette lighter refill canisters and as an aerosol propellant; raw material is used for synthetic rubber and high octane liquid fuels; manufacture of ethylene; as a solvent
- Commercial butane contains a mixture of propane, methane, ethane, *n*-butane and isobutane in varying proportions[1]
- Since 1989, butane has replaced CFCs as the propellant in approximately 90% of domestic aerosols
- Butane is slightly soluble in water and soluble in ethanol, ether and chloroform[2]
- It is highly flammable and produces carbon dioxide and carbon monoxide on combustion; contact with strong oxidising agents can increase the fire and explosion risk

Summary of Human Toxicity

- Butane is quickly absorbed through the lungs, absorbed by the fatty tissue then slowly released back into the blood stream which may be the cause of the acute, direct 'postponed' deaths
- 30–45% of butane is absorbed on inhalation[2]
- Butane is rapidly cleared from the body in exhaled air and is excreted largely unchanged[2]
- Elimination half-life is approximately 0.13 hour
- Spraying directly into the throat is the predominant method of abusing butane; other methods are *bagging* (inhaling from a plastic bag or crisp packet) and *huffing* (inhaling from a cloth or handkerchief)
- Four modes of death can be recognised: anoxia, vagal inhibition of the heart leading to bradycardia and cardiac arrest, respiratory depression and the initiation of cardiac dysrhythmias
- Butane is said to sensitise the myocardium to adrenaline and other endogenous catocholamines

Toxicity of Butane[3]

Concentration	Clinical effects
10,000 ppm	Drowsiness but no systemic effects after 10 min
2.9 – 14.6 mg.m^{-3}	Odour detection

- Occupational Exposure Standards:
  Long-term exposure limit: 600 ppm (1450 mg.m^{-3})
  Short-term exposure limit: 750 ppm (1810 mg.m^{-3})

Acute Clinical Effects

Inhalation Effects

- *Initial effects:* Euphoria, excitation, blurred vision, slurred speech, nausea, vomiting, coughing, sneezing, flushing, headache and increased salivation[2]
- *As dose increases:* Disinhibition, confusion, perceptual distortion, hallucinations (ecstatic or terrifying), delusions

(which may lead to aggressive or risk-taking behaviour), tinnitus, ataxia and apnoea[4]

- *Large doses:* Nystagmus, dysarthria, tachycardia, hypotension, convulsions, cyanosis, CNS depression, drowsiness, coma and death

- Inhalation of butane can 'sensitise' the myocardium to endogenous catecholamines which may cause cardiac arrhythmias and sudden death

- Myocardial infarction has been reported following butane abuse[5]

- Wheezing and coughing may lead to shortness of breath and a chemical or aspiration pneumonitis

Dermal Effects

- Liquid butane may cause burns or frostbite to the skin

- Dermal penetration by butane is not expected to any large extent, as skin contact would be transient because of volatility[6]

Eye Effects

- Butane gas is not irritating

- Liquid butane may cause burns or frostbite to the eyes

Oral Effects

- Butane is a gastric irritant leading to nausea, abdominal pain and vomiting

- Hypokalaemia has been observed after exposure to butane[2]

- Ingestion of butane can 'sensitise' the myocardium to endogenous catecholamines which may cause cardiac arrhythmias and sudden death

Chronic Clinical Effects

Chronic ingestion of butane may cause a numbing sensation in the mouth, abdominal pain and haemorrhagic oesophagitis.

Local chronic sequelae after chronic inhalation of butane includes recurrent epistaxis, chronic rhinitis, increased expectoration, conjunctivitis, halitosis and oral and nasal ulceration. Systemic toxicity of chronic inhalation of butane may result in anorexia, thirst, weight loss, and fatigue. Loss of concentration, depression, irritability, hostility and paranoia have been observed in chronic abusers of butane.[2]

An eczematous rash around the mouth and nose may occur following chronic inhalation of butane.

Management

Measurement of metabolites may be performed but has no clinical relevance other than confirming exposure when this is in doubt.

Inhalation Management

- Maintain a clear airway, give humidified oxygen and ventilate if necessary

- If respiratory irritation occurs assess respiratory function and if necessary perform chest X-rays to check for chemical pneumonitis

- Vagal inhibition of the heart leads to bradycardia or cardiac arrest, treat conventionally

- See systemic management below

Dermal Management

- *If frostbite has occurred:* remove clothing carefully, these may need to be soaked off with tepid water; irrigate the area

- Surgical referral may be necessary

- *If frostbite has not occurred:* remove any remaining contaminated clothing

- Irrigate with copious amounts of water

- Treat burns symptomatically

- Place any contaminated clothes in double, sealed, clear bags and label; store in a secure area away from patients and staff

Eye Management

- Irrigate thoroughly with running water or saline for 15 minutes

- Stain with fluorescein and refer to an ophthalmologist if there is any uptake of the stain

Oral Management

- Encourage oral fluids

- DO NOT INDUCE EMESIS due to risk of aspiration

- Consider gastric lavage (using a cuffed ET tube to ensure the airway is protected) within 1 hour of a substantial ingestion

- Contact a poisons information service for further guidance on gut decontamination

Systemic Management

- Recovery is normally quick once exposure has ceased but support of the respiratory and cardiovascular systems may be needed

- Patients should be kept at rest

- Diazepam may be used for convulsions

- With severe exposure the patient should be kept on a cardiac monitor for 4 hours, avoiding the use of all stimulants except for resuscitation

- Arrhythmias may respond to beta-blockers (e.g. atenolol)

- Respiratory arrest may require assisted ventilation

- Treat pulmonary oedema with PEEP or CPAP ventilation

- Symptomatic and supportive care

Summary of Environmental Hazards

- Butane undergoes rapid photo-oxidation in air

- Butane is not harmful to aquatic life; it does not bioconcentrate in the food chain[4]

- Drinking Water Standards:
    hydrocarbon total: $10 \, \mu g.l^{-1}$ (UK max)

- Soil Guidelines: no data available

- Air Quality Standards: no data available

REFERENCES

1. Ramsey J, Anderson HR, Bloor K & Flanagan RJ, 1989. An introduction to the practice, prevalence and chemical toxicology of volatile substance abuse. *Hum Toxicol;* 8 (4): 261–9

2. Flanagan RJ, Ruprah M, Meredith TJ & Ramsey JD, 1990. An introduction to the clinical toxicology of volatile substances. *Drug Saf;* 5(5): 359–83

3. Clayton GD & Clayton FE (eds.) *Patty's Industrial hygiene and toxicology*, 4th edn. John Wiley & Sons, Inc. New York, 1994

4. Hall AH & Rumack BH (eds.) *TOMES System®* Micromedex, Englewood, Colorado. CD ROM. vol 41 (exp. 31 July 1999)

5. Bauman JE, Dean BS & Krenzelok EP, 1991. Myocardial infarction and neurodevastation following butane inhalation (abs). *Vet Hum Toxicol*; 33(4):389

6. Hathaway GJ, Proctor NH & Hughes JP. *Proctor and Hughes' Chemical Hazards of the Workplace*, 4th edn. Van Nostrand Reinhold, New York, 1996

Henrietta Wheeler

CADMIUM

Key Points

- Cadmium is found as a silver-white, ductile, lustrous metal or as a grey, granular powder
- Inhalation and ingestion of cadmium pose the greatest risk of absorption leading to toxic effects
- Dermal absorption is rare but exposure may cause skin eruptions and pruritus
- Following acute inhalation the lung is the target organ
- Symptoms following acute inhalation range from nausea and diarrhoea to pulmonary oedema and death
- The main target organ following acute ingestion of cadmium is the gastrointestinal tract
- Chronic exposure may result in renal damage
- *In the event of a large spill, stay upwind and out of low areas. Ventilate closed spaces. Protective clothing, eye protection and breathing apparatus should be worn*

FIRST AID

- Terminate exposure and support vital functions
- The casualty should be moved to an uncontaminated area
- Rescuers should, ideally, be trained personnel and must be careful *not to put themselves at risk* and *so wear appropriate protective clothing and, if available, breathing apparatus*
- If the casualty is unconscious a clear airway should be established and maintained; give 100% oxygen if available
- **Inhalation Exposure:** If the patient stops breathing, expired air resuscitation should be started immediately using a pocket mask with a one way valve, if available. It is important where the face is contaminated that expired air resuscitation is NOT attempted unless an airway with rescuer protection is used
- **Dermal Exposure:** Remove contaminated clothing, if possible under a shower and place in double, sealed, clear bags and label; store the bags in a secure area and away from patients and staff
- Wash the skin thoroughly with copious amounts of water
- **Eye exposure:** Irrigate the eyes thoroughly with water or saline for 15 minutes
- **Oral exposure:** Encourage small quantities of oral fluids (no more than 50–100 ml in total)

Detailed Information

- Cadmium is found as a silver-white, ductile, lustrous metal which can easily be cut with a knife; or as a grey, granular powder
- *Common synonym* colloidal cadmium
- CAS 7440-43-9
- UN number not available
- NIOSH/RTECS EU 9800000
- Atomic symbol Cd
- Atomic weight 112.41

- Cadmium is used in fungicides, electroplating, addition to tinning solutions, dyeing, printing and photocopying; in the manufacture of electronic vacuum tubes, special mirrors, as a chemical intermediate for cadmium sulphide and in pyrotechnics
- It is soluble in acids, especially nitric acid and in ammonium nitrate solutions
- Cadmium can ignite spontaneously in air; it is flammable and explosive when exposed to heat or flame
- Cadmium reacts violently with acids giving off hydrogen, and cadmium powder reacts with strong oxidants

Summary of Human Toxicity

- Between 10 and 50% of inhaled cadmium is absorbed, the absorption being greater for smaller particles and fumes than for large particle dust; absorption through the skin is negligible[1]
- The half-life of cadmium in the human body is thought to be around 30 years and it has no known biological function

Cadmium Concentrations

Normal concentration		Hazardous concentration	
Blood	$<2\,\mu g \cdot l^{-1}$ (Non smoker)	Blood	$>20\,\mu g \cdot l^{-1}$
	$<5\,\mu g \cdot l^{-1}$ (Smoker)		
Urine	$<2\,\mu g \cdot l^{-1}$ (Non smoker)	Urine	$>20\,\mu g \cdot l^{-1}$
	$<5\,\mu g \cdot l^{-1}$ (Smoker)		

- Elevated blood concentration indicates exposure within recent past to 18 months prior to sample
- Elevated urine concentration indicates exposure in past 10 years (Note: cadmium concentrations in smokers are elevated due to cadmium absorption from cigarette smoke)
- Exposure to cadmium produces a wide variety of effects involving many organs and biological systems[2]
- Inhalation damages the lungs with the respiratory distress due to pneumonitis and oedema; symptoms are often confused with metal fume fever
- Food poisoning sometimes occurs when acidic foods and liquids are consumed after being stored in containers coated in cadmium glazes; rapid onset of symptoms occurs, mainly gastrointestinal irritation
- The lowest reported lethal concentration by inhalation is 39 $mg \cdot m^{-3}$ for 20 minutes; inhalation of 5 $mg \cdot m^{-3}$ for 8 hours was also fatal[1,2]
- 0.01–0.15 $mg \cdot m^{-3}$ for 9 hours lead to acute gastro-enteritis and 0.50–2.50 $mg \cdot m^{-3}$ for 3 days to pneumonitis
- Ingestion of 150g has been reported as fatal due to either shock due to fluid loss, or acute renal failure[1]
- Ingestion of 15 $mg \cdot l^{-1}$ of cadmium in water resulted in vomiting[1]

- Maximum Exposure Limits: Cadmium and cadmium compounds except cadmium oxide fume, cadmium sulphide and cadmium sulphide pigments:
  - Long-term exposure limit: 0.025 mg.m^{-3}
  - Short-term exposure limit: no data available

Acute Clinical Effects

Inhalation Effects

- Inhalation may cause hypersalivation, a metallic taste, cough, dyspnoea and chest pain[2]

- Pneumonitis and pulmonary oedema may develop over one to four days, with death due to respiratory failure in severe cases[3]

Dermal Effects

- Due to the very low level of absorption of cadmium through the skin, toxic effects are rare but mild irritation may occur

Eye Effects

- Eye discomfort and excessive lacrimation may follow cadmium oxide fume exposure in humans, generally without actual injury[4]

Oral Effects

- Irritation of the gastrointestinal tract usually occurs after small doses; recovery is usually complete

- Larger doses may affect calcium and zinc metabolism

- Other disorders include pulmonary and facial oedema; death may result from shock, fluid loss or renal failure

- Cadmium is a known carcinogen

Chronic Clinical Effects

Patients who inhale cadmium fumes over a long period of time can develop emphysema and are at risk of developing lung cancer. Chronic inhalation may also lead to proteinuria and kidney damage and stones.[1] Osteomalacia, yellow staining of the teeth, anaemia, anosmia and prostate cancer have also been reported following chronic exposure.[5]

Cadmium is known to cause Itai–Itai (ouch–ouch) disease – a bone disease with kidney malfunction.

Management

Blood and urine cadmium concentrations may be determined.

Normal blood cadmium concentrations:
  <27 nmol.l^{-1} (<3 µg.l^{-1}), non-smokers
  <54 nmol.l^{-1} (6 µg.l^{-1}), smokers

Hazardous blood cadmium concentrations:
  >180 nmol.l^{-1} (>20 µg.l^{-1})

Normal urine cadmium concentrations:
  0.4–1.3 nmol/mmol creatinine

Hazardous urine cadmium concentrations:
  >10 nmol/mmol creatinine

Inhalation Management

- If respiratory irritation occurs assess respiratory function and if necessary perform chest X-rays to check for chemical pneumonitis

- Consider the use of steroids to reduce the inflammatory response

- Treat pulmonary oedema with PEEP or CPAP ventilation

- Monitor blood cadmium concentrations in severe cases

- Chelation (with sodium calcium EDTA) may be effective if used within the first 24 hours post exposure

- Symptomatic and supportive care

Dermal Management

- Remove any remaining contaminated clothing, place in double sealed, clear bags and label; store in a secure area away from patients and staff

- Irrigate with copious amounts of water

- Treat symptomatically

Eye Management

- Irrigate thoroughly with running water or saline for 15 minutes

- Stain with fluorescein and refer to an ophthalmologist if there is any uptake of the stain

Oral Management

- Encourage oral fluids

- Although not demonstrably efficacious, chelation therapy (with sodium calcium EDTA) together with gastric lavage may be of some benefit within 1 hour of ingestion

- Contact a poisons information service for further guidance on gut decontamination

- Monitor blood cadmium concentrations in severe cases

- Symptomatic and supportive care

- Charcoal haemoperfusion is ineffective

Summary of Environmental Hazards

- Significant compartments for soluble cadmium salts are likely to be water and soil

- Cadmium (II) is known to adsorb to sediments

- Once in the aquatic environment, cadmium is highly mobile; its dissolved species are extremely labile, and are the first to be released, e.g., when salinity increases in estuaries; in fresh water, cadmium toxicity is influenced by water hardness (the harder the water, the lower the toxicity)

- Cadmium can bioaccumulate and does bioconcentrate in the food chain

- Biomagnification in terrestrial food chains is not expected[1]

- Drinking Water Standards:
  - 5 µg.l^{-1} (UK max)
  - 3 µg.l^{-1} (WHO guideline)

- Soil Guidelines: Dutch Criteria:
  - 0.8 mg.kg^{-1} (target)
  - 12 mg.kg^{-1} (intervention)

- Air Quality Standards:
    < 1–5 ng.m⁻³, rural areas;
    10–20 ng.m⁻³, urban and industrial areas (WHO guidelines)

REFERENCES

1. Hall AH & Rumack BH (eds.) *TOMES System®* Micromedex, Englewood, Colorado. CD ROM. vol 41 (exp. 31 July 1999)

2. Friberg L, Nordberg GF & Vouk VB. *Handbook on the Toxicology of Metals*. vol II. Elsevier, Amsterdam, 1986

3. Barnhart S & Rosenstock L, 1984. Cadmium chemical pneumonitis. *Chest;* 86: 789

4. Rusch G, O'Grodnick JS & Rinehert WE, 1986. Acute inhalation study in the rat and comparative uptake, distribution and excretion for different cadmium containing materials. *Am Ind Hyg Assoc J*; 47(12): 754–63

5. Goldfrank LR, Flomenbaum NE, Lewin NA, Weisman RS, Howland MA & Hoffman RS. *Goldfrank's Toxicologic Emergencies*, 5th edn. Appleton & Lange, New York, 1994

Marie Pickford

CALCIUM PEROXIDE

Key Points

- Calcium peroxide is a white or yellowish, odourless powder which is almost tasteless
- It is a strong oxidiser
- Calcium peroxide powder is a respiratory and dermal irritant; when in contact with hot water or acid it decomposes to hydrogen peroxide
- *The following information is mainly about hydrogen peroxide except where stated*
- Ingestion of hydrogen peroxide causes gastric irritation and occasionally gas embolism
- Hydrogen peroxide is irritating to the skin and respiratory tract
- *In the event of a large spill, stay upwind and out of low areas. Ventilate closed spaces. Protective clothing, eye protection and breathing apparatus should be worn*

FIRST AID

- Terminate exposure and support vital functions
- The casualty should be moved to an uncontaminated area
- Rescuers should, ideally, be trained personnel and must be careful *not to put themselves at risk* and *so wear appropriate protective clothing and, if available, breathing apparatus*
- If the casualty is unconscious a clear airway should be established and maintained; give 100% oxygen if available
- **Inhalation Exposure:** If the patient stops breathing, expired air resuscitation should be started immediately using a pocket mask with a one way valve, if available. It is important where the face is contaminated that expired air resuscitation is NOT attempted unless an airway with rescuer protection is used
- **Dermal Exposure:** Remove contaminated clothing, if possible under a shower and place in double, sealed, clear bags and label; store the bags in a secure area and away from patients and staff
- Wash the skin thoroughly with copious amounts of water
- **Eye exposure:** Irrigate the eyes thoroughly with water or saline for 15 minutes
- **Oral exposure:** Encourage small quantities of oral fluids (no more than 50–100 ml in total)

Detailed Information

Calcium peroxide

- Calcium peroxide is a white or yellowish, odourless powder which is almost tasteless
- *Common synonyms* calcium superoxide, calcium dioxide
- CAS 1305-79-9
- UN 1457
- NIOSH/RTECS EW 3865000
- Molecular formula CaO_2
- Molecular weight 72.08

- Calcium peroxide is used as a rubber stabiliser, seed disinfectant, antiseptic, and in bleaching of oils, modification of starches and high temperature oxidation
- It sinks in cold water; it is soluble in acid with the formation of hydrogen peroxide
- Calcium peroxide is a strong oxidation agent, it is a dangerous fire risk when in contact with organic materials

Summary of Human Toxicity

Calcium peroxide

- Calcium peroxide is an irritant to the mucous membranes, respiratory tract and skin in high concentrations
- Occupational Exposure Standards: no data available

Hydrogen peroxide

- Hydrogen peroxide acts on exposed tissues by releasing oxygen; for each volume of 3% solution, 10 volumes of oxygen may be produced[1]
- Exposure to 3%, household strengths, of hydrogen peroxide by ingestion, dermally or in the eye does not normally give rise to toxic effects
- Inhalation of 90% hydrogen peroxide causes nasal irritation, increased saliva, a scratchy feeling of the throat and respiratory tract inflammation[2]
- Solutions of >10% on the skin may cause burns
- Ingestion is the main route of exposure and causes gastrointestinal irritation
- There is also a risk of gas embolism which although not common from ingestion has occurred after the use of hydrogen peroxide for irrigation of body cavities
- Several deaths from ingestion are reported in the literature; in most cases the exposures were to concentrated solutions of 30–40%. One case reported a 49-year-old female ingesting 240 ml of a 35% solution; she died 78 hours later[3]
- Cerebral infarction, believed to have resulted from gas embolization of the cerebral vasculature, has been reported in an 84-year-old man who took 30 ml of 35% hydrogen peroxide diluted in 100–300 ml of water[4]
- Multiple brain emboli occurred in a 63-year-old who ingested 120 ml of 35% solution; he recovered[5]
- Occupational Exposure Standards:
  Long-term exposure limit: 1 ppm (1.4 mg.m^{-3})
  Short-term exposure limit: 2 ppm (2.8 mg.m^{-3})

Acute Clinical Effects

Inhalation Effects

Calcium peroxide

- Respiratory tract irritant with coughing, wheezing, shortness of breath and potential chemical pneumonitis

Hydrogen peroxide

- Transient dyspnoea and cough; with concentrated solutions there may be more severe irritation and inflammation of the respiratory tract

Dermal Effects

Calcium peroxide

- Skin irritation, blistering and burns may occur

Hydrogen peroxide

- Skin irritation with paraesthesia, blistering and whitening; solutions >10% may cause burns

Eye Effects

Calcium peroxide

- Irritating to eyes

Hydrogen peroxide

- Irritation with a burning sensation, conjunctival hyperaemia, lacrimation and severe pain which should resolve in a few hours, but with more concentrated solutions resolution may by up to 24 hours

- There are rare cases of temporary corneal injury resulting from application of 3% solution to the eye[6]

Oral Effects

Calcium peroxide

- Irritant to gastrointestinal tract; nausea and vomiting may be experienced

Hydrogen peroxide

- Irritation of the gastrointestinal tract, the severity of which depends on the concentration of the solution

- Vomiting is common, and the vomitus may be frothy due to the liberation of oxygen (risk of aspiration)

- Haematemesis and gastric distension, due to the release of oxygen, may occur

- Lethargy, coma, convulsions and respiratory arrest have been reported[7]

- Gas embolism has been reported in several cases of ingestion of hydrogen peroxide

- In severe cases ischaemic ECG changes and EMD (electromechanical dissociation) may be seen because of embolisation of the heart, restricting blood flow[8]

- Cerebral infarction and multiple brain embolism may occur in severe cases

Chronic Clinical Effects

Hydrogen peroxide

Chronic exposure to an aerosol of hydrogen peroxide in a concentration of 41 mEq.m^{-3} (1 ppm = 1.41 mg.m^{-3}) resulted in chronic diffuse interstitial lung disease.[9] Although no cases have been reported, chronic ingestion could cause gastritis.

Management

For both hydrogen and calcium peroxide

Inhalation Management

- Maintain a clear airway, give humidified oxygen and ventilate if necessary

- If respiratory irritation occurs assess respiratory function and if necessary perform chest X-rays to check for chemical pneumonitis

- Patients should be kept at rest

- Treat pulmonary oedema with PEEP or CPAP ventilation

- Symptomatic and supportive care

Dermal Management

- Remove any remaining contaminated clothing, place in double, sealed, clear bags and label; store in a secure area away from patients and staff

- Irrigate with copious amounts of water

- Treat irritation symptomatically

- Bleaching of the skin usually resolves within a few hours

Eye Management

- Irrigate thoroughly with running water or saline for 15 minutes

- Stain with fluorescein and refer to an ophthalmologist if there is any uptake of the stain

Oral Management

- Gastric decontamination is not worthwhile for ingestion of hydrogen peroxide due to its rapid dissociation.

- Encourage oral fluids unless there is evidence of severe injury

- If gastric distension is severe a fine bore gastric tube may be passed to aid the release of gas

- In cases with severe clinical effects patients should be X-rayed to check for the presence of gas emboli

- If facilities are available endoscopic evaluation should be undertaken within 12 hours of the event in any patient with haematemesis, persistent vomiting or other evidence of gastric burns to assess the extent and severity of the injury

- Monitor ECG in severe cases

- Hyperbaric oxygen therapy has been suggested for patients with evidence of cerebral embolism due to hydrogen peroxide[4]

- Contact a poisons information service for further guidance

Summary of Environmental Hazards

Calcium peroxide

- The effects of low concentrations of calcium peroxide on aquatic life is unknown

- It may be dangerous if it enters water supplies

Hydrogen peroxide

- Gaseous hydrogen peroxide is a common air contaminant; it may concentrate in cloud water

- Persistence is unlikely because of the strong oxidising activity of this chemical

- Hydrogen peroxide does not concentrate in the food chain

- It decomposes to water and oxygen

- Drinking Water Standards: no data available

- Soil Guidelines: no data available

- Air Quality Standards: no data available

REFERENCES

1. Gosselin RE, Smith RP & Hodge HC. *Clinical Toxicology of Commercial Products*, 5th edn. Williams & Wilkins, Baltimore, 1984

2. Oberst FW, Comstock CC & Hackley EB, 1954. Inhalation toxicity of ninety percent hydrogen peroxide vapor. *Indust Hyg Occup Med*; 10: 319–27

3. Litovitz TL, Felberg L, Soloway RA, Ford M & Geller R, 1995. 1994 Annual report of the AAPCC toxic exposure surveillance system. *Am J Emerg Med*; 13(5): 551–97

4. Sherman SJ, Boyer LV & Sibley WA, 1994. Cerebral infarction immediately after ingestion of hydrogen peroxide. *Stroke;* 25(5): 1065–7

5. Ijichi I, Itoh T, Sakai R, Nakaji K, Miyauchi T, Takahashi R, Kadosaka S, Hirata M, Yoneda S, Kajita Y & Fujita Y, 1997. Multiple brain embolism after ingestion of concentrated hydrogen peroxide. *Neurology*; 48(1): 277–9

6. Grant MW & Schuman JS. *Toxicology of the Eye*, 4th edn. Charles C Thomas, Springfield, 1993

7. Giberson TP, Kern JD, Pettigrew DW, Eaves CC & Haynes JF, 1989. Near-fatal hydrogen peroxide ingestion. *Ann Emerg Med*; 18(7): 778–9

8. Christensen DW, Faught WE, Black RE, Woodward GA & Timmons OD, 1992. Fatal oxygen embolization after hydrogen peroxide ingestion. *Crit Care Med*; 20(4): 543–4

9. Ellenhorn MJ, Schonwalds S, Ordog G & Wasserberger J. *Ellenhorn's Medical Toxicology - Diagnosis and Treatment of Human Poisoning*, 2nd edn. Williams & Wilkins, London, 1997

Henrietta Wheeler

CARBAMATES

Key Points

- Carbamates are esters of methyl carbamic acid
- Carbamates are absorbed by all routes of exposure
- They cause reversible cholinesterase inhibition resulting in cholinergic effects similar to those caused by organophosphates
- Carbamates are less dangerous than organophosphates due to the rapid reversibility of the acetylcholinesterase inhibition
- They are commonly formulated in organic hydrocarbon solvents
- *In the event of a large spill, stay upwind and out of low areas. Ventilate closed spaces. Protective clothing, eye protection and breathing apparatus should be worn*

First Aid

- Terminate exposure and support vital functions
- The casualty should be moved to an uncontaminated area
- Rescuers should, ideally, be trained personnel and must be careful *not to put themselves at risk* and *so wear appropriate protective clothing and, if available, breathing apparatus*
- If the casualty is unconscious a clear airway should be established and maintained; give 100% oxygen if available
- **Inhalation Exposure:** If the patient stops breathing, expired air resuscitation should be started immediately using a pocket mask with a one way valve, if available. It is important where the face is contaminated that expired air resuscitation is NOT attempted unless an airway with rescuer protection is used
- **Dermal Exposure:** Remove contaminated clothing, if possible under a shower and place in double, sealed, clear bags and label; store the bags in a secure area and away from patients and staff
- Wash the skin thoroughly with copious amounts of water
- **Eye exposure:** Irrigate the eyes thoroughly with water or saline for 15 minutes
- **Oral exposure:** Encourage small quantities of oral fluids (no more than 50–100 ml in total)

Detailed Information

Carbamates and Common Synonyms

Compound	Synonyms and Trade Names
Aldicarb	2-methyl-2-(methylthio)propanal O-[(methylamino)carbonyl]oxime; 2-methyl-2-(methylthio)propionaldehyde O-(methylcarbamoyl)oxime; aldecarb; carbamyl; carbanolate; sulfone aldoxycarb; Temik
Aminocarb	4-(Dimethylamino)-3-methylphenolmethylcarbamate (ester); methylcarbamic acid 4-(dimethylamino)-m-tolyl ester; 4-dimethylamino-m-tolyl methylcarbamate

Compound	Synonyms and Trade Names
Bendiocarb	2,2-Dimethyl-1,3-benzodioxol-4-ol methylcarbamate; methylcarbamic acid 2,3-(isopropylidenedioxy)phenyl ester; Ficam
Bufencarb	3-(1-Ethylpropyl)phenol methylcarbamate mixture with 3-(1-methylbutyl)phenyl methylcarbamate (1:3); methyl carbamic acid m-(1-ethylpropyl)phenyl ester mixture with m-(1-methylbutyl)phenyl ester; metalkamate
Carbaryl	1-Naphthalenol methylcarbamate; methyl carbamic acid 1-naphthyl ester; 1-naphthyl N-methylcarbamate; Arylam; Carylderm; Sevin
Carbofuran	2,3-dihydro-2,2-dimethyl-7-benzofuranol methylcarbamate; methyl carbamic acid 2,3-dihydro-2,2-dimethyl-7-benzofuranyl ester; 2,2-dimethyl-2,3-dihydro-7-benzofuranyl-N-methylcarbamate; 2,2-dimethyl-7-coumaranyl N-methylcarbamate
Dimetan	dimethylcarbamic acid 5,5-dimethyl-3-oxo-1-cyclohexen-1-yl ester; 5,5-dimethyldihydroresorcinol dimethylcarbamate
Dimetilan	dimethylcarbamic acid 1-[(dimethylamino)carbonyl]-5-methyl-1H-pyrazol-3-yl ester; dimethylcarbamic acid ester with 3-hydroxy-N,N,5-trimethylpyrazole-1-carboxamide; 2-dimethylcarbamoyl-3-methyl-5-pyrazolyl dimethylcarbamate
Methiocarb	3,5-dimethyl-4-(methylthio)phenol methylcarbamate; methylcarbamic acid 4-(methylthio)-3,5-xylyl ester; 4-(methylthio)-3,5-xylyl methylcarbamate; 4-methylthio-3,5-dimethylphenyl N-methylcarbamate; mercaptodimethur; metmercapturon
Methomyl	N-[[(methylamino)carbonyl]oxy]-ethanimidothioic acid methyl ester; N-[(methylcarbamoyl)-oxy]thioacetimidic acid methyl ester; S-methyl N-[(methylcarbamoyl)oxy]thioacetimidate; methyl O-(methylcarbamoyl)thiolacetohydroxamate
Oxamyl	2-(dimethylamino)-N-[[(methylamino)-carbonyl]oxy]-2-oxoethanimidothioic acid methyl ester; N',N'-dimethyl-N-[(methylcarbamoyl)oxy]-1-thiooxamimidic acid methyl ester; N,N-dimethyl-α-methylcarbamoyloxyimino-α-(methylthio)acetamide; methyl 1-(dimethylcarbamoyl)-N-(methylcarbamoyloxy)thioformimidate; thioxamyl
Pirimicarb	Dimethylcarbamic acid 2-(dimethylamino)-5,6-dimethyl-4-pyrimidinyl ester; 2-(dimethylamino)-5,6-dimethyl-4-pyrimidinyl dimethylcarbamate; 5,6-dimethyl-2-dimethylamino-4-dimethylcarbamoyloxypyrimidine
Propoxur	2-(1-Methylethoxy)phenol methylcarbamate; o-isopropoxyphenyl N-methylcarbamate; aprocarb; Bay 39007; Bay 9010

Identification Numbers of Carbamates

Compound	CAS	UN	NIOSH/RTECS
Aldicarb	116–06–3	n.a.	UE 2275000
Aminocarb	2032–59–9	n.a.	FC 0175000
Bendiocarb	22781–23–3	n.a.	FC 1140000
Bufencarb	2282–34–0	n.a.	FC 3500000
Carbaryl	63–25–2	2757	FC 5950000
Carbofuran	1563–66–2	2757	FB 9450000
Dimetan	112–15–6	n.a.	FA 1500000
Dimetilan	644–64–4	n.a.	EZ 9084000
Methiocarb	2032–65–7	n.a.	FC 5775000
Methomyl	16752–77–5	n.a.	AK 2975000
Oxamyl	23135–22–0	n.a.	RP 2300000
Pirimicarb	23103–98–2	n.a.	EZ 9100000
Propoxur	114–26–1	n.a.	FC 3150000

n.a. = specific number not available

Molecular Formula and Molecular Weight of Carbamates

Compound	Molecular formula	Molecular weight
Aldicarb	$C_7H_{14}N_2O_2S$	190.27
Aminocarb	$C_{11}H_{16}N_2O_2$	208.26
Bendiocarb	$C_{11}H_{13}NO_4$	223.23
Bufencarb	$C_{13}H_{19}NO_2$	221.30
Carbaryl	$C_{12}H_{11}NO_2$	201.22
Carbofuran	$C_{12}H_{15}NO_3$	221.26
Dimetan	$C_{11}H_{17}NO_3$	211.26
Dimetilan	$C_{10}H_{16}N_4O_3$	240.26
Methiocarb	$C_{11}H_{15}NO_2S$	225.31
Methomyl	$C_5H_{10}N_2O_2S$	162.21
Oxamyl	$C_7H_{13}N_3O_3S$	219.26
Pirimicarb	$C_{11}H_{18}N_4O_2$	238.29
Propoxur	$C_{11}H_{15}NO_3$	209.25

Chemical Properties and Appearance of Carbamates

Compound	Appearance	Properties
Aldicarb	White crystalline solid with a slight sulphurous odour	Highly water soluble
Aminocarb	Crystals	Slightly soluble in water
Bendiocarb	White solid	
Bufencarb	Yellow-amber solid	Stable in neutral or acidic solutions; increases in pH or temperature increases rate of hydrolysis
Carbaryl	White or greyish odourless crystals	Stable to light, heat, and acids; hydrolysed in alkalis; non-corrosive
Carbofuran	White, odourless crystalline solid	Unstable in alkaline media
Dimetan	Crystals	
Dimetilan	Colourless solid or yellow to reddish-brown solid	Hydrolysed by acid and alkali
Methiocarb	White, crystalline powder with a mild odour	Unstable in alkaline media
Methomyl	White, crystalline solid with a slight sulphurous odour	
Oxamyl	Crystalline solid, slight sulphurous odour	
Pirimicarb	Colourless crystalline solid	Soluble in most organic solvents. Decomposes by prolonged boiling wih acids or alkali. Aqueous solutions are unstable to light
Propoxur	White, minute crystals/powder with a faint odour	Decomposes at high temperature forming methyl isocyanate; unstable in alkaline media

- Carbamates are esters of methyl carbamic acid

- They are used as systemic and contact insecticides, acaricides, nematocides, ectoparaciticides, aphicides, molluscicides and bird repellents, in the soil, on crops, and mammals

Summary of Human Toxicity

- Most carbamates are absorbed from all routes of exposure[1]

- They reversibly inhibit acetylcholinesterase to varying degrees resulting in the accumulation of acetylcholine at nerve synapses

- This leads to overstimulation and subsequent suppression of normal nerve impulse transmission throughout the body[2,3]

- Clinical effects may develop within minutes of exposure and last for several hours[1]

- Onset and severity of effects is dose-related and influenced by the individual carbamate, the vehicle and route of exposure[4]

- Once absorbed, carbamates are distributed rapidly to the tissues

- The chemical bond between acetylcholinesterase and carbamates is more unstable than with organophosphate insecticides and the duration of effects, therefore, is shorter[1]

- Symptoms are similar to or indistinguishable from organophosphate poisoning; carbamate-induced effects are usually less severe, but deaths have been reported

- The toxic to fatal dose ratio is substantially larger for carbamates than organophosphates

- Occupational Exposure Standards

  Carbofuran:
  Long-term exposure limit: $0.1 \, mg.m^{-3}$
  Short-term exposure limit: No data available

Methomyl:

      Long-term exposure limit: 2.5 mg.m^{-3}

      Short-term exposure limit: No data available

Propoxur:

      Long-term exposure limit: 0.5 mg.m^{-3}

      Short-term exposure limit: 2 mg.m^{-3}

Acute Clinical Effects

Inhalation Effects

- Laryngeal irritation, violent coughing, diaphoresis, and tachypnoea occur frequently following inhalation of carbamate dusting powders and may not necessarily be associated with systemic toxicity

- Systemic effects may develop; see below

Dermal Effects

- Skin irritation ranging from slight to moderate has been reported, depending on duration of contact and the vehicle used[1]

- Carbamates are absorbed across skin to varying degrees, the rate of which is temperature dependent; systemic effects may develop,[4] see below

Eye Effects

- Splash contact has caused miosis (constricted pupils)

- Slight to moderate irritation with mild hyperaemia, chemosis and discharge; transient corneal epithelial injury and blepharitis also have been reported, possibly due in part to the hydrocarbon vehicle or other ingredients in formulations tested

- Recovery is usually rapid and complete

Oral Effects

- Nausea, vomiting and diarrhoea, with abdominal pain and cramps may occur

- Systemic effects may develop, onset of effects is usually within 15 to 30 minutes; see below

- Aspiration pneumonitis may occur after ingestion of carbamates in hydrocarbon vehicles

Systemic Effects

- Increased salivation, sweating, lacrimation and urinary and faecal incontinence

- Moderate effects include miosis and blurred vision, hypertension and tachycardia or bradycardia

- Chest tightness, bronchoconstriction/spasm and increased pulmonary secretions may occur

- Central nervous system effects include headache, dizziness, confusion

- Muscle twitching, muscle fasciculation and acute muscle weakness may occur

- More severely, central nervous system depression, stupor, coma, convulsions and hypotonicity

- Cardiorespiratory depression may develop

- Death is usually due to respiratory failure as a result of respiratory muscle weakness and central depression of the respiratory centre

- Death or signs of recovery generally occur within one to several hours of exposure; systemic effects subside over five or six hours, and recovery is usually complete 24 hours post exposure

- Disseminated intravascular coagulation following exposure to propoxur has been reported

- Delayed neurotoxicity is rare but has been reported; effects included peripheral axonal neuropathy,[5] and sensorimotor polyneuropathy with axonal degeneration with only partial recovery[6]

Chronic Clinical Effects

Acetylcholinesterase inhibition is the only consistently observed effect in long-term and acute studies.

Contact dermatitis has been reported following occupational exposure to carbamates

Current evidence does not show that chronic occupational exposure to carbamates is associated with neurological defects.[7] No clear indication of carcinogenicity has been attributed to carbamates as a group in long-term animal studies (not including carbamate derivatives)[4]

Management

Blood or urine carbamate concentrations are not routinely measured. Early and follow-up red cell or plasma cholinesterase concentrations are good indicators of exposure and toxicity.

Inhalation Management

- Maintain a clear airway, give humidified oxygen and ventilate if necessary

- If respiratory irritation occurs assess respiratory function and if necessary perform chest X-rays to check for chemical pneumonitis

- Consider the use of steroids to reduce the inflammatory response

- Treat pulmonary oedema with PEEP or CPAP ventilation

- For systemic management, see below

Dermal Management

- Remove any remaining contaminated clothing, place in double, sealed, clear bags and label; store in a secure area away from patients and staff

- Wash thoroughly with soap and water

- For systemic management, see below

Eye Management

- Irrigate thoroughly with running water or saline for 15 minutes

- Stain with fluorescein and refer to an ophthalmologist if there is any uptake of the stain

Oral Management

- DO NOT INDUCE EMESIS

- Consider gastric lavage (using a cuffed ET tube if an organic solvent is involved) followed by 50 g activated charcoal within 1 hour of ingestion

- Contact a poisons information service for further guidance on gut decontamination

- For systemic management, see below

Systemic Management

- Maintain respiration, give oxygen if required, ventilate if necessary

- Excess bronchial secretions should be removed by suction

- If symptoms are moderate or severe, observe for a minimum of 4 hours post exposure with ECG monitoring and the patient kept at complete rest

- Monitor red blood cell cholinesterase concentrations in every symptomatic case until clinical condition improves

- Symptomatic and supportive care

- For moderate or severe symptoms, see antidotes below

Antidotes

- Contact a poisons information service for further guidance and paediatric doses

- Antidotes are seldom required due to the rapid reactivation of cholinesterase and therefore the short duration of clinical effects

- **Diazepam** may have an overall benefit as well as controlling twitching and convulsions, 5–10 mg IV

- *In severe carbamate poisoning:*

- **Atropine** *Note hypoxia must be corrected before atropine is given.* Dose: 2 mg SC or IV repeatedly until atropinisation is achieved and maintained (atropinisation is characterised by decreased bronchial secretions, heart rate >100 bpm, dry mouth, dilated pupils)

- **Pralidoxime** *should not* be given (unless also exposed to organophosphate) as it does not antagonise carbamates and may increase carbamate toxicity

Summary of Environmental Hazards

- Significant compartments involved in the environmental fate of carbamates are likely to be water, soil and sediments

- Carbamates may undergo hydrolysis, biodegradation, microbial degradation, photodegradation and photodecomposition

- Certain carbamates are found as ground water contaminants and may consequently contaminate drinking water[1]

- Carbamates may be rapidly broken down or persist in water and soil for weeks to months

- The rate of degradation is increased with increased pH of the soil or water

- There is no evidence of significant bioaccumulation of carbamates

- Drinking Water Standards:
    Aldicarb: 10 μg.l^{-1} (WHO guideline)
    Carbofuran: 5 μg.l^{-1} (WHO guideline)
    pesticides: 0.1 μg.l^{-1} (UK max)

- Soil Guidelines: Dutch Criteria:
    Carbaryl: 5 mg.kg^{-1} (intervention)
    Carbofuran: 2 mg.kg^{-1} (intervention)

- Air Quality Standards: no data available

REFERENCES

1. International Programme on Chemical Safety, 1986; *Environmental Health Criteria 64. Carbamate Pesticides: A General Introduction.* WHO, Geneva

2. Lima JS & Reis CAG, 1995. Poisoning Due to Illegal Use of Carbamates as a Rodenticide in Rio De Janeiro. *Clinical Toxicology*; 33(6): 687–90

3. Bardin PG, van Eeden SF, Moolman JA, Foden AP, Joubert JR, 1994. Organophosphate and Carbamate Poisoning. *Arch Intern Med*; 154: 1433–41

4. Hayes WJ, Laws ER (eds.) 1991. *Handbook of Pesticide Toxicology*, Volume 2. Academic Press, Ltd., London

5. Dickoff DJ, Gerber O & Turovsky Z, 1987. Delayed neurotoxicity after ingestion of carbamate pesticide. *Neurology*; 37:122–31

6. Umehara F, Izumo S, Arimura K, Osame M, 1991. Polyneuropathy induced by m-tolyl methyl carbamate intoxication. *J Neurol*; 238:47–8

7. Ballantyne B, Marrs TC, 1992. *Clinical and Experimental Toxicology of Organophosphates and Carbamates*, Chapter 8. Butterworth-Heinemann Ltd., Oxford

Catherine Farrow

CARBON DIOXIDE

Key Points

- Carbon dioxide is a colourless, odourless gas or liquid, or white, extremely cold crystalline solid
- It causes asphyxiation, CNS and respiratory stimulation and, at high doses, CNS depression and vasodilation
- Dermal contact with compressed, liquid or dry ice phase may lead to frostbite
- It is heavier than air; accumulating and persisting in unventilated pits or tanks, or close to ground level
- *In the event of a large spill, stay upwind and out of low areas. Ventilate closed spaces. Protective clothing, eye protection and breathing apparatus should be worn*

FIRST AID

- Terminate exposure and support vital functions
- The casualty should be moved to an uncontaminated area
- Rescuers should, ideally, be trained personnel and must be careful *not to put themselves at risk* and *so wear appropriate protective clothing and, if available, breathing apparatus*
- If the casualty is unconscious a clear airway should be established and maintained; give 100% oxygen if available
- **Inhalation Exposure:** If the patient stops breathing, expired air resuscitation should be started immediately using a pocket mask with a one way valve, if available. It is important where the face is contaminated that expired air resuscitation is NOT attempted unless an airway with rescuer protection is used
- **Dermal Exposure:** If frostbite occurs *do not* remove clothing, flush skin with water
- **Eye exposure:** If the eye tissue is frozen seek medical advice as soon as possible
- If the eye tissue is not frozen, irrigate thoroughly with water or saline for 15 minutes

Detailed Information

- Carbon dioxide is a colourless, odourless gas or liquid, or white, extremely cold crystalline solid
- *Common synonyms* carbonic acid gas, carbonic anhydride, dry ice
- CAS 124-38-9
- UN 1013 (compressed)
- UN 1845 (dry ice)
- UN 2187 (refrigerated liquid)
- NIOSH/RTECS FF 6400000
- Molecular formula CO_2
- Molecular weight 44.01
- Carbon dioxide is used for the carbonation of beverages, as an aerosol propellant, as a fire extinguisher and for refrigeration as dry ice

- It is a by-product of ammonia production, lime kiln operations, fermentation or burning of carbonaceous material
- Carbon dioxide gas is about 1.5 times heavier than air; vapours from liquefied gas are initially heavier than air and may spread along the ground away from the source
- Pockets of high concentrations of the gas may persist in open pits or tanks if unventilated, and accumulate close to ground level
- Carbon dioxide is soluble in water, forming a weak acid, carbonic acid; it has no caustic effect and does not support combustion
- At atmospheric pressure the solid form sublimes to the gaseous phase without liquefaction
- A number of metals will burn in a carbon dioxide atmosphere
- Carbon dioxide reacts violently at high temperatures with anhydrous ammonia and various amines

Summary of Human Toxicity

- Carbon dioxide is absorbed by inhalation
- It causes asphyxiation by displacing atmospheric oxygen[1]
- It is a physiologically active gas, causing CNS and respiratory stimulation as well as CNS depression[2]
- Carbon dioxide is a vasodilator and is cardiotoxic, causing diminished contractile force[3]
- The toxicity of carbon dioxide is concentration-related but also depends on concurrent atmospheric oxygen concentration, duration of exposure, and if strenuous exercise is undertaken during exposure[1]
- Concentrations greater than 100,000 ppm can be fatal[1]

Toxicity of Carbon Dioxide[1,2,3,4]

Concentration	Clinical Effects
5500 ppm for 6 h	No effects
10,000 ppm	Considered dangerous to life
30,000 ppm	Mild narcosis, reduced hearing, increased blood pressure and tachycardia
50,000 ppm	Headache, shortness of breath, stimulation of the respiratory centre, dizziness, confusion and laboured breathing
60–100,000 ppm for 5–10 min	Severe headache, sweating, visual disturbances, tremor and paraesthesia
100,000 ppm	Vomiting, hypertension, transient and slight increase in intraocular pressure
>100,000 ppm	Eye, nose and throat irritation. Neurological effects. Death can occur
120,000 ppm for 8–23 min	Unconsciousness
200,000–300,000 ppm for 1 min	Convulsions and coma

- Occupational Exposure Standards :
  Long-term exposure limit: 5000 ppm (9150 mg.m^{-3})
  Short-term exposure limit: 15000 ppm (27400 mg.m^{-3})

Acute Clinical Effects

Inhalation Effects

- As carbon dioxide concentration increases, displacing atmospheric oxygen, asphyxiation occurs[1]

- Inhalation of vapours from liquefied carbon dioxide gas may cause dizziness or asphyxiation without warning

- Inhalation of carbon dioxide may lead to systemic effects; see below

Dermal Effects

- Carbon dioxide at room temperature will cause no injuries to skin[2]

- Contact with compressed or liquefied gas or dry ice may cause burns, blisters, severe injury and/or frostbite

Eye Effects

- Only a brief, transient stinging sensation and local vasodilation is expected from a splash of a carbonated drink[4]

- Contact of dry ice with the eye is not likely to cause serious adverse effect

Oral Effects

- Not applicable

Systemic Effects

- At a low concentration or early stages of exposure, carbon dioxide is a stimulant, causing increased depth and rate of respiration, blood pressure and pulse

- Vasodilation causes flushed skin and increased cerebral blood flow

- At increasing concentrations and duration of exposure, a depressive phase develops culminating in cardiorespiratory failure

- Effects may progress through headache, drowsiness, mental confusion, decreased co-ordination and judgement, hyporeflexia, giddiness, tinnitus, weakness, lassitude, tremors, nausea, diarrhoea, flaccid paralysis, convulsions, coma, pulmonary oedema and death

- At high concentrations in air carbon dioxide causes stinging sensation of the eyes, nose and throat and may produce respiratory acidosis

- Significant acidosis with hypocapnia have also been reported

- Post-hypoxic injury causing acute renal failure has been reported

- Other organ systems could also be affected depending on the degree and length of hypoxia

- Severe damage to the CNS and retinal ganglion cells has been reported, resulting in constriction of visual fields, enlargement of blind spots, photophobia, loss of convergence and accommodation, deficient dark adaptation, headache, insomnia and personality changes

Chronic Clinical Effects

Headache and dyspnoea may occur on exertion following prolonged inhalation of lower concentrations. Long-term exposure can cause acidosis and adverse effects on calcium-phosphorus metabolism resulting in calcium deposits in soft tissues.[1] Chronic exposure may cause constant respiratory stimulation, which can stress the renal cortex.[1]

Prolonged exposure to carbon dioxide resulted in a progressive central nervous system depression following a brief period of stimulation. Blood flow to the skin and rate of blood flow were increased, while other circulatory function indexes, such as blood pressure, core temperature and respiratory rate decreased, and mental function became impaired.[3]

Management

Arterial blood gases and pH are useful indicators of exposure and toxicity.

Inhalation Management

- Maintain a clear airway, give humidified oxygen and ventilate if necessary

- Treat pulmonary oedema with PEEP or CPAP ventilation

- See systemic management below

Dermal Management

- *If frostbite has occurred*: remove clothing carefully, these may need to be soaked off with tepid water; irrigate the area

- Surgical referral may be necessary

- *If frostbite has not occurred*: remove any remaining contaminated clothing

- Irrigate with copious amounts of water

- Treat burns symptomatically

- Place any contaminated clothes in double, sealed, clear bags and label; store in a secure area away from patients and staff

Eye Management

- Irrigate thoroughly with running water or saline for 15 minutes

- Stain with fluorescein and refer to an ophthalmologist if there is any uptake of the stain

Oral Management

- Not applicable

Systemic Management

- In the depressive phase, the most immediate threat is hypoxia; maintain a clear airway, give humidified oxygen and ventilate if necessary

- Monitor arterial blood gases and pH

- Correct metabolic acidosis with sodium bicarbonate if necessary

- Check renal function and other organs susceptible to hypoxia if necessary

- Symptomatic and supportive care

Summary of Environmental Hazards

- Carbon dioxide is utilised by all photosynthesising plants and therefore forests are important for global carbon dioxide reduction

- The gradual rise in average global temperature is due to the absorption of infrared radiation by increasing amounts of CO_2 in the air; this retards dissipation of heat from the earth's surface

- This phenomenon is ascribed to the burning of fossil fuels, especially coal, aided by aerosols and other contaminants[5]

- Drinking Water Standards: no data available

- Soil Guidelines: no data available

- Air Quality Standards: no data available

REFERENCES

1. Hall AH & Rumack BH (eds.) *TOMES System®* Micromedex, Englewood, Colorado. CD ROM. vol 41 (exp. 31 July 1999)

2. Hathaway GJ, Proctor NH & Hughes JP. *Proctor and Hughes' Chemical Hazards of the Workplace*, 4th edn. Van Nostrand Reinhold, New York, 1996

3. Clayton GD & Clayton FE (eds.) *Patty's Industrial hygiene and toxicology*, 4th edn. John Wiley & Sons, Inc., New York, 1994

4. Grant MW & Schuman JS. *Toxicology of the Eye*, 4th edn. Charles C Thomas, Springfield, 1993

5. Sax NI & Lewis RJ. *Hawley's Condensed Chemical Dictionary*, 11th edn. Van Nostrand Reinhold Company, New York, 1987

Catherine Farrow

CARBON DISULPHIDE

Key Points

- Carbon disulphide is a colourless to light yellow liquid with a sweet ether–like odour
- Exposure primarily affects the central nervous system, resulting in headache, delirium, dizziness, tremor, insomnia, convulsions and coma
- Carbon disulphide causes skin irritation and potentially severe chemical burns
- *Vapours are heavier than air and may spread along the ground or confined spaces. In the event of a large spill, stay upwind and out of low areas. Ventilate closed spaces. Protective clothing, eye protection and breathing apparatus should be worn*

FIRST AID

- Terminate exposure and support vital functions
- The casualty should be moved to an uncontaminated area
- Rescuers should, ideally, be trained personnel and must be careful *not to put themselves at risk* and *so wear appropriate protective clothing and, if available, breathing apparatus*
- If the casualty is unconscious a clear airway should be established and maintained; give 100% oxygen if available
- **Inhalation Exposure:** If the patient stops breathing, expired air resuscitation should be started immediately using a pocket mask with a one way valve, if available. It is important where the face is contaminated that expired air resuscitation is NOT attempted unless an airway with rescuer protection is used
- **Dermal Exposure:** Remove contaminated clothing, if possible under a shower and place in double, sealed, clear bags and label; store the bags in a secure area and away from patients and staff
- Wash the skin thoroughly with copious amounts of water
- **Eye exposure:** Irrigate the eyes thoroughly with water or saline for 15 minutes
- **Oral exposure:** Encourage small quantities of oral fluids (no more than 50–100 ml in total)

Detailed Information

- Carbon disulphide is a colourless to light yellow liquid with a sweet ether-like odour
- *Common synonyms* carbon bisulphide, carbon sulphide
- CAS 75-15-0
- UN 1131
- NIOSH/RTECS FF 6650000
- Molecular formula CS_2
- Molecular weight 76.14
- Carbon disulphide is used for the preparation of rayon viscose fibres; solvent for lipids, sulphur, rubber, phosphorous, oils, resins and waxes; insecticide and fumigant for grain; used for production of Cellophane, carbon tetrachloride, soil disinfectant and optical glass[1,2]
- Vapour may form explosive mixtures in air; vapours may travel back to the source of ignition and flash back
- Carbon disulphide is slightly soluble in water; soluble in benzene, carbon tetrachloride, chloroform, ethanol, methanol and oils
- Carbon disulphide reacts with strong oxides and reacts violently with aluminium and chlorine; it is incompatible with halogens, amines, air, azides, rust and metals[3]
- It is highly flammable and easily ignited by heat, sparks or flames; containers may explode when heated
- When burnt it produces oxides of sulphur and carbon monoxide

Summary of Human Toxicity

- Carbon disulphide is primarily absorbed by inhalation; it is highly tissue–bound and is rapidly distributed to the liver, kidney, and brain[4]
- Exposure to carbon disulphide is generally by inhalation but dermal and ingestion exposures may occur
- It is one of the strongest organic skin irritants
- The mechanism of toxicity involves reaction with the free amine and sulphydryl group of amino acids, proteins and other biological molecules; it is also metabolised by microsomal enzymes to a reactive intermediate which binds covalently to macromolecules[1]
- Exposure causes damage to the central and peripheral nervous systems; chronic exposure may accelerate the development of, or worsen, coronary heart disease
- The lowest lethal oral dose in humans is $14 \, mg.kg^{-1}$ [5]

Toxicity of Carbon Disulphide[1,2]

Concentration	Clinical effects
30–50 ppm	Headache, fatigue, anorexia, psychiatric changes
1,000 ppm	Acute psychosis
2,000 ppm for 5 min	Fatal

- Maximum Exposure Limits:
  Long-term exposure limit: 10 ppm ($32 \, mg.m^{-3}$)
  Short-term exposure limit: no data available

Acute Clinical Effects

Inhalation Effects

- Carbon disulphide is an irritant to the respiratory tract resulting in coughing, wheezing and dyspnoea
- See systemic effects, below

Dermal Effects

- Mild exposure to a dilute form may cause irritation, burning, erythema and peeling of the skin

- Severe burns may occur, particularly with concentrated forms

- Systemic effects from cutaneous exposure are usually preceded by severe local irritation; see systemic effects, below.

Eye Effects

- Ocular exposure results in immediate irritation which may be severe

- Nyastagmus and diplopia have also been observed

Oral Effects

- Ingestion of carbon disulphide may lead to irritation of the mucous membranes, nausea and vomiting

- Oral exposure to large quantities may lead to systemic toxicity; see below

Systemic Effects

- Exposure may lead to headache, dizziness, nausea, vomiting, nervousness and tremor

- Drowsiness, convulsions and coma may occur in severe exposure

- Respiratory failure may result from exposure to high concentrations; death usually occurs from respiratory paralysis

- Acute exposure may cause psychosis

Chronic Clinical Effects

The central and peripheral nervous systems are the major target organs for toxicity, of which psychosis, tremor and polyneuritis are common manifestations. Polyneuritis, seen in up to 88% of poisoned patients, includes lower extremity weakness and paraesthesias. Extrapyramidal signs, chorea and athetosis may occur.[4]

Behavioural changes may manifest themselves as personality changes, memory impairment and irritability; manic-depressive type psychoses with psychomotor excitation, delirium and hallucinations may also be seen. Other manifestations include gastric disturbances and visual disturbances, including retinal microaneurysms and discrete pigmentary changes in the posterior pole.[1] Colour blindness and minimal visual field constriction have been noted; pallor of the temporal portion of the optic nerve heads and extensive atrophy of the optic nerve may occur.[1] Blindness reported rarely.

Chronic exposure has been associated with development of atherosclerosis and ischaemic heart disease. The death rate from coronary artery disease in workers in the viscose rayon manufacturing industry was found to be 2.5 times that of non-exposed individuals.[4]

Acute and prolonged exposure may produce adverse reproductive effects in males and females. In females these including menstrual abnormalities, spontaneous abortions and premature births. In males, exposure can inhibit sperm production and loss of libido; changes can occur in prostate, seminal vesicle, Cowper's glands and accessory glands.

Management

Inhalation Management

- Maintain a clear airway, give humidified oxygen and ventilate if necessary

- If respiratory irritation occurs assess respiratory function and if necessary perform chest X-rays to check for chemical pneumonitis

- See systemic management, below

Dermal Management

- Remove any remaining contaminated clothing, place in double, sealed, clear bags and label; store in a secure area away from patients and staff

- Irrigate with copious amounts of water

- An emollient may be required or treat burns symtomatically

- For large or prolonged exposure, see systemic management below

Eye Management

- Irrigate thoroughly with running water or saline for 15 minutes

- Stain with fluorescein and refer to an ophthalmologist if there is any uptake of the stain

Oral Management

- NO GASTRIC LAVAGE OR EMETIC

- Encourage oral fluids

- Symptomatic and supportive care

- For large exposures, see systemic management below

Systemic Management

- Diazepam can be given for convulsions

- Ventilate if respiratory depression occurs

- Symptomatic and supportive care

Summary of Environmental Hazards

- Carbon disulphide decomposes rapidly in air

- In the atmosphere it reacts with atomic oxygen with a half-life of 9 days and the photochemically produced hydroxyl radicals have a half-life of 5–9 hours to 8 days[1]

- In water volatilisation of carbon disulphide is the significant process; the half-life is 206 hours in a model river

- Most carbon disulphide spilt on to soil will volatilise

- Adsorption to sediment and concentration in the food chain are not thought to be important routes

- Drinking Water Standards: no data available

- Soil Guidelines: no data available

- Air Quality Standards: $100\,\mu g.m^{-3}$ averaging time 24 hours (WHO guideline)

REFERENCES

1. Hall AH & Rumack BH (eds.) *TOMES System®* Micromedex, Englewood, Colorado. CD ROM. vol 41 (exp. 31 July 1999)

2. Hathaway GJ, Proctor NH & Hughes JP. *Proctor and Hughes' Chemical Hazards of the Workplace*, 4th edn. Van Nostrand Reinhold, New York, 1996

3. Sax NI. *Dangerous Properties of Industrial Materials*, 6th edn. Van Nostrand Reinhold, New York, 1984

4. Haddad LM & Winchester JF (eds.) *Clinical Management of Poisoning and Drug Overdose*, 2nd edn. WB Saunders Co, Philadelphia, 1990

5. Clayton GD & Clayton FE (eds.) *Patty's Industrial Hygiene and Toxicology*, 4th edn. John Wiley & Sons, Inc., New York, 1994

Henrietta Wheeler

CARBON MONOXIDE

Key Points

- Carbon monoxide is a colourless, odourless gas
- Severe exposure causes coma, convulsions, and cardiovascular collapse
- Low-level exposures cause non-specific symptoms which can often be mistaken for other illnesses
- Delayed neurological sequelae can occur
- Contact with compressed gas may lead to frostbite
- *In the event of a large spill, stay upwind and out of low areas. Ventilate closed spaces. Protective clothing, eye protection and breathing apparatus should be worn*

FIRST AID

- Terminate exposure and support vital functions
- The casualty should be moved to an uncontaminated area
- Rescuers should, ideally, be trained personnel and must be careful *not to put themselves at risk* and *so wear appropriate protective clothing and, if available, breathing apparatus*
- If the casualty is unconscious a clear airway should be established and maintained; give 100% oxygen if available
- **Inhalation Exposure:** If the patient stops breathing, expired air resuscitation should be started immediately using a pocket mask with a one way valve, if available. It is important where the face is contaminated that expired air resuscitation is NOT attempted unless an airway with rescuer protection is used
- **Dermal Exposure:** If frostbite occurs *do not* remove clothing, flush skin with water
- If frostbite is not present remove contaminated clothing, if possible under a shower and place in double, sealed, clear bags and label; store the bags in a secure area away from patients and staff
- Wash the skin thoroughly with copious amounts of water
- **Eye Exposure:** If the eye tissue is frozen seek medical advice as soon as possible
- If eye tissue is not frozen, irrigate thoroughly with water or saline for 15 minutes

Detailed Information

- Carbon monoxide is a naturally occurring colourless, odourless gas which is sparingly soluble in water

- *Common synonyms* carbonic oxide; common names include coal gas, town gas, flue gas

- CAS 630-08-0

- UN 1016

- NIOSH/RTECS FG 3500000

- Molecular formula CO

- Molecular weight 28.01

- Carbon monoxide is used as a reducing agent in metallurgic operations especially in the Mond process for the recovery of nickel; in organic synthesis especially the Fischer-Tropsch processes for petroleum-type products and in the oxo reaction; in the manufacture of metal carbonyls

- It is produced by the incomplete combustion of organic materials

- Carbon monoxide decomposes into carbon and carbon dioxide

- Mixes readily with air forming explosive mixtures

Summary of Human Toxicity

- Absorption of carbon monoxide occurs exclusively through the lungs

- CO combines reversibly with the oxygen carrying sites of the haemoglobin (Hb) molecule with an affinity 200–300 times greater than oxygen itself

- The carboxyhaemoglobin (COHb) thus formed is unavailable for oxygen transportation; in addition the partial saturation of the Hb molecule results in tighter oxygen binding and thus impaired delivery to tissues

- Permanent neurological sequelae can occur following recovery from CO poisoning

- Carbon monoxide toxicity can result from inhalation or ingestion of dichloromethane (methylene chloride) due to hepatic metabolism of the chemical to CO[1]

- Exposure to CO at levels of 500 ppm would be expected to cause mild symptoms only, exposure to levels of 4000 ppm would be rapidly fatal

- Normal COHb levels due to endogenous CO production are 0.4–0.7%; in non-smokers in urban areas this level may be raised to 1–2% as a result of environmental exposure, smokers may have a COHb level of 5–6%

- The half-life of COHb in blood is approximately:
  5 hours in a subject breathing air
  80 minutes in a subject breathing 100% oxygen
  20–25 minutes in 100% oxygen at 3 ATM[2]

- Occupational Exposure Standards:
  Long-term exposure limit: 50 ppm (58 mg.m⁻³)
  Short-term exposure limit: 300 ppm (349 mg.m⁻³)

Acute Clinical Effects

Inhalation Effects

- Mild to moderate exposures cause headache, weakness, fatigue, nausea, vomiting, irritability, dizziness, drowsiness, disorientation, inco-ordination, visual disturbances, hypotension, tachycardia and hyperventilation

- Severe exposures can rapidly cause coma, convulsions, severe hypotension, respiratory depression, cardiovascular collapse, cerebral oedema, death

- Low-level exposure to carbon monoxide causes non-specific symptoms which are often mistaken for other illnesses e.g. viral illness or food poisoning

- Permanent neurological sequelae following recovery from acute symptoms of poisoning include dementia, amnestic syndromes, psychosis, paralysis, apraxias and agnosias[3]

Dermal Effects

- Rapid release of the compressed gas may cause local frostbite

Eye Effects

- Rapid release of the compressed gas may cause local frostbite

Oral Effects

Not applicable

Chronic Clinical Effects

Repeated exposure to low levels of carbon monoxide can cause nausea, diarrhoea, abdominal pain, headache, fatigue, dizziness, paraesthesiae, chest pain and palpitations.[4] The symptoms are protean in nature and can be mistakenly attributed to other illnesses, e.g. food poisoning or viral illness. Signs that may indicate chronic carbon monoxide poisoning as a possible cause for unexplained illness include the fact that symptoms tend to occur in bad weather (due to heating systems being switched on), recurrent episodes are common, several people are affected simultaneously (e.g. a whole family) and spontaneous recovery occurs when the patient is outside the home.[2]

Management

Inhalation Management

Blood carboxyhaemoglobin concentration should be determined as soon as possible and monitored.

- Ensure an adequate airway and give oxygen, ideally 100% high flow oxygen using a leak-tight mask with a circuit which minimises rebreathing[2]

- Fire victims should be evaluated for airway obstruction from thermal or chemical injury, pulmonary oedema, trauma and concomitant inhalation of other toxic gases (e.g. cyanide and methaemoglobin-inducers)

- Control convulsions with diazepam but be aware of the risk of respiratory depression; hypotensive patients should be given IV fluids if necessary

- In severe poisoning where focal neurological signs are present give mannitol: dosage $1g.kg^{-1}$ of a 20% solution over 20 minutes to prevent or reduce cerebral oedema

- Metabolic acidosis should not be treated aggressively unless the acidosis itself contributes to toxicity; acidosis increases the unloading of oxygen in tissues and may shift the oxygen dissociation curve to the right; ideally the pH should be corrected to 7.2 using sodium bicarbonate

- Monitor ECG, respiration, blood gases and electrolytes

- Obtain baseline cardiac enzymes in all patients; patients with a history of cardiovascular disease should have serial cardiac enzymes performed

- Observe for development of cerebral oedema with serial neurological examinations, CT scans and fundoscopy

- Perform a baseline chest X-ray to assess for inhalation injury or aspiration of vomit

- Hyperbaric oxygen therapy may be considered in cases where a patient has a COHb concentration of > 20%, any history of unconsciousness, any neurological signs, cardiac arrhythmias, pregnancy; contact a poisons information service for further guidance and paediatric information

Dermal Management

- *If frostbite has occurred*: remove clothing carefully, these may need to be soaked off with tepid water; irrigate the area

- Surgical referral may be necessary

- *If frostbite has not occurred*: remove any remaining contaminated clothing

- Irrigate with copious amounts of water

- Treat burns symptomatically

- Place any remaining contaminated clothing in double, sealed, clear bags and label; store in a secure area away from patients and staff

Eye Management

- If the eye tissue is frozen refer to an ophthalmologist

Oral Management

- Not applicable

Summary of Environmental Hazards

- The residence time of carbon monoxide in the atmosphere is believed to be approximately 0.2 years

- Oxidation in the atmosphere and take-up by the soil, vegetation and inland fresh waters have been identified as the major removal mechanisms[5]

- The oceans act as reservoirs for carbon monoxide, since considerable quantities are dissolved in water; because of the equilibrium that exists, carbon monoxide is dissolved or released according to conditions depending on the partial pressure of carbon monoxide in the atmosphere and on the water temperature[5]

- Drinking Water Standards: no data available

- Soil Guidelines: no data available

- Air Quality Standards:
    10 ppm for 8 hours (UK)
    $60 mg.m^{-3}$ for 30 min; $30 mg.m^{-3}$ for 1 hour;
    $10 mg.m^{-3}$ for 8 hours (time weighted averages)
    (WHO guidelines)

REFERENCES

1. Hathaway GJ, Proctor NH & Hughes JP. *Proctor and Hughes' Chemical Hazards of the Workplace*, 4th edn. Van Nostrand Reinhold, New York, 1996

2. Crawford R, Campbell DGD & Ross J, 1990. Carbon monoxide poisoning in the home: recognition and treatment. *Br Med J*; 301: 977–9

3. Ginsberg MD, 1985. Carbon monoxide intoxication: clinical features, neuropathology and mechanisms of injury. *J Toxicol-Clin Toxicol*; 23(4–6): 281–8

4. Meredith T & Vale A, 1988. Carbon monoxide poisoning. *Br Med J*; 296(6615): 77–9

5. International Programme on Chemical Safety. *Environmental Health Criteria 13: Carbon monoxide*. WHO, Geneva, 1979

Nicola Scott

CHLORINE

Key Points

- Chlorine is a greenish-yellow gas, the odour of which is familiar from household bleach and swimming pools
- Chlorine exposure can result from occupational or domestic use
- Chlorine is a respiratory irritant and may cause coughing, choking, hypoxia and pulmonary oedema and in higher concentrations may be fatal
- Contact with compressed liquid gas may cause frostbite or burns to skin and eyes
- *Chlorine is heavier then air and may accumulate in low or confined areas; in the event of a large spill, stay upwind and out of low areas. Ventilate closed spaces. Protective clothing, eye protection and breathing apparatus should be worn*

FIRST AID

- Terminate exposure and support vital functions
- The casualty should be moved to an uncontaminated area
- Rescuers should, ideally, be trained personnel and must be careful *not to put themselves at risk* and *so wear appropriate protective clothing and, if available, breathing apparatus*
- If the casualty is unconscious a clear airway should be established and maintained; give 100% oxygen if available
- **Inhalation Exposure:** If the patient stops breathing, expired air resuscitation should be started immediately using a pocket mask with a one way valve, if available. It is important where the face is contaminated that expired air resuscitation is NOT attempted unless an airway with rescuer protection is used
- **Dermal Exposure:** If frostbite occurs *do not* remove clothing, flush skin with water
- If frostbite is not present remove contaminated clothing, if possible under a shower and place in double, sealed, clear bags and label; store the bags in a secure area away from patients and staff
- Wash the skin thoroughly with copious amounts of water
- **Eye Exposure:** If the eye tissue is frozen seek medical advice as soon as possible
- If eye tissue is not frozen, irrigate thoroughly with water or saline for 15 minutes

Detailed Information

- Chlorine is a greenish-yellow compound with a melting point of $-101°C$ and boiling point of $-34.5°C$, thus most exposures are to chlorine gas; the odour of which is familiar from household bleach and swimming pools

- *Common synonyms* chlorine mol., dichlorine, molecular chlorine

- CAS 7782-50-5

- UN 1017

- NIOSH/RTECS FO 2100000

- Molecular formula Cl_2

- Molecular weight 70.91

- Chlorine is used as a disinfectant, for sterilising and purifying water and in the manufacturing of bleaching agents

- Chlorine can support combustion and is a serious fire risk

- Chlorine is extremely reactive, it reacts violently with many combustible materials and other chemicals, including water when heated; it reacts with most organic compounds and many inorganic compounds

Summary of Human Toxicity

- Chlorine reacts with tissue water to form hydrochloric and hypochlorous acids, and thus is a potent irritant of the eyes, skin and mucous membranes

- Injury is proportional to the concentration of the gas, duration of exposure and water content of exposed tissues

- Evidence exists suggesting that patients with pre-existing respiratory disease, such as asthma, may be at greater risk from chlorine exposure

- *Industrial exposure*: Chlorine gas is corrosive to eyes, skin, respiratory tract and mucous membranes. Severe exposure may cause laryngospasm, airway obstruction, respiratory arrest, pulmonary oedema and cardiovascular collapse

- *Household exposure*: The mixing of household cleaning agents (for example bleach and acids) may liberate chlorine gas. Single acute exposure commonly produces coughing, lacrimation, conjunctivitis and tachycardia. A few cases may develop vomiting, sweating and headache

- One or two breaths of gas accumulating above swimming pool chlorinator tablets has caused marked respiratory distress and hypoxaemia in children

Toxicity of Chlorine[1,2]

Concentration	Clinical effects
1–3 ppm for 1 h	Mild mucous membrane irritation
5–15 ppm	Moderate irritation of upper respiratory tract, stinging and burning eyes
30 ppm	Immediate chest pain, vomiting, and coughing
40–60 ppm	Toxic pneumonitis and pulmonary oedema
430 ppm for 30 min	Lethal
1000 ppm for few min	Fatal

- Occupational Exposure Standards:
  Long-term exposure limit: 0.5 ppm (1.5 mg.m^{-3})
  Short-term exposure limit: 1 ppm (2.9 mg.m^{-3})

Acute Clinical Effects

Inhalation Effects

- Initially, symptoms of exposure include irritation of the eyes, nose and throat which occur very rapidly, and are followed by coughing and wheezing, dyspnoea, sputum production and chest pain

- Nausea and vomiting may occur

- During this time the large and small airways may become constricted

- Metabolic acidosis with hyperchloraemia has been reported following large exposures via inhalation

- At very high concentrations anoxia may occur, leading to cardiac and/or respiratory arrest

- In severe cases chemical pneumonitis and pulmonary oedema may occur after a latent phase of 12–24 hours

Dermal Effects

- Skin irritation may occur with burns at higher concentrations

- Contact with compressed liquid gas may cause frostbite or burns

Eye Effects

- Stinging and burning sensation of the eyes, with associated blepharospasm, redness, and watering

- With continued exposure sensitivity may decrease and signs and symptoms diminish[1]

- Contact with compressed liquid gas can cause frostbite

Oral Effects

- Not applicable

Chronic Clinical Effects

Green hair coloration has been noted in blonde or grey-haired individuals following regular swimming in chlorinated water. Dental enamel erosion has been noted in competitive swimmers.[3]

Management

There is no analysis available to measure blood chlorine concentration.

Inhalation Management

Immediate management:

- Patients should be kept at rest and assessed for respiratory difficulty, using baseline lung function tests as appropriate

- Give oxygen for dyspnoea

- Give bronchodilators (e.g. salbutamol; orally or inhaled) for bronchospasm

Patients with no immediate symptoms require no treatment, a record of their peak flow may be of use in assessing any subsequent respiratory effects.

Admission criteria

- All patients showing immediate effects should be assessed in hospital where lung sounds should be checked, and baseline lung function tests and chest X-rays taken

- Patients with mild effects only may be discharged, but should be advised to return if symptoms recur or develop over the following 24–36 hours

- All patients showing **immediate** moderate or severe effects should be admitted for at least 24 hours

- Assess and consider admitting all patients with pre-existing respiratory disease for at least 24 hours

- All patients who have developed moderate or severe clinical effects, even if pulmonary oedema did not occur, should be reviewed for lung function tests

Further management

- The airway should be assessed regularly and chest X-rays taken if indicated by the clinical condition of the patient

- Pulmonary oedema should be treated with PEEP or CPAP

- Corticosteroids may inhibit the inflammatory response and early use should be considered in severe cases

Dermal Management

- *If frostbite has occurred*: remove clothing carefully, these may need to be soaked off with tepid water; irrigate the area

- Surgical referral may be necessary

- *If frostbite has not occurred*: remove any remaining contaminated clothing

- Irrigate with copious amounts of water

- Treat burns symptomatically

- Place any contaminated clothes in double, sealed, clear bags and label; store in a secure area away from patients and staff

- Skin burns should be treated as a thermal injury

Eye Management

- Irrigate thoroughly with running water or saline for 15 minutes

- Stain with fluorescein and refer to an ophthalmologist if there is any uptake of the stain

Oral Management

- Not applicable

Summary of Environmental Hazards

- Free chlorine released into water is extremely unstable; it will oxidise inorganic compounds rapidly and oxidise organic compounds at a slower rate[4]

- Chlorine is harmful to many forms of aquatic life in concentrations below 0.1 ppm; it is not expected to accumulate in the food chain

- Drinking Water Standards: 5 mg.l[-1] (WHO guideline)

- Soil Guidelines: no data available

- Air Quality Standards: no data available

REFERENCES

1. Grant MW & Schuman JS. *Toxicology of the Eye*, 4th edn. Charles C Thomas, Springfield, 1993

2. Hathaway GJ, Proctor NH & Hughes JP. *Proctor and Hughes' Chemical Hazards of the Workplace*, 4th edn. Van Nostrand Reinhold, New York, 1996

3. Centerwall BS, Armstrong CW & Funkhouser LS, 1986. Erosion of dental enamel among competitive swimmers at a gas–chlorinated swimming pool. *Am J Epidemiol*; 123(4): 641–7

4. Hall AH & Rumack BH (eds.) *TOMES System*® Micromedex, Englewood, Colorado. CD ROM. vol 41 (exp. 31 July 1999)

Henrietta Wheeler

CHLOROFLUOROCARBONS (CFCs)

Key Points

- Relatively inert non-flammable, odourless, colourless liquids or gases
- On inhalation CFCs can sensitise the myocardium to endogenous catecholamines
- Dermal contact may cause burns and frostbite
- *In the event of a large spill, stay upwind and out of low areas. Ventilate closed spaces. Protective clothing, eye protection and breathing apparatus should be worn*

First Aid

- Terminate exposure and support vital functions
- The casualty should be moved to an uncontaminated area
- Rescuers should, ideally, be trained personnel and must be careful *not to put themselves at risk* and *so wear appropriate protective clothing and, if available, breathing apparatus*
- If the casualty is unconscious a clear airway should be established and maintained; give 100% oxygen if available
- **Inhalation Exposure:** If the patient stops breathing, expired air resuscitation should be started immediately using a pocket mask with a one way valve, if available. It is important where the face is contaminated that expired air resuscitation is NOT attempted unless an airway with rescuer protection is used
- **Dermal Exposure:** If frostbite occurs *do not* remove clothing, flush skin with water
- If frostbite is not present remove contaminated clothing, if possible under a shower and place in double, sealed, clear bags and label; store the bags in a secure area away from patients and staff
- Wash the skin thoroughly with copious amounts of water
- **Eye Exposure:** If the eye tissue is frozen seek medical advice as soon as possible
- If eye tissue is not frozen, irrigate thoroughly with water or saline for 15 minutes
- **Oral Exposure:** Encourage small quantities of oral fluids (no more than 50–100 ml in total) unless perforation is suspected

Detailed Information

- Relatively inert, non-flammable, odourless, colourless liquids or gases

- CFCs are coded by the 'rule of 90'; for example for CFC 12 add 90 to 12 to give 102 indicating 1 carbon, 0 hydrogen and 2 fluorine giving the formula CCl_2F_2

- Used as refrigerants and aerosol propellants

- High thermal stability, but when pyrolysis occurs in the presence of humidity, products usually include hydrofluoric and hydrochloric acid and, in the presence of either water or oxygen, phosgene

- Their use has been restricted by the Montreal Protocol on Substances that Deplete the Ozone Layer (1988), and also by EC Regulations 3093/94

Common Synonyms for CFCs

Trichlorofluoromethane	CFC 11; fluorocarbon 11; Freon 11; fluorotrichloromethane; monofluorotrichloromethane; trichloromonofluoromethane
Dichlorodifluoromethane	CFC 12; Freon 12; fluorocarbon 12; difluorochloromethane; propellant 12; refrigerant 12
Chlorotrifluoromethane	CFC 13; R-13; Freon 13; monochlorotrifluoromethane; trifluoromethylchloride; trifluoromonochlorocarbon
1,1,2,2-tetrachloro-1,2-difluoroethane	CFC 112; Freon 112; 1,2-difluoro-1,1,2,2-tetrachloroethane
1,1,1,2-tetrachloro-2,2-difluoroethane	CFC 112a; Freon 112a
1,1,2-trichloro-1,2,2-trifluoroethane	CFC 113; Freon 113; trichlorotrifluoroethane; 1,2,2-trifluoro-1,1,2-trichloroethane
1,1,1-trichloro-2,2,2-trifluoroethane	CFC 113a; Freon 113a
1,2-dichloro-1,1,2,2-tetrafluoroethane	CFC 114; Freon 114; fluorocarbon 114; propellant 114; R 114; tetrafluorodichloroethane; sym-dichlorotetrafluoroethane; 1,1,2,2-tetrafluoro-1,2-dichloroethane; cryofluorane
1,1-dichloro-1,2,2,2-tetrafluoroethane	CFC 114a; Freon 114a
1-chloro-1,1,2,2,2-pentafluoroethane	CFC 115; Freon 115; chloropentafluoroethane; monochloropentafluoroethane

Identification Numbers

CFC number	CAS	UN	NIOSH/RTECS
CFC 11	75-69-4	n.a.	PB 6125000
CFC 12	75-71-8	1028	PA 8200000
CFC 13	75-72-9	1022	PA 6410000
CFC 112	76-12-0	n.a.	KI 1420000
CFC 112a	76-11-9	n.a.	KI 1425000
CFC 113	76-13-1	n.a.	KJ 4000000
CFC 113a	354-58-5	n.a.	KJ 3975000
CFC 114	76-14-2	n.a.	KI 1101000
CFC 114a	374-07-2	n.a.	n.a.
CFC 115	76-15-3	1020	n.a.

n.a. = specific number not available

Chemical Formulae and Molecular Weights

CFC number	Chemical formula	Molecular weight
CFC 11	CCl_3F	137.37
CFC 12	CCl_2F_2	120.92
CFC 13	$CClF_3$	104.46
CFC 112	$CCl_2F.CCl_2F$	203.82
CFC 112a	$Cl_3C.CClF_2$	203.82
CFC 113	$CCl_2F.CClF_2$	187.38
CFC 113a	$CCl_3.CF_3$	187.38
CFC 114	$CCF_2.CClF_2$	170.92
CFC 114a	$CCl_2F.CF_3$	170.92
CFC 115	$CClF_2.CF_3$	154.47

Summary of Human Toxicity

- Inhalation is the most common route of entry, although oral, dermal or eye exposure is possible

- Irrespective of route of absorption, CFCs are almost exclusively excreted through the lungs

- CFCs cause myocardial depression and sensitise the myocardium to endogenous catecholamines; this cardiac sensitisation is unpredictable in terms of exposure duration and quantity

- Sudden sniffing death: this is described following abuse of CFCs and is associated with agitation and physical exertion and sudden collapse (presumably associated with an increase in catecholamines)

- Inhalation of 4 to 5% of bromochlorofluoromethane led to dizziness and tingling of the fingers after 1 minute

Occupational Exposure Standards

CFC number	Long-term exposure limit	Short-term exposure limit
CFC 11	1,000 ppm (5710 mg.m^{-3})	1,250 ppm (7140 mg.m^{-3})
CFC 12	1,000 ppm (5030 mg.m^{-3})	1,250 ppm (6280 mg.m^{-3})
CFC 13	no data available	no data available
CFC 112	100 ppm (847 mg.m^{-3})	100 ppm (847 mg.m^{-3})
CFC 113	1,000 ppm (7790 mg.m^{-3})	1,250 ppm (9740 mg.m^{-3})
CFC 114	1,000 ppm (7110 mg.m^{-3})	1,250 ppm (8890 mg.m^{-3})
CFC 115	1,000 ppm (6420 mg.m^{-3})	no data available

Acute Clinical Effects

Inhalation Effects

- Nausea, headache, disorientation, respiratory irritation with cough, sore throat and dyspnoea

- Euphoria with intentional abuse

- Frostbite of lips, tongue, buccal mucosa, hard palate, trachea, bronchi, larynx and oesophageal burns following intentional abuse[1]

- Ventricular fibrillation and ventricular tachycardia may occur, usually following deliberate abuse,[2] industrial use or in poorly ventilated areas; patients often present with cardiorespiratory arrest

- Pulmonary oedema and cerebral oedema (both usually found at post-mortem)

- Jaundice and deranged LFTs may occur

Dermal Effects

- Defatting and erythema of the skin

- Contact dermatitis

- Frostbite with severe tissue damage may occur from direct contact[3]

Eye Effects

- Mildly irritant with inflammation and conjunctivitis

- Freon 11 in rabbits eyes has caused mild inflammation of the eyelids and hyperaemia lasting for several hours[4]

Oral Effects

- Nausea and vomiting may be expected

- Necrosis and perforation of the stomach has been reported following accidental ingestion of Freon 11[5]

Chronic Clinical Effects

Palpitations and light-headedness have been reported in some workers, no neurological or electro-neurophysiological abnormalities were detected.[6] Occupational overexposure to Freons may precipitate ventricular arrhythmias.[7]

Management

Plasma CFC concentrations are not clinically useful.

Inhalation Management

- Maintain a clear airway, give humidified oxygen and ventilate if necessary

- If respiratory irritation occurs assess respiratory function and if necessary perform chest X-rays to check for chemical pneumonitis

- Monitor ECG

- Patients should be kept at complete bedrest, the use of stimulants (including adrenaline and noradrenaline) is best avoided except for resuscitation because of the risk of sensitisation

- Consider endoscopic evaluation in patients with evidence of burns

Dermal Management

- *If frostbite has occurred*: remove clothing carefully, these may need to be soaked off with tepid water; irrigate the area

- Surgical referral may be necessary

- *If frostbite has not occurred*: remove any remaining contaminated clothing

- Irrigate with copious amounts of water

- Treat burns symptomatically

- Place any contaminated clothes in double, sealed, clear bags and label; store in a secure area away from patients and staff

Eye Management

- Irrigate thoroughly with running water or saline for 15 minutes

- Stain with fluorescein and refer to an ophthalmologist if there is any uptake of the stain

Oral Management

- Encourage oral fluids

- If facilities are available, early gastro-oesophagoscopy should be considered within 12 hours of the event, if the patient is symptomatic, to assess the extent and severity of the injury

Summary of Environmental Hazards

- Releases of CFCs into the atmosphere do not degrade in the troposphere but diffuse into the stratosphere with a half-life of 20 years

- In the stratosphere CFCs react slowly with oxygen free radicals and release chlorine atoms; these chlorine atoms catalytically destroy ozone and irreversibly damage the ozone layer

- Their use has been restricted by the Montreal Protocol on Substances that Deplete the Ozone Layer (1988), and also by EC Regulation 3093/94

- CFCs do not degrade readily in the ambient atmosphere, some are considered to have a half-life of >100 years for the photochemical reaction producing hydroxy radicals

- If released into water, CFCs are expected to be lost rapidly by volatilisation

- If released into soil, CFCs volatilise rapidly from surfaces or leach through soil

- Drinking Water Standards:
        hydrocarbon total: $10 \, \mu g \, . \, l^{-1}$ (UK max)

- Soil Guidelines: no data available

- Air Quality Standards: no data available

REFERENCES

1. Elliott DC, 1991. Frostbite of the mouth: a case report. *Milt Med*; 156: 18–9

2. Brady WJ, Stremski E, Eljaiek L & Aufderheide TP, 1994. Freon inhalational abuse presenting with ventricular fibrillation. *Am J Emerg Med*; 12: 533–6

3. Wegener EE, Barraza KR & Das S, 1991. Severe frostbite caused by freon gas. *South Med J*; 84(9): 1143–6

4. Grant MW & Schuman JS. *Toxicology of the Eye*, 4th edn. Charles C Thomas, Springfield, 1993

5. Haj M, Burstein Z, Horn E & Stambler B, 1980. Perforation of the stomach due to trichlorofluoromethane (Freon 11) ingestion. *Isr J Med Sci*; 16: 392–4

6. Campbell DD, Lockey JE, Petajan J, Gunter BJ & Rom WN, 1986. Health effects among refrigeration repair workers exposed to fluorocarbons. *Br J Indust Med*; 43: 107–11

7. Kaufman JD, Silverstein MA & Moure-Eraso R, 1994. Atrial fibrillation and sudden death related to occupational solvent exposure. *Am J Indust Med*; 25: 731–5

Nicola Bates

CHLOROFORM

Key Points

- Chloroform is a clear, colourless, non-flammable highly volatile liquid, with a sweet odour
- Toxicity may occur from ingestion, inhalation, injection or prolonged skin contact
- Chloroform is a potent CNS depressant and may cause hepatic and renal damage; it is cardiotoxic
- Chemical pneumonitis and pulmonary oedema may also occur
- *In the event of a large spill, stay upwind and out of low areas. Ventilate closed spaces. Protective clothing, eye protection and breathing apparatus should be worn*

FIRST AID

- Terminate exposure and support vital functions
- The casualty should be moved to an uncontaminated area
- Rescuers should, ideally, be trained personnel and must be careful *not to put themselves at risk* and *so wear appropriate protective clothing and, if available, breathing apparatus*
- If the casualty is unconscious a clear airway should be established and maintained; give 100% oxygen if available
- **Inhalation Exposure:** If the patient stops breathing, expired air resuscitation should be started immediately using a pocket mask with a one way valve, if available. It is important where the face is contaminated that expired air resuscitation is NOT attempted unless an airway with rescuer protection is used
- **Dermal Exposure:** Remove contaminated clothing, if possible under a shower and place in double, sealed, clear bags and label; store the bags in a secure area away from patients and staff
- Wash the skin thoroughly with copious amounts of water
- **Eye Exposure:** Irrigate thoroughly with water or saline for 15 minutes
- **Oral Exposure:** Encourage small quantities of oral fluids (no more than 50–100 ml in total)

Detailed Information

- Clear, colourless, non-flammable highly volatile liquid, with a sweet odour
- *Common synonyms* methane trichloride, methenyl chloride, methyl trichloride, trichloromethane, trichloroform
- CAS 67-66-3
- UN 1888
- NIOSH/RTECS FS 9100000
- Molecular formula $CHCl_3$
- Molecular weight 119.39
- Chloroform was used as an anaesthetic, as a dry cleaning agent and in cough medicines and toothpastes; it is now used as a solvent

- Miscible with alcohol, ether, benzene, cabon disulphide, carbon tetrachloride, fixed and volatile oils; slightly soluble in water
- Chloroform may react explosively with fluorine, dinitrogen trioxide, triisopropylphosphine, aluminium, lithium, sodium, sodium/methanol, sodium hydroxide/methanol and sodium methoxide
- It is not affected by acids but reacts with alkali metals; attacks many plastics
- On heating, chloroform breaks down to form hydrochloric acid, phosgene and chlorine

Summary of Human Toxicity

- The initial effects of chloroform toxicity are those of CNS depression; these effects come on rapidly following ingestion or inhalation
- Hepatotoxicity may occur 10–48 hours post-exposure with the peak in liver enzyme elevation at 3–4 days post-exposure; these usually return to normal within 6–8 weeks
- Renal failure is usually evident within 24–48 hours
- Death may occur early from arrhythmias
- Death has occurred following ingestion of 114 ml, although ingestion of as little as 10 ml has also been reported to cause CNS depression and death[1,2]
- Chloroform can produce deep anaesthesia at a concentration of as little as 0.75%
- During chloroform anaesthesia blood concentrations are 71 mg.l^{-1} (plane 1), 106 mg.l^{-1} (plane 2), 122 mg.l^{-1} (plane 3) and 165 mg.l^{-1} (plane 4) [2]

Toxicity of Chloroform[3]

Concentration	Clinical effects
205–307 ppm	Lowest amount that can be detected by smell
389 ppm	Endured for 30 minutes without effects
1024 ppm	Dizziness, nausea; fatigue and headache may be felt later
1475 ppm	Dizziness, salivation
4096 ppm	Vomiting, sensation of fainting
14,336–16,384 ppm	Narcotic limiting concentration

Occupational Exposure Standards:
    Long-term exposure limit: 2 ppm (9.9 mg.m^{-3})
    Short-term exposure limit: no data available

Acute Clinical Effects

Inhalation Effects

- See systemic effects below

Dermal Effects

- Irritant to the skin causing a defatting action

- Urticaria, erythema, blistering and burns may occur in larger exposures

- Dermal absorption may occur but is probably only significant following prolonged contact

Eye Effects

- Stinging sensation, pain and hyperaemia of the conjunctiva

- The epithelium may be damaged but recovery is rapid and usually complete within 1–3 days

Oral Effects

- See systemic effects below

Systemic Effects

- Dizziness, headache, disorientation and anorexia may occur. Nausea and vomiting are common following inhalation or ingestion

- Coma may be rapid in onset and last for many hours

- Chloroform may cause hypotension and arrhythmias. It sensitises the myocardium to endogenous catecholamines

- Respiratory depression, chemical pneumonitis and pulmonary oedema may occur. Respiratory arrest may occur rarely

- Hepatotoxicity is common and renal damage may occur

- After exposure the breath usually has a chloroform odour

Chronic Clinical Effects

Dizziness, fatigue, drowsiness, memory impairment, increased dreams, anorexia and palpitations have been reported in chronically exposed workers. There is evidence of slight liver damage with higher concentrations of serum prealbumin and transferrin levels. Neurobehavioural testing revealed dose-related negative changes and increased scores in passive mood states.[4]

Chronic ingestion of 1.6–2.6 g of chloroform daily for 10 years led to hepatitis and nephrosis.[5] Chronic abuse of chloroform may cause psychotic behaviour.

Management

Chloroform concentrations may be determined in blood.

Inhalation Management

- Maintain a clear airway, give humidified oxygen and ventilate if necessary

- If respiratory irritation occurs, assess function and, if necessary, perform chest X-rays to check for chemical pneumonitis

- Consider the use of steroids to reduce the inflammatory response

- Treat pulmonary oedema with PEEP or CPAP ventilation

- Symptomatic and supportive care; see systemic management below

Dermal Management

- Remove any remaining contaminated clothing, place in double, sealed, clear bags and label; store in a secure area away from patients and staff

- Irrigate with copious amounts of water

- Treat irritation symptomatically

Eye Management

- Irrigate thoroughly with running water or saline for 15 minutes

- Stain with fluorescein and refer to an ophthalmologist if there is any uptake of the stain

Oral Management

- Chloroform is radiopaque and an X-ray may confirm ingestion

- DO NOT INDUCE EMESIS because of the rapid onset of CNS depression and the risk of aspiration

- Consider gastric lavage within 1 hour of ingestion because of very rapid absorption of chloroform (using a cuffed ET tube to protect the airway)

- Contact a poisons information service for further guidance on gut decontamination

- See systemic management below

Systemic Management

- All patients initially require at least 24 hours observation with ECG monitoring

- Patients should be kept at complete bedrest, the use of stimulants (including adrenaline and noradrenaline) is best avoided except for resuscitation because of the risk of sensitisation of the myocardium

- In symptomatic patients the hepatic and renal function should be monitored for at least 3 days post-exposure

- Chest X-rays will be necessary to monitor development of any respiratory complications

- Chloroform is known to deplete glutathione stores and N-acetylcysteine (used in the treatment of paracetamol overdose) has been suggested as a possible antidote for hepatotoxic organic solvents. It has been used with success in patients with carbon tetrachloride poisoning. The regimen is the same as that for paracetamol poisoning but a longer duration of therapy is recommended. There is limited data on the use of N-acetylcysteine in chloroform poisoning, but its use should be considered to minimise hepatic damage

- Contact a poisons information service for further guidance

Summary of Environmental Hazards

- Environmental releases from industrial uses are to the atmosphere

- Releases to water will be lost primarily by evaporation and will end up in the atmosphere

- Release to the atmosphere may be transported long distances and will photo-oxidise with a half-life of 80 days

- Spills and other releases of chloroform on land will evaporate rapidly or, due to poor adsorption to soil, leach into the groundwater where it will reside for long periods of time

- Chloroform is not expected to concentrate in the food chain

- Drinking Water Standards:
  hydrocarbon total: $10 \, \mu g.l^{-1}$ (UK max)
  chloroform: $200 \, \mu g.l^{-1}$ (WHO guideline)

- Soil Guidelines: Dutch criteria:
  $0.001 \, mg.kg^{-1}$ (target)
  $10 \, mg.kg^{-1}$ (intervention)

- Air Quality Standards: no data available

REFERENCES

1. Boardman EC, 1866. Death from the imbibition of chloroform. *Br J Med*; 1: 541

2. Baselt RC & Cravey RH. *Disposition of Toxic Drugs and Chemicals in Man*, 4th edn. Chemical Toxicology Institute, 1995

3. Clayton GD & Clayton FE (eds.) *Patty's Industrial hygiene and toxicology*, 4th edn. John Wiley & Sons, Inc., New York, 1994

4. Li L-H, Jiang X-Z, Liang Y-X, Chen Z-Q, Zhou Y-F & Wang Y-L, 1993. Studies on the toxicity and maximum allowable concentration of chloroform. *Biomed Environ Sci*; 6: 179–86

5. Wallace CJ, 1950. Hepatitis and nephrosis due to cough syrup containing chloroform. *Calif Med*; 73:442

Nicola Bates

CHLOROPICRIN

Key Points

- Chloropicrin is a colourless, slightly oily, non-flammable liquid with an intense pungent odour
- Inhalation is the main route of exposure; it is a potent lacrimator and respiratory irritant
- Chloropicrin is irritating to the skin, eyes and mucous membranes
- Exposure may cause pulmonary oedema which is the most frequent cause of rapid death
- *In the event of a large spill, stay upwind and out of low areas. Ventilate closed spaces. Protective clothing, eye protection and breathing apparatus should be worn*

First Aid

- Terminate exposure and support vital functions
- The casualty should be moved to an uncontaminated area
- Rescuers should, ideally, be trained personnel and must be careful *not to put themselves at risk* and *so wear appropriate protective clothing and, if available, breathing apparatus*
- If the casualty is unconscious a clear airway should be established and maintained; give 100% oxygen if available
- **Inhalation Exposure:** If the patient stops breathing, expired air resuscitation should be started immediately using a pocket mask with a one way valve, if available. It is important where the face is contaminated that expired air resuscitation is NOT attempted unless an airway with rescuer protection is used
- **Dermal Exposure:** Remove contaminated clothing, if possible under a shower and place in double, sealed, clear bags and label; store the bags in a secure area away from patients and staff
- Wash the skin thoroughly with copious amounts of water
- **Eye Exposure:** Irrigate thoroughly with water or saline for 15 minutes
- **Oral Exposure:** Encourage small quantities of oral fluids (no more than 50–100 ml in total)

Detailed Information

- Chloropicrin is a colourless, slightly oily, non-flammable liquid with an intense pungent odour
- *Common synonyms* acquinite, aquinite, dolochlor, nitro-chlorform, nitrotrichloromethane, trichloronitromethane
- CAS 76-06-2
- UN 1580
- UN 1583 (absorbed)
- NIOSH/RTECS PB 6300000
- Molecular formula CCl_3NO_2
- Molecular weight 164.37
- Chloropicrin is used to disinfect cereals and grains, as a fumigant, soil insecticide, and as a war gas; also used as a chemical intermediate

- Chloropicrin vapours are heavier than air and may collect in low areas
- Chloropicrin is not readily soluble in water but is miscible with benzene, absolute alcohol, and carbon disulfide[1]
- Strong oxidant which reacts violently with combustible substances and reducing agents
- Reacts violently with alkaline-earth and alkali metals and metal powders, with a risk of fire and explosion
- Decomposes when heated or exposed to light, giving off corrosive hydrochloric acid and nitrous vapours

Summary of Human Toxicity

- Chloropicrin is a severe irritant to the eyes, skin, lungs, and mucous membranes and it is a sensitiser[2]
- Development of irritation may be delayed for 2–5 hours depending on the concentration[3]
- Chloropicrin reacts with sulphydryl groups in haemoglobin resulting in impaired oxygen transport[3]
- It is more injurious to the small and medium bronchi than to the trachea and large bronchi
- Causes pulmonary oedema, which is the most frequent cause of early deaths[2]
- An 18-year-old girl died of pulmonary oedema within 4 hours of exposure to chloropicrin in a confined space[4]

Toxicity of Chloropicrin[2,5,6]

Concentration	Clinical effect
0.3–0.37 ppm for 3–30 sec	Closing of eyelids, painful irritation in the eyes, depending on individual susceptibility
1.1 ppm	Odour detectable, irritation to eyes
4 ppm	Incapacitant, renders an individual unfit for combat
15 ppm	Intolerable, respiratory tract injury
119 ppm for 30 min	Estimated lethal exposure
297.6 ppm for 10 min	Lowest lethal concentration

- Occupational Exposure Standards:
  Long-term exposure limit: 0.1 ppm (0.68 mg.m^{-3})
  Short-term exposure limit: 0.3 ppm (2.1 mg.m^{-3})

Acute Clinical Effects

Inhalation Effects

- Headache, nausea, and vomiting are common
- Irritation of mucous membranes with a burning sensation in the mouth, rhinorrhoea, rhinitis, and swelling of the throat may occur
- Leading to tightness of chest with coughing, sneezing, and increased bronchial secretions

- Irregular respiration with periods of apnoea have been reported

- Severe exposure can result in pulmonary oedema

- Deaths may occur from secondary infection, bronchopneumonia, or bronchitis obliterans

- Severe exposures may also cause coma, hepatic necrosis, renal damage, and cardiac necrosis[3]

- Vertigo, fatigue and headache may also occur with exacerbation of postural hypotension from exposures producing lacrimation

- Anaemia and cardiac arrhythmias have been reported

- Individuals exposed to chloropicrin may be more susceptible to subsequent exposure due to sensitisation

Dermal Effects

- Exposure may lead to irritation with a burning sensation and erythema

- Blistering and dermatitis may also occur

Eye Effects

- Lacrimation, discomfort, blepharospasm and pain are common

Oral Effects

- Irritant to mucous membranes with a burning sensation in the mouth

- Ingestion causes severe gastro-enteritis with, nausea, vomiting, colic and diarrhoea

Chronic Clinical Effects

Chloropicrin is a sensitiser and has been reported to induce recurrent asthma attacks. Persons exposed to chloropicrin may be more susceptible to subsequent exposure.[6]

Chronic occupational exposure may lead to eye irritation, unpleasant taste, nausea and loss of appetite.

Management

Inhalation Management

- Maintain a clear airway, give humidified oxygen and ventilate if necessary

- If respiratory irritation occurs assess respiratory function and if necessary perform chest X-rays to check for chemical pneumonitis

- Consider the use of steroids to reduce inflammation response

- Treat pulmonary oedema with PEEP or CPAP ventilation

- In severe cases monitor renal, liver and cardiac function

- Symptomatic and supportive care

Dermal Management

- Remove any remaining contaminated clothing, place in double, sealed, clear bags and label; store in a secure area away from patients and staff

- Irrigate with copious amounts of water

- Topical steroids and antiseptic solutions may be used for dermatitis and erythema

- Treat burns symptomatically

Eye Management

- Irrigate thoroughly with water or saline for 15 minutes

- Stain with fluoroscein and refer to an ophthalmologist if there is any uptake of the stain

Oral Management

- NO GASTRIC LAVAGE OR EMETIC

- Encourage oral fluids

- Symptomatic and supportive care

Summary of Environmental Hazards

- In air chloropicrin undergoes photolysis to phosgene, nitrosyl chloride, chlorine, and nitric oxide with a half life of approximately 20 days [3]

- In water chloropicrin is expected to readily volatilise with a half-life in a model river of 4.3 hours and a model lake 5.2 days[3]

- It undergoes photolysis in the surface layers of water with a half-life of about 3 days

- Chloropicrin is not expected to adsorb to sediment or bioconcentrate in fish

- Drinking Water Standards:
  pesticide: $0.1 \, \mu g.l^{-1}$ (UK max)

- Soil Guidelines: no data available

- Air Quality Standards: no data available

REFERENCES

1. Budavari S, O'Neil MJ, Smith A, Heckelman PE & Kinneary JF (eds.) *The Merck Index*, 12th edn. Merck & Co., Inc., Whitehouse Station, 1996

2. Clayton GD & Clayton FE (eds.) *Patty's Industrial Hygiene and Toxicology*, 4th edn. vol.2A John Wiley & Sons, Inc., New York, 1994

3. Hall AH & Rumack BH (eds.) *TOMES System®* Micromedex, Englewood, Colorado. CD ROM. vol 41 (exp. 31 July 1999)

4. Gonmori K, Muto H, Yamamoto T & Takahashi K, 1987. A case of homicidal intoxication by chloropicrin. *Am J Foren Med Pathol*; 8(2): 135–8

5. Ellenhorn MJ, Schonwalds S, Ordog G & Wasserberger J. *Ellenhorn's Medical Toxicology – Diagnosis and Treatment of Human Poisoning*, 2nd edn. Williams & Wilkins, London, 1997

6. Hathaway GJ, Proctor NH & Hughes JP. *Proctor and Hughes' Chemical Hazards of the Workplace*, 4th edn. Van Nostrand Reinhold, New York, 1996

Jennifer Butler

COPPER

Key Points

- Copper is a lustrous, ductile, odourless solid metal and a distinct reddish-brown colour; it is non-magnetic and a good conductor of heat and electricity
- It may be absorbed by ingestion, across the skin or by inhalation of fumes or dust
- Severe systemic toxicity is much less likely following exposure to metallic copper than copper salts

FIRST AID

- Terminate exposure and support vital functions
- The casualty should be moved to an uncontaminated area
- If the casualty is unconscious a clear airway should be established and maintained; give 100% oxygen if available
- **Dermal Exposure:** Remove contaminated clothing, if possible under a shower and place in double, sealed, clear bags and label; store the bags in a secure area away from patients and staff
- Wash the skin thoroughly with copious amounts of water
- **Eye Exposure:** Irrigate thoroughly with water or saline for 15 minutes
- **Oral Exposure:** Encourage small quantities of oral fluids (no more than 50–100 ml in total)

Detailed Information

- Copper is a lustrous, ductile, odourless solid metal of a distinct reddish-brown colour; it is non-magnetic and a good conductor of heat and electricity
- *Common synonyms* copper metallic
- CAS 7440-50-8
- UN specific number not available
- NIOSH/RTECS GL 5325000
- Atomic symbol Cu
- Atomic weight 63.57
- Copper is used primarily in electrical equipment, also in plating, plumbing and heating, and as an alloy component
- When exposed to moisture, copper gradually develops a coating of green basic carbonate
- Copper is insoluble in water and is not combustible
- It decomposes on heating giving off toxic fumes
- Copper reacts violently with a wide variety of substances; it reacts with acetylene, other alkenes and azides to form shock-sensitive compounds
- Mixtures of finely dispersed copper powder and various strong oxidants heated or subjected to shock or friction are an explosion hazard

Summary of Human Toxicity

- Copper is absorbed by ingestion, across the skin and by inhalation of fumes or dust
- Acute copper poisoning can occur from industrial exposure by inhalation of fumes or dusts, or the contamination of food with copper cooking utensils
- The major route of copper excretion is via the bile and faeces[1]
- Almost all of the copper absorbed is stored in the liver, brain, heart, kidney and muscle[1]
- Excess copper can cause liver cell necrosis; copper is then released into the serum and taken up by erythrocytes[2]
- If the release of copper is large enough, haemolytic crisis can occur
- The use of copper piping in household plumbing systems may result in copper water pollution; acidity increases the leaching of copper into tap water; concentrations may also increase in soft and chlorinated water
- Water containing 1 mg.l^{-1} of copper is associated with acute gastrointestinal effects[1]
- Occupational Exposure Standards :
  Long-term exposure limit: 0.2 mg.m^{-3} fume;
  1 mg.m^{-3} dust/mist

  Short-term exposure limit: 2 mg.m^{-3} dust/mist

Acute Clinical Effects

Inhalation Effects

- Metal fume fever may develop following exposure to metallic copper fumes, characterised by fever, chills, headache, muscle pain and vomiting which may be delayed in onset for several hours
- Copper dust irritates the mucous membranes of the respiratory tract causing nasal discharge, headache, sore throat, burning sensation, cough and shortness of breath
- Prolonged or repeated exposure can lead to systemic effects; see below

Dermal Effects

- Contact dermatitis has been reported[3]
- Prolonged or repeated exposure can lead to systemic effects, see below

Eye Effects

- Dust irritates the eyes causing redness
- Copper causes mechanical damage to the eye, especially if driven into the eye by explosion
- The degree of destruction to the eye depends on the size and location of the foreign body
- Intraocular foreign bodies of copper may cause haemorrhage, rapid abscess formation, destruction of the vitreous body, and detached retina[4]

- Copper foreign body in the cornea, conjunctiva or sclera causes a purulent inflammatory reaction and may cause fibrinous iritis [4]

Oral Effects

- Abdominal pain, nausea, vomiting, epigastric burns and diarrhoea

- Systemic effects are possible but are more likely following chronic exposure or ingestion of copper salts, see below

Systemic Effects

- Systemic effects are much more likely following exposure to copper salts (e.g. copper sulphate, see p. 79) than metallic copper

- Metabolic acidosis, cirrhosis and methaemoglobinaemia have been reported following ingestion of metallic copper-contaminated food or water

- Haemolysis has been reported[1]

Chronic Clinical Effects

Industrial chronic copper poisoning is associated with anorexia, nausea, vomiting and hepatomegaly. [5] Exposure to copper-contaminated tap water has been reported to cause green pigmentation of blonde hair. [3] Green discoloration of the skin may occur following chronic dermal exposure. Chronic ingestion of copper coins has been reported to cause hepatic necrosis.[3]

Management

Serum copper concentrations may be determined. Normal serum concentrations are 11–20 μmol.l^{-1} (0.7–1.3 mg.l^{-1})

Inhalation Management

- Maintain a clear airway, give humidified oxygen and ventilate if necessary

- If respiratory irritation occurs assess respiratory function and if necessary perform chest X-rays to check for chemical pneumonitis

- Consider the use of steroids to reduce the inflammatory response

- Treat pulmonary oedema with PEEP or CPAP ventilation

- If systemic effects develop, see below

Dermal Management

- Remove any remaining contaminated clothing, place in double, sealed, clear bags and label; store in a secure area away from patients and staff

- Irrigate with copious amounts of water

- Irritation should be treated symptomatically

Eye Management

- Irrigate thoroughly with running water or saline for 15 minutes

- Stain with fluorescein and refer to an ophthalmologist if there is any uptake of the stain; surgical removal of foreign bodies if appropriate

Oral Management

- Encourage oral fluids

- Consider gastric lavage within 1 hour of a substantial ingestion

- Contact a poisons information service for further guidance on gut decontamination

- Give plasma expanders, blood or IV fluids for shock and analgesics for pain

- Consider the use of steroids to reduce the inflammatory response

- Take abdominal X-ray to check for perforation

- If systemic effects develop, see below

Systemic Management

- Monitor and correct electrolytes, arterial pH and hepatic function tests

- Haemodialysis may be of benefit in the early stages of poisoning when the metal is still present in the circulation as free copper [2]

- In severe cases, chelation therapy using calcium sodium edetate, D-penicillamine or DMPS should be instituted; contact a poisons information service for further guidance

Summary of Environmental Hazards

- Copper is not likely to accumulate in the atmosphere due to the short residence time for airborne copper aerosols; however, airborne copper may be transported over long distances

- Copper accumulates significantly in the food chain

- Drinking Water Standards:
  3000 μg.l^{-1} (UK max)
  2000 μg.l^{-1} (WHO provisional guideline)
  1000 μg.l^{-1} (WHO level where customers may complain)

- Soil Guidelines: Dutch Criteria:
  36 mg.kg^{-1} (target)
  190 mg.kg^{-1} (intervention)

- Air Quality Standards: no data available

REFERENCES

1. Friberg L, Nordberg GF & Vouk VB. *Handbook on the Toxicology of Metals*. Elsevier, Amsterdam, 1989

2. Ellenhorn MJ, Schonwalds S, Ordog G & Wasserberger J. *Ellenhorn's Medical Toxicology – Diagnosis and Treatment of Human Poisoning*, 2nd edn. Williams & Wilkins, London, 1997

3. Hall AH & Rumack BH (eds.) *TOMES System* ® Micromedex, Englewood, Colorado. CD ROM. vol 41 (exp. 31 July 1999)

4. Grant MW & Schuman JS. *Toxicology of the Eye*, 4th edn. Charles C Thomas, Springfield, 1993

5. Baselt RC & Cravey RH. *Disposition of Toxic Drugs and Chemicals in Man*, 4th edn. Chemical Toxicology Institute, 1995

Elizabeth Schofield

COPPER SULPHATE

Key Points

- Copper sulphate is a highly water soluble blue crystal, crystalline granule or powder
- It is irritant to the mucous membranes and skin
- Ingestion of copper sulphate causes moderate gastrointestinal distress and may lead to hepatic and renal failure and death
- *In the event of a large spill, stay upwind and out of low areas. Ventilate closed spaces. Protective clothing, eye protection and breathing apparatus should be worn*

FIRST AID

- Terminate exposure and support vital functions
- The casualty should be moved to an uncontaminated area
- Rescuers should, ideally, be trained personnel and must be careful *not to put themselves at risk* and *so wear appropriate protective clothing and, if available, breathing apparatus*
- If the casualty is unconscious a clear airway should be established and maintained; give 100% oxygen if available
- **Dermal Exposure:** Remove contaminated clothing, if possible under a shower and place in double, sealed, clear bags and label; store the bags in a secure area away from patients and staff
- Wash the skin thoroughly with copious amounts of water
- **Eye Exposure:** Irrigate thoroughly with water or saline for 15 minutes
- **Oral Exposure:** Encourage small quantities of oral fluids (no more than 50–100 ml in total)

Detailed Information

- Copper sulphate is a highly water soluble blue crystal, crystalline granule or powder
- *Common synonyms* blue copper, blue stone, blue vitriol, copper II sulphate, copper II sulphate pentahydrate
- CAS 7758-99-8
- UN 9109
- NIOSH/RTECS GL 8900000
- Molecular formula $CuSO_4.5H_2O$
- Molecular weight 249.7
- Copper sulphate is used in agriculture as a fungicide, algicide, in animal nutrition and in some fertilisers
- Copper sulphate is readily soluble in water forming an acidic solution
- Copper sulphate is not combustible

Summary of Human Toxicity

- Copper is absorbed by ingestion, by inhalation and across the skin
- The more highly water soluble, the more toxic the copper salt
- The major route of copper excretion is via the bile and faeces[1]
- Almost all of the copper absorbed is stored in the liver, brain, heart, kidney and muscle[1]
- Excess copper or copper salts can cause liver cell necrosis; copper is then released into the serum and taken up by erythrocytes[2]
- If the release of copper is large enough, haemolytic crisis can occur
- Copper also causes focal necrosis of the proximal tubules[3]
- Kidney damage may occur as a direct toxic effect of copper and/or associated haemolysis
- Clinical effects develop following ingestion of at least 15 mg of copper as a copper salt[1]
- Blood concentration of 3 mg.l^{-1} is associated with gastrointestinal effects after ingestion of large amounts of copper sulphate[1]
- Haemoglobinaemia has been reported after ingestion of about 175 g copper sulphate[3]
- Renal failure and death may follow ingestion of 1 g of copper sulphate[3]
- Death usually occurs within 1 to 7 days post ingestion of 10 to 20 g of copper sulphate[4]
- Blood concentration of 8 mg.l^{-1} is associated with renal and hepatic damage[1]
- Occupational Exposure Standards:
  Long-term exposure limit: 1 mg.m^{-3} (as Cu) dust/mist
  Short-term exposure limit: 2 mg.m^{-3} (as Cu) dust/mist

Acute Clinical Effects

Inhalation Effects

- Copper sulphate irritates the mucous membranes of the respiratory tract causing nasal discharge initially
- Sore throat, cough, shortness of breath
- Prolonged or repeated exposure can lead to systemic effects; see below

Dermal Effects

- Copper salts irritate the skin causing pain and erythema
- Prolonged or repeated exposure, or exposure to broken skin can lead to systemic effects; see below
- Allergic contact dermatitis has been reported[2]

Eye Effects

- Copper sulphate solution irritates the eyes

- Redness, pain, impaired vision, inflammation and purulence

- The severity of the reaction depends on the concentration and duration of exposure

- Copper sulphate crystals may cause local inflammation and necrosis, corneal opacity and symblepharon (adhesion of the lid to the conjunctival membrane)

Oral Effects

- Sore throat, epigastric burns, moderate gastrointestinal distress with prompt (violent) vomiting (vomitus may be blue or green in colour) and diarrhoea with intense abdominal pain

- More severely, haematemesis, and melaena, with ulceration of stomach and small intestine mucosae, and collapse

Systemic Effects

- Metallic taste, myalgia, acidosis, pancreatitis, methaemoglobin formation, haemolysis, haemolytic anaemia, hypotension, oliguria, anuria and jaundice

- Tremor and toxic psychosis have been reported

- Focal necrosis of the proximal tubules can lead to tubular proteinuria, haematuria generalised amino aciduria, phosphaturia, uricosuria and hypercalciuria

- More severe effects may take 1 to 7 days to manifest

- In serious cases, hepatic centrilolobular and renal tubule necrosis, leading to hepatic and renal failure, coma and death

Chronic Clinical Effects

Vineyard sprayer's lung disease, a histiocytic granulomatous lung and liver disease, has been described following occupational exposure to copper sulphate for 2 to 15 years.[1,5]

Chronic use of copper sulphate-containing eye drops (concentration not specified) has resulted in discoloration of the cornea, causing slight or no visual disturbance.

Management

Serum copper concentrations may be determined. Normal serum copper conentrations are: $11–20 \mu mol.l^{-1}$ ($0.7–1.3 mg.l^{-1}$)

Inhalation Management

- Maintain a clear airway, give humidified oxygen and ventilate if necessary

- If respiratory irritation occurs assess respiratory function and if necessary perform chest X-rays to check for chemical pneumonitis

- Consider the use of steroids to reduce the inflammatory response

- Treat pulmonary oedema with PEEP or CPAP ventilation

- For systemic effects, see below

Dermal Management

- Remove any remaining contaminated clothing, place in double, sealed, clear bags and label; store in a secure area away from patients and staff

- Irrigate with copious amounts of water

- Skin burns should be treated as a thermal injury

Eye Management

- Irrigate thoroughly with running water or saline for 15 minutes

- Stain with fluorescein and refer to an ophthalmologist if there is any uptake of the stain

Oral Management

- Encourage oral fluids, unless perforation is suspected

- Consider gastric lavage within I hour of a substantial ingestion

- Contact a poisons information service for further guidance on gut decontamination

- Give plasma expanders, blood or IV fluids for shock and analgesics for pain

- Consider the use of steroids to reduce the inflammatory response

- Take abdominal X-ray to check for perforation

- For systemic effects, see below

Systemic Management

- Check whole blood copper concentration

- Monitor and correct electrolytes, arterial pH, hepatic and renal function tests

- Haemodialysis may be of benefit in the early stages of poisoning when the metal is still present in the circulation as free copper[3]

- In severe cases, chelation therapy using calcium sodium edetate, D-penicillamine or DMPS should be instituted; contact a poisons information service for further guidance

Summary of Environmental Hazards

- Copper is not likely to accumulate in the atmosphere due to the short residence time for airborne copper aerosols; however, airborne copper may be transported over long distances

- The mobility of copper in rocks is likely to be enhanced by acidic conditions, while the precipitation of copper is favoured by alkaline conditions in the soil and surface water

- Copper accumulates significantly in the food chain

- Drinking Water Standards: copper:
      $3000 \mu g.l^{-1}$ (UK max)
      $2000 \mu g.l^{-1}$ (WHO provisional guideline)
      $1000 \mu g.l^{-1}$ (WHO level where customers may complain)
      sulphate: $250 mg.l^{-1}$ (UK max)

- Soil Guidelines: Dutch Criteria
      copper: $36 mg.kg^{-1}$ (target)
      $190 mg.kg^{-1}$ (intervention)

- Air Quality Standards: no data available

REFERENCES

1. Friberg L, Nordberg GF & Vouk VB. *Handbook on the Toxicology of Metals*. Elsevier, Amsterdam, 1989

2. Hall AH & Rumack BH (eds.) *TOMES System* ® Micromedex, Englewood, Colorado. CD ROM. vol 41 (exp. 31 July 1999)

3. Ellenhorn MJ, Schonwalds S, Ordog G & Wasserberger J. *Ellenhorn's Medical Toxicology – Diagnosis and Treatment of Human Poisoning*, 2nd edn. Williams & Wilkins, London, 1997

4. Baselt RC & Cravey RH. *Disposition of Toxic Drugs and Chemicals in Man*, 4th edn. Chemical Toxicology Institute, 1995

5. Clayton GD & Clayton FE (eds.) *Patty's Industrial hygiene and toxicology*, 4th edn. John Wiley & Sons, Inc., New York, 1994

Catherine Farrow

CROWD CONTROL AGENTS

Key Points

- Crowd control agents are irritant to the skin, eyes and upper respiratory tract
- They are chemical activators of lacrimal glands, hence the name tear gas
- In the majority of cases effects are short-lived and self-limiting
- Some individuals develop reversible skin reactions, e.g. cellulitis, bullae
- Severe cases are only likely to occur following exposure to high concentrations in confined spaces

FIRST AID

- In the majority of cases effects resolve spontaneously within 15–30 minutes after cessation of exposure and treatment in hospital is often not required
- Rescuers and medical personnel should wear gloves and ensure area is well ventilated
- **Dermal Exposure:** Remove contaminated clothing, if possible dry, and place in double, sealed, clear bags and label; store the bags in a secure area away from patients and staff
- If water comes into contact with clothing this may cause the gas to evaporate further and affect the skin
- Place casualties in a well ventilated area, preferably where there is a free flow of air to ensure rapid evaporation and dispersal of the gas

Detailed Information

- The most common agents are:

CN

- *Common synonyms* 2-chloroacetophenone; 1-chloroacetophenone; tear gas; alpha-chloroacetophenone; monochloroacetone; 2-chloro-1-phenylethanone; phenyl chloromethyl ketone; Mace
- CAS 532-27-4
- UN 1697
- NIOSH/RTECS AM 6300000
- Molecular formula C_7H_7OCl
- Molecular weight 154.60

CS

- *Common synonyms* o-chlorobenzylidine malononitrile; super tear gas; malononitrile; ortho-chlorobenzylidene; beta, beta-dicyano-ortho-chlorostyrene
- CAS 2698-41-1
- UN number not available
- NIOSH/RTECS OO 3675000
- Molecular formula $C_{10}H_5ClN_2$
- Molecular weight 188.62

CR

- *Common synonyms* dibenzoxazepine
- CAS 257-07-8
- UN number not available
- NIOSH/RTECS HQ 3950000
- Molecular formula $C_{13}H_9NO$
- Molecular weight 195.23

- Note CS, CN and CR are code names and are not derived from the chemical names or formulae[1]
- These chemicals are all solids in their natural states but are used as liquid aerosols for easy and relatively safe projection and dispersal, however the presence of solvents or propellants may modify toxicity
- These chemicals are used as self-defence sprays and crowd control agents

Summary of Human Toxicity

- Crowd control agents are irritant to the skin, eyes and upper respiratory tract
- They are chemical activators of lacrimal glands, hence the name tear gas
- In the majority of cases effects are short-lived and self-limiting[2]
- Severe cases are only likely to occur following exposure to high concentrations in confined spaces
- They differ in their relative toxicity:

    dibenzoxazepine (CR): the most potent lacrimator with the least systemic effects

    2-chloroacetophenone (CN): most toxic of these agents, deaths from pulmonary injury and/or asphyxia have been reported.[1] Constituent of Mace[3]

    o-chlorobenzylidine malononitrile (CS): ten times more potent lacrimator than CN, but less toxic

Occupational Exposure Standards

Chemical	Long-term exposure limit	Short-term exposure limit
CN gas	0.05 ppm (0.32 mg . m^{-3})	no data available
CS gas	no data available	no data available
CR gas	no data available	no data available

Acute Clinical Effects

Onset and Duration

- Immediate onset, effects usually settle within 15–30 minutes after removal from exposure

- Occasionally ocular and mucous membrane effects can last for up to 24 hours

Inhalation Effects

- Nasal discomfort, pain and rhinorrhoea

- Sore throat, tight chest, coughing, sneezing and increased secretions

- Bronchospasm and laryngospasm may occur

- Pulmonary oedema may occur 12–24 hours later following excessive exposure

- Patients with pre-existing respiratory disease (e.g. asthma, bronchitis) may be more at risk of severe effects

Dermal Effects

- Burning sensation and erythema which usually settles within 24 hours

- Prolonged exposure, particularly when clothing is wet, can produce chemical burns

- CS gas can cause erythematous dermatitis and allergic contact dermatitis with vesicles, blisters and crusts, in both localised areas and other parts of the body;[4] this is often accompanied by marked oedema and onset is between 12 hours to 3 days post exposure[5,6]

- Skin exposed to CR gas may become painful on contact with water up to 48 hours later

- CN gas is a skin sensitiser and can produce allergic contact dermatitis (pruritus, weeping, papulovesicular rash) within 72 hours of exposure

Eye Effects

- Lacrimation, pain, blepharospasm, conjunctival erythema and periorbital oedema

Oral Effects

- Stinging or burning sensation, possibly nausea and vomiting

Management

Medical personnel should wear gloves and ensure the area is well ventilated. When dealing with heavily contaminated casualties, staff are advised to use respirators.

In the majority of cases effects resolve spontaneously within 15–30 minutes after cessation of exposure and treatment in hospital is often not required.

Inhalation Management

- Patients with persistent respiratory symptoms should be admitted for observation

- Humidified oxygen may provide symptomatic relief

- Patients with severe respiratory effects should be treated supportively

Dermal Management

- The most important first line treatment is removal from exposure and removal of contaminated clothing (dry if possible). Place clothing in double, sealed, clear bags and label; store in a secure area away from patients and staff

- If water comes into contact with clothing this may cause the gas to further evaporate and may affect the skin

- Casualties should be placed in a well ventilated area, preferably where there is a free flow of air to ensure rapid evaporation and dispersal of the gas

- The skin may be washed with soap and water if necessary; further treatment is unlikely to be required in the majority of cases

- Any chemical burns should be treated as thermal burns

- Topical steroids and antiseptic solutions may be used for dermatitis and erythema

- Clothing may be decontaminated by washing in a conventional washing machine with a normal powder or liquid; the clothing should be washed several times before wearing to ensure all the tear gas is removed

- Some police forces in the UK now use a 'CS Gas Incapacitant' which contains a 5% solution of CS in the solvent methyl iso-butyl ketone, propelled by nitrogen

- Methyl iso-butyl ketone is an irritant to the eyes; on contact with the skin it may cause a 'tingling' effect, irritation, erythema, drying and flaking, and blistering

- The onset of these effects may be delayed for up to 8 hours and they can persist for up to 1 week; treatment is symptomatic, with the use of an emollient cream if the skin is dry (see page 168 for further details)

Eye Management

- Usually tear secretions are sufficient to remove the chemical from the eye, but where ocular effects persist, eye irrigation should be undertaken using saline or water

- Referral to ophthalmologist is indicated for patients with severe ocular effects

- Note: CS hydrolyses rapidly on contact with water, CR does not and therefore CR in the hair can be washed into the eyes during showering and produce a second eye exposure

Oral Management

- Symptomatic and supportive care

Summary of Environmental Hazards

CN

- CN sinks in water and may smother benthic life[7]

- In air the half-life is about 9.2 days

- It volatilises slowly in water and is not likely to bioaccumulate

CS and CR

- No data available

- Drinking Water Standards: hydrocarbon total: $10 \, \mu g \, .l^{-1}$ (UK max)

- Soil Guidelines: no data available

- Air Quality Standards: no data available

REFERENCES

1. Beswick FW, 1983. Chemical agents used in riot control and warfare. *Hum Toxicol*; 2(2): 247–56

2. Ballantyne B & Swanston DW, 1978. The comparative acute mammalian toxicity of 1- chloroacetophenone (CN) and 2-chlorobenzylidine malononitrile (CS). *Arch Toxicol*; 40(2): 75–95

3. Hu H, Fine J, Epstein P, Kelsey K, Reynolds P & Walker B, 1989. Tear gas – harassing agent or toxic chemical weapon? *J Am Med Assoc*; 262(5): 660–3

4. Parneix-Spake A, Theisen A, Roujeau JC & Revuz J, 1993. Severe cutaneaous reactions to self-defense sprays. *Arch Dermatitis*; 129(7): 913

5. Leenutaphong V & Goerz G, 1989. Allergic contact dermatitis from chloroacetophenone (tear gas). *Contact Dermatitis*; 20(4): 316

6. Ro YS & Lee CW, 1991. Tear gas dermatitis. Allergic contact sensitization due to CS. *Int J Dermatol*; 30(8): 576–7

7. Hall AH & Rumack BH (eds.) *TOMES System* ® Micromedex, Englewood, Colorado. CD ROM.vol 41 (exp. 31 July 1999)

Henrietta Wheeler

1,2-DIBROMOETHANE

Key Points

- 1,2-Dibromoethane is a colourless, viscous liquid with a chloroform-like odour
- It is rapidly absorbed orally, dermally and by inhalation; severe symptoms can be delayed for 24 hours post exposure
- On inhalation, shortness of breath, bronchospasm, laryngeal oedema, chemical pneumonitis and pulmonary oedema may occur
- Dermal exposure results in erythema and blistering, first or second degree burns may take 24 hours to develop
- Systemic effects can occur from inhalation, ingestion and dermal exposure
- 1,2-Dibromoethane may cause CNS depression, renal and hepatic failure or skeletal muscle necrosis
- *In the event of a large spill, stay upwind and out of low areas. Ventilate closed spaces. Protective clothing, eye protection and breathing apparatus should be worn*

FIRST AID

- Terminate exposure and support vital functions
- The casualty should be moved to an uncontaminated area
- Rescuers should, ideally, be trained personnel and must be careful *not to put themselves at risk* and *so wear appropriate protective clothing and, if available, breathing apparatus*
- If the casualty is unconscious a clear airway should be established and maintained; give 100% oxygen if available
- **Inhalation Exposure:** If the patient stops breathing, expired air resuscitation should be started immediately using a pocket mask with a one way valve, if available. It is important where the face is contaminated that expired air resuscitation is NOT attempted unless an airway with rescuer protection is used
- **Dermal Exposure:** Remove contaminated clothing, if possible under a shower and place in double, sealed, clear bags and label; store the bags in a secure area away from patients and staff
- Wash the skin thoroughly with copious amounts of water
- **Eye Exposure:** Irrigate thoroughly with water or saline for 15 minutes
- **Oral Exposure:** Encourage small quantities of oral fluids (no more than 50–100 ml in total)

Detailed Information

- 1,2-Dibromoethane is a colourless, viscous liquid with a sweet, pleasant, chloroform-like odour
- *Common synonyms* DBE, ethylene bromide, 1,2-ethylene dibromide, glycol dibromide
- CAS 106-93-4
- UN 1605
- NIOSH/RTECS KH 9275000
- Molecular formula $C_2H_4Br_2$

- Molecular weight 187.88
- 1,2-Dibromoethane has been used as a fumigant; in leaded petrol as a lead scavenger (anti-knock agent) and as a chemical intermediate in the industrial synthesis of other brominated compounds; it may also be found as a solvent in paints, varnish and finish removers
- It is slightly soluble in water (0.4%) and is soluble in most organic solvents
- 1,2-Dibromoethane is stable under normal conditions of handling and storage; exposing dibromoethane to heat or light in the presence of moisture should be avoided as it may hydrolyse to produce hydrogen bromide
- It is incompatible with strong oxidisers, chemically active metals such as sodium, potassium, calcium, zinc, hot aluminium and hot magnesium, and liquid ammonia
- It is non-flammable and non-combustible; at high temperature it decomposes to release hydrogen bromide, bromine, carbon monoxide and carbon dioxide

Summary of Human Toxicity

- 1,2-Dibromoethane is systemically absorbed through ingestion, inhalation and intact skin
- It is a extremely irritating to mucous membranes, eyes and skin
- Due to its low vapour pressure and relative stability, the number of serious human exposures has been very low
- Symptoms following acute exposure may also be delayed up to 24–48 hours; deaths from acute exposure to high concentrations of 1,2-dibromoethane are usually due to pulmonary oedema or pneumonia following lung damage
- Acute inhalation may also cause varying degrees of renal or hepatic impairment[1]
- Ingestion of 140 mg.kg[-1] body weight was fatal,[2] although ingestion of 3 ml (6840 mg) has also been reported to cause death[3]
- Maximum Exposure Limits:
  Long-term exposure limit: 0.5 ppm (3.9 mg.m[-3])
  Short-term exposure limit: no data available

Acute Clinical Effects

Inhalation Effects

- 1,2-Dibromoethane is a severe respiratory tract irritant, resulting in coughing, wheezing, shortness of breath and chest pain
- Prolonged exposure or exposure to high concentrations may cause bronchial or laryngeal oedema, chemical pneumonitis and pulmonary oedema
- Systemic effects may be expected; see below

Dermal Effects

- Dilute solutions would be expected to cause slight irritation characterised by erythema and exfoliation

- Concentrated solutions may cause erythema, blistering, oedema and necrosis; superficial and deep burns can take 24 hours to develop

- Systemic effects may be expected; see below

Eye Effects

- 1,2-Dibromoethane in the eye results in conjunctival irritation or corneal abrasions, but no permanent damage is expected

Oral Effects

- Concentrated 1,2-dibromoethane is absorbed rapidly following ingestion and causes burns and blisters to the buccal mucosa and gastrointestinal tract

- Systemic effects may be expected; see below

Systemic Effects

- Mild: nausea, vomiting, diarrhoea, abdominal cramps or pain, weakness and headache

- Moderate: tachycardia (or bradycardia), mild hypotension (or severe hypertension), oliguria, jaundice, agitation, confusion, delirium and coma, which may be intermittent

- Severe: metabolic acidosis and shock; cardiac, renal and hepatic failure occur within 12–48 hours; rarely reported effects include skeletal muscle necrosis, cerebral oedema and intracerebral oedema

Chronic Clinical Effects

Chronic inhalation may cause bronchitis and/or shortness of breath which can progress to pulmonary oedema or pulmonary fibrosis. Impairment of renal and hepatic function may also occur. 1,2-Dibromoethane is a skin irritant and can cause sensitisation.

Although human epidemiological studies have proved inconclusive, the extensive evidence for carcinogenicity in animal studies indicates that 1,2-dibromoethane is a potential human carcinogen.[2]

Chronic exposure of five years, at a mean concentration of 0.68 mg.m^{-3} in the breathing zone, significantly decreased sperm counts and motility in occupationally exposed workers.[4]

Management

Plasma 1,2-dibromoethane concentrations are not useful

Inhalation Management

- Maintain a clear airway, give humidified oxygen and ventilate if necessary

- If respiratory irritation occurs assess respiratory function and if necessary perform chest X-rays to check for chemical pneumonitis

- Consider the use of steroids to reduce the inflammatory response

- Treat pulmonary oedema with PEEP or CPAP ventilation

- All patients should be observed for systemic effects; see systemic management below

Dermal Management

- Remove any remaining contaminated clothing and place in double, sealed, clear bags and label; store in a secure area away from patients and staff

- Irrigate with copious amounts of water paying attention to the nail beds, skin folds and hair

- Treat burns symptomatically

- All patients should be observed for systemic effects; see systemic management below

Eye Management

- Irrigate thoroughly with running water or saline for at least 15 minutes

- Stain with fluorescein and refer to an ophthalmologist if there is any uptake of the stain

Oral Management

- NO GASTRIC LAVAGE OR EMETIC

- Encourage oral fluids

- Consider nasogastric aspiration of stomach contents within 1 hour of ingestion

- Activated charcoal may be given, although there is no evidence of its ability to adsorb 1,2-dibromoethane

- Contact a poisons information service for further guidance on gut decontamination

- If facilities are available, early gastro–oesophagoscopy should be undertaken within 12 hours of the event to assess the extent and severity of the injury

- All patients should be observed for systemic effects; see systemic management below

Systemic Management

- All patients with any symptoms should be observed for 24–48 hours

- Treatment is symptomatic and supportive; there is no specific antidote for 1,2-dibromoethane

- Check and correct urea and electrolytes, blood gases and pH

- Monitor ECG, WBC count, haematocrit, respiration, renal and hepatic function

- Raised serum levels of uric acid, skeletal muscle enzymes, hepatic aminotransferases, creatinine phosphokinase and creatinine have occurred in severe poisoning[4]

- Haemodialysis has been used to treat renal failure and acidosis in two patients, serum bromide levels were not recorded

- Contact a poisons information service for further guidance

Summary of Environmental Hazards

- 1,2-Dibromoethane degrades in the atmosphere by reaction with photochemically produced hydroxyl radicals with a half-life of 32 days

- When spilled on land 1,2-dibromoethane will partially evaporate

- In water 1,2-dibromoethane will evaporate; the estimated half-life is between 1 and 5 days

- Persistence in soil can vary greatly from several days to several months, depending on the soil

- Due to potential slow biodegradation rate in the soil it will potentially leach into groundwater

- 1,2-Dibromoethane will not bioconcentrate in the food chain

- Drinking water standards:
    hydrocarbon total: $10\,\mu g.l^{-1}$ (UK max)

- Soil guidelines: no data available

- Air Quality Standards: no data available

REFERENCES

1. Hathaway GJ, Proctor NH & Hughes JP. *Proctor and Hughes' Chemical Hazards of the Workplace*, 4th edn. Van Norstrand Reinhold, New York, 1996

2. International Programme on Chemical Safety. *Environmental Health Criteria 177: 1,2 Dibromoethane*. WHO, Geneva, 1996

3. Singh S, Chaudry D, Garg M & Sharma BK, 1993. Fatal ethylene dibromide ingestion. *J Assoc. Physicians of India*. 41 (9): 608

4. Hall AH & Rumack BH (eds.) *TOMES System* ® Micromedex, Englewood, Colorado. CD ROM. vol 41 (exp. 31 July 1999)

Robie Kamanyire

1,2-DICHLOROETHANE

Key Points

- 1,2-Dichloroethane is a colourless, oily liquid with a chloroform-like odour and is sweet to taste
- It is irritating to both the respiratory system and gastrointestinal tract
- 1,2-Dichloroethane causes CNS depression and can sensitise the myocardium to endogenous catecholamines
- 1,2-Dichloroethane can cause liver and renal damage
- It is heavier than air and can travel great distances to an ignition source to flash back and cause fire or explosion
- *1,2-Dichloroethane vapours are heavier than air and may accumulate in low or confined areas; in the event of a large spill, stay upwind and out of low areas. Ventilate closed spaces. Protective clothing, eye protection and breathing apparatus should be worn*

FIRST AID

- Terminate exposure and support vital functions
- The casualty should be moved to an uncontaminated area
- Rescuers should, ideally, be trained personnel and must be careful *not to put themselves at risk* and *so wear appropriate protective clothing and, if available, breathing apparatus*
- If the casualty is unconscious a clear airway should be established and maintained; give 100% oxygen if available
- **Inhalation Exposure:** If the patient stops breathing, expired air resuscitation should be started immediately using a pocket mask with a one way valve, if available. It is important where the face is contaminated that expired air resuscitation is NOT attempted unless an airway with rescuer protection is used
- **Dermal Exposure:** Remove contaminated clothing, if possible under a shower and place in double, sealed, clear bags and label; store the bags in a secure area away from patients and staff
- Wash the skin thoroughly with copious amounts of water
- **Eye Exposure:** Irrigate thoroughly with water or saline for 15 minutes
- **Oral Exposure:** Encourage small quantities of oral fluids (no more than 50–100 ml in total)

Detailed Information

- 1,2-Dichloroethane is a clear, colourless, oily liquid that has a chloroform-like odour and sweet taste
- *Common synonyms* 1,2-bichloroethane, dichloroethane, dichloroethylene, EDC, ethylene dichloride, glycol dichloride
- CAS 107-06-2
- UN 1184
- NIOSH/RTECS KI 0525000
- Molecular formula $C_2H_4Cl_2$
- Molecular weight 98.96

- 1,2-Dichloroethane is used in the synthesis of vinyl chloride, as an insecticide, fumigating agent, anti-knocking agent in petrol, textile and polyvinyl chloride cleaning agent; in paints, coatings and adhesives; metal degreasing agent, solvent in printing inks, resins and rubber
- 1,2-Dichloroethane is heavier than air and is slightly soluble in water, miscible in alcohol, chloroform, ether and most organic solvents
- It reacts violently with aluminium, dinitrogen tetroxide, ammonia and dimethylaminopropylamine[1]
- 1,2-Dichloroethane is flammable and on heating decomposes to chloride fumes, vinyl chloride and phosgene gas

Summary of Human Toxicity

- 1,2-Dichloroethane is absorbed through the lungs, gastrointestinal tract and skin
- May depress and sensitise the myocardium to endogenous catecholamines after inhalation and ingestion; this cardiac sensitisation is unpredictable in terms of exposure duration and quantity
- 1,2-Dichloroethane is metabolised in the liver and eliminated via the kidney
- Alcohol is thought to increase the toxicity of 1,2-dichloroethane
- 1,2-Dichloroethane is a potential carcinogenic risk to humans[2]
- 1,2-Dichloroethane odour is not a dependable guide to exposure as olfactory warning properties are limited by the development of tolerance[3]
- Ingestion of 8–200 ml has been reported as fatal, death is usually due to circulatory or respiratory collapse[2]

Toxicity of 1,2-dichloroethane[2,4,5]

Concentration	Clinical effects
0.05–0.15 mg.l⁻¹	Long-term repeated exposure may lead to neurological changes, anorexia, irritation of mucous membranes, liver and kidney impairment
6 ppm	Odour detectable
356 mg.m⁻³	Odour threshold in air
7 mg.l⁻¹	Odour threshold in water
40 ppm	Impairment of CNS, irritability, increased morbidity including disease of liver and bile duct
10–200 ppm	Anorexia, dizziness, insomnia, vomiting, lacrimation, constipation, epigastric pain, tender liver on palpation and elevated uribilogen concentration

- Maximum Exposure Limits:
  Long-term exposure limit: 5 ppm (21 mg.m⁻³)
  Short-term exposure limit: no data available

Acute Clinical Effects

Inhalation Effects

- 1,2-Dichloroethane is irritating to the respiratory tract leading to coughing and wheezing

- Nausea, vomiting, diarrhoea and cramp-like epigastric pain can occur

- Metallic taste in the mouth has been noted

- Respiratory depression, dyspnoea, cyanosis and pulmonary oedema which may be immediate or delayed

- Inhalation may cause CNS depression and systemic effects; see below

Dermal Effects

- 1,2-Dichloroethane is a strong irritant causing defatting of the skin

- Prolonged exposure may lead to severe irritation, oedema and necrosis

- 1,2-Dichloroethane is readily absorbed through the skin and can cause systemic effects; see below

Eye Effects

- 1,2-Dichloroethane vapour causes lacrimation

- At higher concentrations and splashes 1,2-dichloroethane causes immediate discomfort, hyperaemia of the conjunctiva and slight corneal epithelial damage

- Eyes normally return to normal within 1–2 days[6]

Oral Effects

- 1,2-Dichloroethane is a gastric irritant leading to nausea, vomiting (which may have blood or bile in it), abdominal pain and diarrhoea[7]

- Constricted cardiac pain and epigastric pain may occur

- Aspiration may lead to cough, wheeze, frothy sputum production, cyanosis and pulmonary oedema

- Shock and circulatory collapse lead to death

- Ingestion may cause CNS depression and systemic effects; see below

Systemic Effects

- Headache, confusion, ataxia, dizziness, lethargy, dilated pupils, convulsions, drowsiness and coma

- Oliguria, anuria, acute tubular necrosis and renal failure

- Liver necrosis and impairment, hypoprothombinaemia, hypo- or hyperglycaemia and hypercalcaemia have all been noted

- Extrapyramidal symptoms are occasionally noted

- Death may occur from ventricular arrhythmias as at high concentrations the solvent sensitises the myocardium to adrenaline and other catecholamines

Chronic Clinical Effects

Chronic exposure may lead to neurological changes, insomnia, anorexia, irritation of the mucous membranes and liver and renal impairment.

Chronic skin exposure may lead to red, raw, dry, shiny skin with a 'cigarette paper'-like appearance. Bluish, purple discoloration has also been reported.[3]

Management

Laboratory analysis may be of use in the diagnosis if exposure is uncertain.

Inhalation Management

- Maintain a clear airway, give humidified oxygen and ventilate if necessary

- If respiratory irritation occurs assess respiratory function and if necessary perform chest X-rays to check for chemical pneumonitis

- See systemic management below

Dermal Management

- Remove any remaining contaminated clothing and place in double, sealed, clear bags and label; store in a secure area away from patients and staff

- Irrigate with copious amounts of water

- An emollient may be required

- Treat irritation and necrosis symptomatically

- See systemic management below

Eye Management

- Irrigate thoroughly with running water or saline for 15 minutes

- Stain with fluorescein and refer to an ophthalmologist if there is any uptake of the stain

Oral Management

- Encourage oral fluids for small ingestion

- DO NOT INDUCE EMESIS due to risk of aspiration

- Consider gastric lavage followed by 50g activated charcoal (ensuring airway is protected) within 1 hour of a substantial ingestion

- Symptomatic and supportive care

- See systemic management below

Systemic Management

- Diazepam should be used for convulsions

- Patients should be kept at rest

- With severe exposures the patient should be kept on a cardiac monitor for 12 hours, avoiding the use of all stimulants except for resuscitation

- Monitor blood sugar, calcium, renal and liver function

- Haemodialysis or peritoneal dialysis are of use in acute renal failure

- Treat pulmonary oedema with PEEP or CPAP ventilation

- Symptomatic and supportive care

Summary of Environmental Hazards

- In the atmosphere 1,2-dichloroethane may be transported long distances and is removed by photo-oxidation, with a half-life of 1 month[8]

- In water it is primarily removed via evaporation, with a half-life of hours to 10 days

- If released on the land 1,2-dichloroethane evaporates to the air or percolates down to the ground water where it may persist for long periods

- 1,2-Dichloroethane has low toxicity to fish and is not expected to concentrate in the food chain

- Drinking Water Standards:
      hydrocarbon total: $10 \,\mu g . l^{-1}$ (UK max)
      1,2-dichloroethane: $30 \,\mu g . l^{-1}$ (WHO guideline)

- Soil Guidelines: Dutch criteria: $4 \,mg . kg^{-1}$ (intervention)

- Air Quality Standards: $0.7 \,mg . m^{-3}$ for continuous exposure, averaging time 24 hours (WHO guideline)

REFERENCES

1. Commission of the European Communities (eds.) *Organo-chlorine solvents. Health risks to workers*. Royal Society of Chemistry, 1986

2. Hathaway GJ, Proctor NH & Hughes JP. *Proctor and Hughes' Chemical Hazards of the Workplace*, 4th edn. Van Nostrand Reinhold, New York, 1996

3. Ellenhorn MJ, Schonwalds S, Ordog G & Wasserberger J. *Ellenhorn's Medical Toxicology – Diagnosis and Treatment of Human Poisoning*, 2nd edn. Williams & Wilkins, London, 1997

4. Gosselin RE, Smith RP & Hodge HC. *Clinical Toxicology of Commercial Products*, 5th edn. Williams & Wilkins, Baltimore, 1984

5. International Programme on Chemical Safety. *Environmental Health Criteria 176: 1,2-Dichloroethane*. WHO, Geneva 1995

6. Grant MW & Schuman JS. *Toxicology of the Eye*, 4th edn. Charles C Thomas, Springfield, 1993

7. Ware GW (ed.) 1,2-Dichloroethane. In *Reviews of Environmental Contamination and Toxicology*, Vol 106. Springer-Verlag, 1988

8. Hall AH & Rumack BH (eds.) *TOMES System* ® Micromedex, Englewood, Colorado. CD ROM.vol 41 (exp. 31 July 1999)

Henrietta Wheeler

DICHLOROMETHANE

Key Points

- Dichloromethane is a clear, colourless, liquid with a sweet, penetrating, ether-like odour
- It has anaesthetic-like properties causing profound CNS depression at high concentration with drowsiness, coma and death
- It is an eye, skin and respiratory tract irritant
- Dichloromethane is extremely volatile and high concentrations can accumulate in unventilated areas
- *In the event of a large spill, stay upwind and out of low areas. Ventilate closed spaces. Protective clothing, eye protection and breathing apparatus should be worn*

FIRST AID

- Terminate exposure and support vital functions
- The casualty should be moved to an uncontaminated area
- Rescuers should, ideally, be trained personnel and must be careful *not to put themselves at risk* and *so wear appropriate protective clothing and, if available, breathing apparatus*
- If the casualty is unconscious a clear airway should be established and maintained; give 100% oxygen if available
- **Inhalation Exposure:** If the patient stops breathing, expired air resuscitation should be started immediately using a pocket mask with a one way valve, if available. It is important where the face is contaminated that expired air resuscitation is NOT attempted unless an airway with rescuer protection is used
- **Dermal Exposure:** Remove contaminated clothing, if possible under a shower and place in double, sealed, clear bags and label; store the bags in a secure area away from patients and staff
- Wash the skin thoroughly with copious amounts of water
- **Eye Exposure:** Irrigate thoroughly with water or saline for 15 minutes
- **Oral Exposure:** Encourage small quantities of oral fluids (no more than 50–100 ml in total)

Detailed Information

- Dichloromethane is a clear, colourless, highly volatile, non-inflammable liquid with a sweet, penetrating, ether-like odour
- *Common synonyms* DCM, methylene bichloride, methylene chloride, methylene dichloride
- CAS 75-09-2
- UN 1593
- NIOSH/RTECS PA 8050000
- Molecular formula CH_2Cl_2
- Molecular weight 84.93
- Dichloromethane is a multipurpose solvent, used as a paint remover and in the manufacturer of photographic film; in aerosol propellants and in urethane foam

- It is extremely volatile therefore high airborne concentrations can accumulate in poorly ventilated areas
- The vapour is not flammable and is not explosive when mixed with air
- Dichloromethane hydrolyses slowly in the presence of moisture, producing small quantities of hydrogen chloride
- Dichloromethane is soluble in alcohol and ether; slightly soluble in water
- Reacts violently with nitric acid; reacts with alkaline-earth and alkali metals and attacks many plastics and rubber[1]
- When heated dichloromethane decomposes releasing hydrogen chloride plus small amounts of phosgene and chlorine

Summary of Human Toxicity

- Dichloromethane is absorbed by the lungs, gastrointestinal tract and to a lesser extent dermally and it is able to cross the blood-brain barrier; inhalation is the most frequent form of exposure
- It is rapidly excreted, mostly via the lungs in exhaled air
- At high concentrations, most of the absorbed dichloromethane is exhaled unchanged; the remainder is metabolised to carbon monoxide, carbon dioxide and inorganic chloride[2]
- Serious poisoning can occur without raised carboxyhaemoglobin (COHb) concentration, although raised concentrations of COHb can persist for several hours[3]
- CNS effects are thought to be due to dichloromethane itself or dichloromethane in combination with other sources of COHb, rather than the COHb metabolite
- The raised COHb concentrations are usually not expected to cause adverse effects in healthy individuals, but caution is advocated in patients with cardiovascular disease[4]
- Inhalation can cause fatigue, weakness, sleepiness, light-headedness, chills, and nausea; pulmonary oedema can develop after several hours; liver and kidney damage may occur, although data are minimal
- Estimated lethal oral dose is 0.5–5 ml.kg^{-1}[3]
- The lowest published lethal concentration is 20,000 ppm for 2 hours[3]

Toxicity of dichloromethane [2,3,5]

Concentration	Clinical effects
>300 ppm	Sweet odour
500–1000 ppm for 1–2 h	Unpleasant odour, slight anaesthetic effects, headache, light-headedness, eye irritation and elevated COHb concentration
2300 ppm for 5 min	Odour strong, intensely irritating and dizziness
7200 ppm for 8–16 min	Paraesthesia and tachycardia
>50,000 ppm	Immediately life threatening

- Maximum Exposure Limits:
  Long-term exposure limit: 100 ppm (350 mg.m^{-3})
  Short-term exposure limit: 300 ppm (1060 mg.m^{-3})

Acute Clinical Effects

Inhalation Effects

- Nausea, vomiting, diarrhoea, abdominal pain, blurred vision and headache

- Respiratory tract irritation, coughing and dyspnoea may occur; pulmonary oedema can occur although this is rare

- See systemic effects below

Dermal Effects

- Dermal irritation, erythema, paraesthesia and exfoliation may occur due to the defatting action of the solvent

- Burning and erosions can occur from prolonged contact

- Elevated liver enzymes have been reported following dermal exposure[3]

- For prolonged or severe exposure, see systemic effects below

Eye Effects

- Dichloromethane is highly irritant to the eyes causing lacrimation and conjunctivitis

- Splashes to the eye may cause immediate pain but significant injury is not expected, although immersion has caused severe corneal burns[5,6]

Oral Effects

- Dichloromethane can cause severe mucous membrane irritation, burns, blisters and erosion

- Aspiration leading to pulmonary oedema may occur

- See systemic effects below

Systemic Effects

- CNS depression may include drowsiness, weakness, dizziness and coma; convulsions may occur

- Raised liver enzymes and renal damage have been reported[3]

- Encephalopathy has been reported following repeated exposure[3]

- Angina, myocardial infarction, cardiac arrhythmias and cardiac arrest have been reported, although the cardiovascular system is not generally a target for dichloromethane toxicity[3]

- Hypotension, shock and metabolic acidosis have been reported[7]

- Respiratory failure may develop secondary to CNS depression in severe cases

Chronic Clinical Effects

Chronic exposure may cause damage to the CNS including confusion, delusions, slurred speech, memory impairment, anxiety, focal seizures, encephalopathy, and visual and auditory hallucinations. These effects may have been due to chronic carbon monoxide poisoning from dichloromethane metabolism.[3]

Dichloromethane is a suspected human carcinogen; the IARC has determined that there is sufficient evidence for carcinogenicity in animals and inadequate evidence in humans.

Management

Dichloromethane concentrations are not of use. Dichloromethane is metabolised to carbon monoxide; carboxyhaemoglobin concentration may be a useful indicator of exposure

Inhalation Management

- Maintain a clear airway, give humidified oxygen and ventilate if necessary

- If respiratory irritation occurs assess respiratory function and if necessary perform chest X-rays to check for chemical pneumonitis

- Consider the use of steroids to reduce the inflammatory response

- Treat pulmonary oedema with PEEP or CPAP ventilation

- Symptomatic and supportive care

- See systemic management below

Dermal Management

- Remove any remaining contaminated clothing, place in double, sealed, clear bags and label; store in a secure area away from patients and staff

- Irrigate with copious amounts of water

- An emollient may be required

- Treat irritation and burns symptomatically

- For prolonged or severe exposure, see systemic management below

Eye Management

- Irrigate thoroughly with running water or saline for 15 minutes

- Stain with fluorescein and refer to an ophthalmologist if there is any uptake of the stain

Oral Management

- Encourage oral fluids for ingestion of small quantities

- DO NOT INDUCE EMESIS due to risk of aspiration

- Activated charcoal may be of use

- Consider gastric lavage within 1 hour of a substantial ingestion, ensuring the airway is protected

- Contact a poisons information service for further guidance on gut decontamination

- Treat pulmonary oedema with PEEP or CPAP ventilation

- Symptomatic and supportive care

- See systemic management below

Systemic Management

- Give oxygen

- Monitor carboxyhaemoglobin, renal and liver functions in symptomatic patients

- Symptomatic and supportive care

Summary of Environmental Hazards

- In the atmosphere dichloromethane degrades by reaction with photochemically produced hydroxyl radicals with a half-life of 6 months[2]

- Due to its high volatility dichloromethane is rapidly transferred from water to the atmosphere; the estimated half-life for volatilisation is 3 – 5.6 hours

- In soil, surface dichloromethane is expected to evaporate into the atmosphere; it may partially leach into groundwater where its fate is unknown

- It is not expected to bioaccumulate or bioconcentrate in the food chain

- Drinking Water Standards:
    hydrocarbon total: $10\,\mu g.l^{-1}$ (UK max)
    dichloromethane: $20\,\mu g.l^{-1}$ (WHO guideline)

- Soil Guidelines: Dutch Criteria: detection threshold (target)
    $20\,mg.kg^{-1}$ (intervention)

- Air Quality Standards:
    $3\,mg.m^{-3}$ averaging time 24 hours (WHO guideline)

REFERENCES

1. Dutch Institute for the Working Environment and the Dutch Chemical Industry Association (ed.) *Chemical Safety Sheets – Working Safely with Hazardous Chemicals.* Samson Chemical Publishers, The Netherlands, 1991

2. International Programme on Chemical Safety. *Environmental Health Criteria 164: Methylene Chloride*, 2nd edn. WHO, Geneva 1996

3. Hall AH & Rumack BH (eds.) *TOMES System®* Micromedex, Englewood, Colorado. CD ROM.vol 41 (exp. 31 July 1999)

4. Hathaway GJ, Proctor NH & Hughes JP. *Proctor and Hughes' Chemical Hazards of the Workplace*, 4th edn. Van Nostrand Reinhold, New York, 1996

5. Hall AH & Rumack BH, 1990. Methylene chloride exposure in furniture-stripping shops: ventilation and respirator use practices. *J Occup Med*; 32(1): 33–7

6. Grant MW & Schuman JS. *Toxicology of the Eye*, 4th edn. Charles C Thomas, Springfield, 1993

7. Bates N, Edwards N, Roper J & Volans G (eds.) *Paediatric Toxicology. Handbook of Poisoning in Children.* Macmillan, London, 1997

Simon Ward

DIOXINS

Key Points

- The term dioxin is applied to a large group of closely related chemicals known as polychlorinated dibenzo-*p*-dioxin and polychlorinated dibenzofurans; the most toxic and thoroughly studied is 2,3,7,8-tetrachlorodibenzo-*p*-dioxin (TCDD)
- Note: For the purpose of this entry TCDD is interchangeable with dioxin throughout
- TCDD (often referred to as dioxin) is one of the most potentially toxic synthetic chemicals yet discovered
- The major route of human exposure is ingestion of dioxin-containing foods, particularly meat, dairy products and fish
- No death due to systemic dioxin toxicity has been reported
- The two clinical effects repeatedly reported in exposed populations are chloracne and transient hepatic effects
- *In the event of a large spill, stay upwind and out of low areas. Ventilate closed spaces. Protective clothing, eye protection and breathing apparatus should be worn*

First Aid

- Terminate exposure and support vital functions
- The casualty should be moved to an uncontaminated area
- Rescuers should, ideally, be trained personnel and must be careful *not to put themselves at risk* and *so wear appropriate protective clothing and, if available, breathing apparatus*
- If the casualty is unconscious a clear airway should be established and maintained; give 100% oxygen if available
- **Inhalation Exposure:** If the patient stops breathing, expired air resuscitation should be started immediately using a pocket mask with a one way valve, if available. It is important where the face is contaminated that expired air resuscitation is NOT attempted unless an airway with rescuer protection is used
- **Dermal Exposure:** Remove contaminated clothing, if possible under a shower and place in double, sealed, clear bags and label; store the bags in a secure area away from patients and staff
- Wash the skin thoroughly with copious amounts of water
- **Eye Exposure:** Irrigate thoroughly with water or saline for 15 minutes
- **Oral Exposure:** Encourage small quantities of oral fluids (no more than 50–100 ml in total)

Detailed Information

- Dioxins exist as colourless to white needles or crystals
- The term dioxin is applied to a large group of closely-related chemicals known as polychlorinated dibenzo-*p*-dioxin and polychlorinated dibenzofurans; the most toxic and thoroughly studied is 2,3,7,8-tetrachlorodibenzo-*p*-dioxin (TCDD)

Synonyms of Common Dioxins

Compound	Synonyms
1-Chlorodibenzo-*p*-dioxin	1-CDD
2,3,7,8-Tetrachlorodibenzo-*p*-dioxin	2,3,7,8-Tetrachlorodibenzo-1,4-dioxin; TCDD
2,7-Dichlorodibenzo-*p*-dioxin	2,7-Dichlorodibenzodioxin; DCDD
2-Chlorodibenzo-*p*-dioxin	2-CDD
Dibenzo-*p*-dioxin	Diphenylene dioxide

Identification Numbers of Common Dioxins

Compound	CAS	UN	NIOSH/RTECS
1-Chlorodibenzo-*p*-dioxin	39227-53-7	–	HP 3095300
2,3,7,8-Tetrachloro dibenzo-*p*-dioxin	1746-01-6	–	HP 3500000
2,7-Dichloro dibenzo-*p*-dioxin	33857-26-0	–	HP 3100000
2-Chlorodibenzo-*p*-dioxin	39227-54-8	–	HP 3095500
Dibenzo-*p*-dioxin	262-12-4	–	HP 3090000

Molecular Formula and Weight of Common Dioxins

Compound	Molecular formula	Molecular weight
1-Chlorodibenzo-*p*-dioxin	$C_{12}H_7O_2Cl$	218.5
2,3,7,8-Tetrachloro dibenzo-*p*-dioxin	$C_{12}H_4Cl_4O_2$	322
2,7-Dichloro dibenzo-*p*-dioxin	$C_{12}H_6Cl_2O_2$	253.08
2-Chlorodibenzo-*p*-dioxin	$C_{12}H_7O_2Cl$	218.5
Dibenzo-*p*-dioxin	$C_{12}H_8O_2$	184.20

Dioxin Content of Herbicides

Compound	Mean dioxin concentration
Agent Orange	1.77–2.11 ppm
Agent Purple	32.8–45 ppm
Agent Pink	65.6 ppm
Agent Green	65.6 ppm
Silvex	1–70 ppm
2,4,5-T	0.01–0.05 ppm

- There are 75 possible congeners and isomers of polyhalogenated dibenzo-*p*-dioxin and 135 possible congeners and isomers of polyhalogenated dibenzofurans[1]
- Dioxins are ubiquitous in the environment, generally at low concentrations; environment concentrations peaked around 1970 and have been declining since then, mainly due to changes in industrial processes
- Dioxins do not occur naturally, they are formed during the production of many chlorinated organic solvents, hexachlorophene, and the herbicide 2,4,5-T; emission from

coal-burning power plants, exhaust from diesel engines, cigarette smoke and the incomplete burning of wastes containing chlorine, such as PVC plastic

- Major sources of atmospheric contamination by dioxins are hospital and municipal waste incinerators and metallurgic processes; industrial waste incineration has lower emissions[2]

- Agent orange, agent pink, agent purple and agent green were herbicides used in Vietnam; see table above for a list of dioxin contents in the various preparations

- TCDD is sparingly soluble in water at room temperature

- When heated to decomposition dioxin emits toxic fumes of chlorides; combustion of TCP-contaminated 2,4,5-T can result in conversion to small amounts of TCDD[3]

Summary of Human Toxicity

- Dioxins are absorbed following ingestion, inhalation and dermal exposure

- The major route of human exposure is ingestion of dioxin-containing foods, particularly meat, dairy products and fish; e.g. contaminated fish from the Baltic Sea are important sources of exposure for regular fish eaters[3]

- Plant crops do not concentrate dioxins

- Gastrointestinal absorption varies with the vehicle used, dioxins are less well absorbed from aqueous soil suspensions than from oil or solvent mixtures; more stable congeners are absorbed to a greater extent than less stable ones

- Metabolism of TCDD in man is unknown; it may be slowly metabolised in man to more water-soluble metabolites[3]

- Dioxins are highly lipid-soluble and tend to accumulate in body fat and may be released systemically with weight loss; high concentrations are found in the pancreas and liver

- Half-life is estimated to be 7 years; excretion is mainly through the faeces

- No death due to systemic dioxin toxicity has been reported; the two main clinical effects repeatedly reported in exposed populations are chloracne and transient hepatic effects[4]

- Cumulative oral doses of 100 $\mu g . kg^{-1}$ are estimated to be the minimal toxic dose; dermal exposure to soil concentrations of > 100 ppm may lead to chloracne, lowest toxic dose via dermal exposure is 107 $\mu g . kg^{-1}$ [3]

- Approximately 1.3 g of TCDD contaminated a populated area of 2 km^2 near Seveso, Italy; many animals died but there were no human fatalities as a result of acute exposure but 134 cases of severe chloracne were reported[3]

- Soil concentrations likely to produce chloracne with daily contact probably exceed 100 ppm

- Women may excrete up to half their accumulated dioxin during lactation as dioxin is concentrated in breast milk[5]

- Average daily intake is estimated at 0.05 ng, 98% of which is from dietary sources[6]

- TCDD is a probable human carcinogen, linked with soft tissue sarcomas

- Maximum Exposure Limits: No data available

Acute Clinical Effects

Note: most exposures are to mixtures containing very low concentrations of dioxins; other components may contribute to the toxicity

Inhalation Effects

- Dyspnoea has been noted

- See systemic effects below

Dermal Effects

- Chemical burns of the skin, mucous membranes; nausea, vomiting and severe muscle pains have been reported

- See systemic effects below

Eye Effects

- Eye irritation and blepharoconjunctivitis may occur

Oral Effects

- Nausea and vomiting may occur

- See systemic effects below

Systemic Effects

- Nausea and vomiting may occur

- Onset of symptoms following acute exposure may take days to weeks and can include skin, eye and respiratory tract irritation; headache, dizziness, blurred vision and nausea

- Effects may be non-specific including abdominal pain, fatigue, irritability and difficulty with muscular and mental processes

- Chloracne can result from inhalation, ingestion or dermal contact and may indicate systemic toxicity; chloracne is the only visible sign of exposure although its absence does not rule out exposure

- Acneform lesions may appear 1 to 3 weeks post exposure, although may take months to years to develop; most cases resolve over 1 to 3 years

- First signs include inflammation, resembling lupus erythematosis or erythema elevatum diutinum; photosensitivity may be present

- Initially comedones, milia or epidermal small, pale yellow cysts develop; acne usually involves the face and upper body

- Hyperpigmentation, hirsutism, increased skin fragility and vesicular eruptions on exposed areas have been reported; the skin is generally dry, in contrast to adolescent acne vulgaris

- A grading system has been developed for chloracne, see below

Chloracne Grading System[7]

Grade	Clinical Effects
Grade 1	No change
Grade 2	Few comedones in specific sites only
Grade 3	More comedones in specific sites, no cysts
Grade 4	Numerous comedones in specific sites with cysts
Grade 5	Numerous comedones, cysts in specific and other sites
Grade 6	The same as Grade 4, with inflammatory changes

- Polyphyria cutanea tarda may occur in moderate or severe exposures

- Polyneuropathy with sensory impairment and lower extremity weakness are common in moderate exposure; mild exposure may result in asymptomatic electromyogram alterations

- Hepatotoxicity is the most consistent finding in acute exposures and manifests with elevated liver enzyme activity; mild fibrosis, fatty changes, haemofuscin deposition, parenchymal-cell degeneration and hepatomegaly; in severe exposures, approximately a third of those exposed develop liver damage[8]

- Prothrombin time prolongation has been reported occasionally, this may be related to liver damage; pancreatitis has been reported[3]

Chronic Clinical Effects

Links between chronic exposure to TCDD and cardiovascular disorders, such as atherosclerosis and myocardial infarction, have been investigated; these studies have not shown these disorders to be conclusively related to TCDD exposure. Enzyme induction is prominent in chronic exposure. Workers exposed to dioxins and who consume alcohol seem to have a greater risk of elevated serum gamma glutamyl transferase concentration; the risk increases with dioxin concentrations.[3,9]

Self-limiting haemorrhagic cystitis has been reported following chronic exposure to TCDD-containing soil.[10]

Dioxin is a probable human carcinogen, linked with soft tissue sarcomas.

Management

NB Care workers must ensure adequate protection to prevent self-contamination when carrying out decontamination and medical treatment.

Inhalation Management

- Maintain a clear airway, give humidified oxygen and ventilate if necessary

- If respiratory irritation occurs assess respiratory function and if necessary perform chest X-rays to check for chemical pneumonitis

- Observe closely for development of systemic symptoms; see systemic management below

Dermal Management

- Remove any remaining contaminated clothing, place in double, sealed, clear bags and label; store in a secure area away from patients and staff

- Irrigate with copious amounts of water

- Emollients may ease irritation

- Symptomatic and supportive care; see systemic management below

Eye Management

- Irrigate thoroughly with running water or saline for 15 minutes

- Stain with fluorescein and refer to an ophthalmologist if there is any uptake of the stain

Oral Management

- Encourage oral fluids

- Symptomatic and supportive care; see systemic management below

Systemic Management

- Monitor LFTs, INR/PT, FBC and uroporphyrins following acute exposure

- Chloracne is refractory and may not respond to usual acne treatment

- Long-term topical use of dilute retinoic acid and oral tetracycline has been used for secondary pustular follicles

- Surgery or dermabrasion may be of use in severe cases

- Symptomatic and supportive care

Summary of Environmental Hazards

- Atmospheric dioxins emitted from all sources are deposited on soil, vegetation and surface water

- Although TCDD has a very low vapour pressure it has been shown to be volatile and to occur in air in both the gas-phase and particulate phase; dioxin is not biodegradable

- Gas-phase dioxin reacts with photochemically produced hydroxyl radicals in air with an estimated half-life of 8.3 days; direct photolysis of gas-phase dioxin may occur at a faster rate than hydroxyl radical reaction[3]

- Particulates may be physically removed from the atmosphere by wet and dry deposition; dioxin may be transported long distances through the atmosphere

- Due to its low solubility, most dioxin in water is expected to be associated with sediments or suspended materials; aquatic sediments may be an important, and the ultimate, environmental sink for all global releases of dioxin

- Photolysis half-life at the water surface is estimated to range from 21 hours (summer) to 118 hours (winter); these rates increase significantly as water depth increases[3]

- Volatilisation half-life from the water column of an environmental pond was estimated to be 46 days; although when adsorption to sediment is considered in the model, removal half-life is estimated to be over 50 years[3]

- Half-life in lakes is estimated to be 1.5 years[3]

- Volatilisation from soil surfaces during warm, summer months may be a major mechanism for dioxin removal from soil; volatilisation during cold, winter months is extremely slow

- The half-life of dioxin on soil surfaces may vary from 1 to 3 years, but half-lives in soil interiors may be as long as 12 years[3]

- Absorption by animals and bioaccumulation make the food chain the most important route to man (>90%); fish, dairy and meat products, and vegetables all being important

- Dioxin has been shown to bioconcentrate in aquatic organisms and may bioaccumulate in animals; plant crops do not concentrate dioxins

- Soil concentrations of 1 ppb are estimated to increase the risk of developing cancer by 1 in 1 million; highest average soil concentrations in Seveso were 584 ppb

- Drinking water standards:
     hydrocarbon total: $10\,\mu g.l^{-1}$ (UK max)

- Soil Guidelines: no data available

- Air Quality Standards: no data available

REFERENCES

1. Clayton GD, Clayton FE, (eds.) *Patty's Industrial Hygiene and Toxicology*, 4th edn vol II, Part E. John Wiley & Sons Inc., New York. 1994

2. European Centre for Ecotoxicology and Toxicology of Chemicals. *Technical Report No. 49. Exposure of Man to Dioxins. A perspective on industrial waste incineration.* 1992

3. Hall AH & Rumack BH (eds.) *TOMES System ®* Micromedex, Englewood, Colorado. CD ROM.vol 41 (exp. 31 July 1999)

4. Ellenhorn MJ, Schonwalds S, Ordog G & Wasserberger J. *Ellenhom's Medical Toxicology – Diagnosis and Treatment of Human Poisoning,* 2nd edn. Williams & Wilkins, London., 1997

5. Koninckx PR, Braet P, Kennedy SH & Barlow DH, 1994. Dioxin pollution and endometriosis in Belgium. *Hum Reprod*; 9(6): 1001–2

6. Sax NI & Lewis RJ. *Hawley's Condensed Chemical Dictionary*, 11th edn. Van Nostrand Reinhold Company, New York, 1987

7. Moses M, Lilis R, Crow KD, Thornton J, Fischbein A, Anderson HA & Selikoff IJ. 1984. Health status of workers with past exposure to 2,3,7,8-tetrachlorodibenzo-*p*-dioxin in the manufacture of 2,4,5-trichlorophenoxyacetic acid: comparison of findings with and without chloracne. *Am J Ind Med;* 5:161–82

8. Martin JV, 1984. Lipid abnormalities in workers exposed to dioxin. *Br J Ind Med*; 41:254–6

9. Calvert GM, Hornung RW, Sweeney MH , Fingerhut MA & Halperin WE, 1992. Hepatic and gastrointestinal effects in an occupational cohort exposed to 2,3,7,8-tetrachlorodibenzo-para-dioxin. *JAMA*; 267: 2209–14

10. Beale MG, Shearer WT, Karl MM & Robson AM, 1977. Long term effects of dioxin exposure. *Lancet*; 1 (8014): 748

Henrietta Wheeler

ETHANOL

Key Points

- Absolute ethanol is a colourless liquid with a pleasant odour and a burning taste
- It is well absorbed orally and may be absorbed via the respiratory tract and dermally following high concentrations or prolonged exposure
- Dose-related CNS depression occurs, ranging from inebriation to anaesthesia, coma, respiratory failure, and death in significant exposures
- Other effects include hypothermia, hypoglycaemia, acidosis, electrolyte imbalances, and gastrointestinal upset
- Ethanol vapours can produce CNS depression, eye and upper respiratory tract irritation
- *In the event of a large spill, stay upwind and out of low areas. Ventilate closed spaces. Protective clothing, eye protection and breathing apparatus should be worn*

FIRST AID

- Terminate exposure and support vital functions
- The casualty should be moved to an uncontaminated area
- Rescuers should, ideally, be trained personnel and must be careful *not to put themselves at risk* and *so wear appropriate protective clothing and, if available, breathing apparatus*
- If the casualty is unconscious a clear airway should be established and maintained; give 100% oxygen if available
- **Inhalation Exposure:** If the patient stops breathing, expired air resuscitation should be started immediately using a pocket mask with a one way valve, if available. It is important where the face is contaminated that expired air resuscitation is NOT attempted unless an airway with rescuer protection is used
- **Dermal Exposure:** Remove contaminated clothing, if possible under a shower and place in double, sealed, clear bags and label; store the bags in a secure area away from patients and staff
- Wash the skin thoroughly with copious amounts of water
- **Eye Exposure:** Irrigate thoroughly with water or saline for 15 minutes
- **Oral Exposure:** Encourage small quantities of oral fluids (no more than 50–100 ml in total)

Detailed Information

- Absolute ethanol has a pleasant odour, burning taste, and is a colourless liquid
- *Common synonyms* absolute ethanol, alcohol, ethyl alcohol, ethyl hydrate, grain alcohol, methyl carbinol
- CAS 64-17-5
- UN 1170
- NIOSH/RTECS KQ 6300000
- Chemical formula C_2H_6O
- Molecular weight 46.07

- Ethanol is used as a solvent for resins, fatty acids, oils and hydrocarbons; in the manufacture of acetaldehyde, acetic acid, ethylene, dyes, pharmaceuticals, detergents, cleaning preparations, cosmetics, explosives; in antifreezes, beverages and as a yeast-growing medium
- Ethanol used in industry is synthesised almost entirely by hydration of ethylene
- Industrial ethanol is one of the largest volume organic chemicals used in industrial and consumer products
- Ethanol vapours mix well with air forming explosive mixtures
- Miscible with water, methanol, ether, chloroform and acetone
- It can react vigorously with oxiders; it has an explosive reaction with oxidising coating around potassium metal
- Ethanol is a flammable liquid when exposed to heat or flame

Summary of Human Toxicity

- Ethanol is rapidly absorbed across the gastrointestinal tract; respiratory and dermal absorption is possible following high concentrations or prolonged exposure
- Intoxication associated with ingestion of alcoholic beverages is well known, industrial exposure to ethanol vapours is generally of no importance providing the correct occupational hygiene standards are in place
- Healthy adults absorb 80–90% ingested ethanol completely in 1 hour; the kidneys and lungs excrete only 5–10% of the absorbed dose unchanged
- Following inhalation of ethanol vapour of 11–20 mg.l^{-1}, 62% absorption would be expected[1]
- The average adult metabolises 7–10 g of ethanol per hour, thereby reducing blood ethanol concentrations by 150–500 mg.l^{-1}.h^{-1}
- The legal blood concentration limit for driving in the UK is currently 0.8 g.l^{-1} (80 mg.100 ml^{-1})

Toxicity of Ethanol

Blood concentration	Clinical effects
1.5 g.l^{-1}	Intoxication even in tolerant adults
3-4 g.l^{-1}	Stupor and coma
≥ 5 g.l^{-1}	Often fatal

- Concentrations of 11 g.l^{-1} have been survived with supportive treatment although fatalities have occurred at levels as low as 1.8 g.l^{-1}
- Fatal dose of absolute alcohol is approximately 6–10 ml.kg^{-1}
- Ethanol ingestion from alcoholic beverages can effect the metabolism of other chemicals; e.g. enhanced hepatoxic effects of carbon tetrachloride in alcoholics
- Tolerance to ethanol vapour has been reported[1]

Toxicity of Inhaled Ethanol[1]

Concentration	Clinical effects
84 ppm	Odour threshold
5,000 – 10,000 ppm	Coughing and smarting of eyes, nose; symptoms disappeared after few min
15,000 ppm	Lacrimation and coughing
20,000 ppm	Impossible to tolerate even for a short time

- Occupational Exposure Standards:
  Long-term exposure limit: 1000 ppm (1920 mg . m^{-3})
  Short-term exposure limit: no data available

Acute Clinical Effects

Inhalation Effects

- High vapour concentrations are irritating to eyes, nose, respiratory tract

- Prolonged exposure or high concentrations may cause systemic effects; see systemic effects below

Dermal Effects

- Ethanol is an astringent to the skin

- Prolonged exposure may result in systemic effects; see systemic effects below

Eye Effects

- Eye exposure to liquid generally causes transient pain, irritation, and reflex lid closure

- A foreign-body sensation may persist for one to two days

- Vapours produce transient stinging and lacrimation, but no apparent adverse effects

Oral Effects

- Nausea, vomiting, gastrointestinal bleeding and abdominal pain are common; diarrhoea may occur

- See systemic effects below

Systemic Effects

Mild (blood concentration <1.5 g . l^{-1})

- Impaired visual acuity, co-ordination and reaction time, emotional lability

Moderate (blood concentration 1.5–3 g . l^{-1})

- Slurred speech, confusion, ataxia, emotional lability, perceptual and sensation disturbances, possible blackout spells, and inco-ordination with impaired objective performance in standardised tests

- There may be diplopia, flushing, tachycardia, sweating and incontinence

- Bradypnoea may occur early, and tachypnoea may develop in cases of metabolic acidosis; hypoglycaemia and hypokalaemia

- CNS depression may progress to coma

Severe (blood concentration 3–5 g . l^{-1})

- Cold clammy skin, hypothermia and hypotension

- Atrial fibrillation and atrioventricular block have been reported[2]

- Respiratory depression may occur; respiratory failure may follow severe intoxication; aspiration of vomit may result in pneumonitis and pulmonary oedema

- Convulsions due to severe hypoglycaemia have been reported

- Acute hepatitis can occur

Fatal (blood concentration >5 g . l^{-1})

- Deep coma, respiratory depression, respiratory arrest and circulatory failure may occur

Chronic Clinical Effects

Effects of chronic alcoholism from excessive use of alcoholic beverages is well known and is not discussed here.

Inhalation exposure to an airborne ethanol is sufficient to produce chemical dependence in rats, there is no human data available.

Management

Blood ethanol concentrations are a good indicator of toxicity.

Inhalation Management

- Symptomatic and supportive care

- For prolonged or large exposure see systemic management below

Dermal Management

- Remove any remaining contaminated clothing, place in double, sealed, clear bags and label; store in a secure area away from patients and staff

- Irrigate the skin with water

- Emollients may ease irritation

- Symptomatic and supportive care

- For prolonged exposure see systemic management below

Eye Exposure

- Irrigate thoroughly with running water or saline for 15 minutes

- Stain with fluorescein and refer to an ophthalmologist if there is any uptake of the stain

Oral Management

- Gastric lavage is rarely necessary as ethanol is rapidly absorbed; activated charcoal does not significantly reduce ethanol absorption

- Symptomatic and supportive care

- For large exposure see systemic management below

Systemic Management

- Monitor blood glucose and arterial pH. Administer glucose for hypoglycaemia; acidosis will usually respond to correction of hypoglycaemia but additional sodium bicarbonate may be required

- Convulsions usually respond to correction of hypoglycaemia but diazepam may be required

- Ventilate for respiratory depression

- Monitor ECG in severe cases

- Haemodialysis should be reserved for life threatening cases; it should be considered if ethanol concentration are >5 g.l⁻¹ or the arterial pH <7

- Symptomatic and supportive care

Summary of Environmental Hazards

- In the atmosphere ethanol should be reduced significantly by rainout or it will photodegrade in hours in a polluted urban atmosphere and in approximately 4 to 6 days in less polluted areas [2]

- In water it will volatilise and probably biodegrade; ethanol is not expected to adsorb to sediment or bioconcentrate in fish

- Spills on land will volatilise, biodegrade and leach into the ground water, but there is no data on the rates of these processes

- Drinking Water Standards:
      hydrocarbon total: $10 \mu g.l^{-1}$ (UK max)

- Soil Guidelines: no data available

- Air Quality Standards: no data available

REFERENCES

1. Clayton GD & Clayton FE (eds.) *Patty's Industrial hygiene and toxicology*, 4th edn. John Wiley & Sons, Inc., New York, 1994

2. Hall AH & Rumack BH (eds.) *TOMES System®* Micromedex, Englewood, Colorado. CD ROM. vol 41 (exp. 31 July 1999)

3. Hathaway GJ, Proctor NH & Hughes JP. *Proctor and Hughes' Chemical Hazards of the Workplace*, 4th edn. Van Nostrand Reinhold, New York, 1996

Henrietta Wheeler

ETHYLENE GLYCOL

Key Points

- Ethylene glycol is a clear, colourless, odourless, sweet tasting, viscous liquid
- Highly toxic by ingestion and injection; dermal absorption is possible but unlikely to produce systemic effects
- At normal temperature and pressure, inhalation is not a significant route of exposure
- Irritant to skin, eyes, and respiratory tract
- Causes CNS depression, metabolic disturbances, and renal damage
- *In the event of a large spill, stay upwind and out of low areas. Ventilate closed spaces. Protective clothing, eye protection and breathing apparatus should be worn*

FIRST AID

- Terminate exposure and support vital functions
- The casualty should be moved to an uncontaminated area
- If the casualty is unconscious a clear airway should be established and maintained; give 100% oxygen if available
- **Dermal Exposure:** Remove contaminated clothing, if possible under a shower and place in double, sealed, clear bags and label; store the bags in a secure area away from patients and staff
- Wash the skin thoroughly with copious amounts of water
- **Eye Exposure:** Irrigate thoroughly with water or saline for 15 minutes
- **Oral Exposure:** Encourage small quantities of oral fluids (no more than 50–100 ml in total)

Detailed Information

- Ethylene glycol is a clear, colourless, odourless, viscous liquid with a sweet taste[1]
- *Common synonyms* 1,2-ethanediol, ethylene alcohol, ethylene dihydrate, glycol alcohol, monoethylene glycol
- CAS 107-21-1
- UN specific number not available
- NIOSH/RTECS KW 2975000
- Molecular formula $C_2H_6O_2$
- Molecular weight 62.07
- Ethylene glycol is used as an antifreeze, an industrial humectant, as a solvent in paints and plastics and in hydraulic brake fluids
- It is soluble in water, alcohol and acetone
- Ethylene glycol interacts with strong oxidisers such as sodium peroxide, potassium permanganate, nitrates, and perchlorates
- It emits acrid smoke and irritating fumes when heated to decomposition

Summary of Human Toxicity

- Ethylene glycol is readily absorbed by ingestion and injection; at normal temperature and pressure, inhalation is not a significant route of exposure but may be possible if heated or misted[2]
- Ethylene glycol, itself, is of a low order of toxicity,[3] toxicity is due mainly to its metabolites
- It is metabolised by alcohol dehydrogenase to glycoaldehyde, this is further metabolised to glycolic acid, which appears to be the principle cause of the acidosis; the aldehydes (glycoaldehyde, glycolic acid and glyoxylate) may inhibit oxidative phosphorylation and respiration, oxalates cause renal damage and hypocalcaemia by binding to calcium oxalate, crystals of which appear in the urine
- **Ingestion:** The minimum lethal dose is reported to be approximately 100 ml in adults, although individuals have survived higher doses[4]
- **Inhalation:** The lowest published toxic concentration for a human has been reported as 10,000 $mg.m^{-3}$ [5]
- Therapy for ethylene glycol is principally aimed at blocking the action of alcohol dehydrogenase to prevent the formation of the toxic metabolites; this is achieved by the administration of ethanol which allows renal excretion of the unmetabolised parent compound
- Fomepizole (4-methylpyrazole) is a new treatment that may be used as an alternative to ethanol therapy. It acts similarly to ethanol by blocking the action of alcohol dehydrogenase, thereby preventing the formation of toxic metabolites. It appears to be well tolerated and without the side-effects of ethanol, thus reducing the need for close supervision[6]
- Occupational Exposure Standards:
  Long-term exposure limit: 10 $mg.m^{-3}$ (particulate)
  60 $mg.m^{-3}$ (vapour)

  Short-term exposure limit: no data available (particulate)
  125 $mg.m^{-3}$ (vapour)

Acute Clinical Effects

Inhalation Effects

- Initial irritation to nose and throat
- If ethylene glycol is heated or misted, see systemic effects below

Dermal Effects

- Irritation at the site of contact
- Rash may develop in sensitive individuals[2]
- Limited absorption may occur although systemic effects are unlikely

Eye Effects

- Irritation, lacrimation, mild temporary conjunctival reaction, but no significant corneal damage expected[7]

Oral Effects

- See systemic effects below

Systemic Effects

- Clinical effects are divided into three main phases:

    Stage I: 20 minutes to 12 hours post ingestion: CNS toxicity predominates. Nausea and vomiting may also occur. Nystagmus, partial or complete paralysis of the eye muscles may be evident. Severe anion gap acidosis and convulsions may develop in severe cases. Hypotonia, hyporeflexia.[3] Formation of calcium oxalate may lead to hypocalcaemia

    Stage II: 12–24 hours post ingestion: Cardiopulmonary symptoms become evident. Tachycardia, tachypnoea occur. Pulmonary oedema and congestive heart failure usual cause of death. ARDS also reported[8]

    Stage III: 24–72 hours post ingestion: Renal effects may occur, and include oliguria, which can persist for up to 50 days, flank pain, acute tubular necrosis, and renal failure. The renal damage may be permanent[9]

Chronic Toxicity

Little is known about the effects in humans of chronic exposure to ethylene glycol. Main complaints in volunteers exposed to ethylene glycol for 4 weeks were throat irritation, headache, and low backache. There may be an increased risk of kidney damage and the development of kidney stones. It may be a sensitising agent.

Management

Blood ethylene concentrations are a good indicator of toxicity.

Inhalation Management

- Maintain a clear airway, give humidified oxygen and ventilate if necessary

- If respiratory irritation occurs assess respiratory function and if necessary perform chest X-rays to check for chemical pneumonitis

- Consider the use of steroids to reduce inflammatory response

- Treat pulmonary oedema with PEEP or CPAP ventilation

- If exposed to heated or misted ethylene glycol, see systemic management below

- Symptomatic and supportive care

Dermal Management

- Remove any remaining contaminated clothing, place in double, sealed, clear bags and label; store in a secure area away from patients and staff

- Wash area thoroughly with water

- Treat irritation symptomatically

Eye Management

- Irrigate thoroughly with running water or saline for 15 minutes

- Stain with fluoroscein and refer to an ophthalmologist if there is any uptake of the stain

Oral Management

- Consider gastric lavage within I hour of a substantial ingestion

- Contact a poisons information service for further guidance on gut decontamination

- Although data is lacking, small, accidental ingestion by adults of diluted ethylene glycol probably requires no treatment. However, it may be wise to give the loading dose of oral ethanol

- See systemic management below

Systemic Management

- Determine ethylene glycol concentration and osmolar gap

- Monitor and correct electrolytes

- Acidosis should be treated aggressively with bicarbonate, large doses may be required

- **Ethanol therapy:**[10] While waiting for lab results, all patients should receive a loading dose of ethanol:

    Give EITHER :

    7.5 ml.kg^{-1} of 10% ethanol in water IV over 30 minutes (solutions stronger than 10% should **NOT** be used for parenteral administration)

    OR

    1 ml.kg^{-1} 100% ethanol (suitably diluted) orally over 15–30 minutes

    OR

    2.5 ml.kg^{-1} of 40% ethanol (most spirits i.e. whisky, gin etc.) suitably diluted orally over 15–30 minutes

- Ethanol therapy is continued if:

    Ethylene glycol level >200 mg.l^{-1}

    Acidosis is present

    Increased osmolar gap (>10 mOsm.kg^{-1} H$_2$O)

    Calcium oxalate crystals in the urine

- Dosages (adjusted to achieve blood ethanol concentration of 1–1.5 g.l^{-1}) IV infusions should be diluted with 5% dextrose (also compatible with 0.9% saline), **monitor blood glucose**

Patient	Amount of Ethanol needed	5% Ethanol (Oral or IV)	10% Ethanol (Oral or IV)	40% Ethanol (Oral ONLY)
Non–drinker/Child	66 mg.kg^{-1}.h^{-1}	1.65 ml.kg^{-1}.h^{-1}	0.825 ml.kg^{-1}.h^{-1}	0.2 ml.kg^{-1}.h^{-1}
Average Adult	110 mg.kg^{-1}.h^{-1}	2.76 ml.kg^{-1}.h^{-1}	1.38 ml.kg^{-1}.h^{-1}	0.3 ml.kg^{-1}.h^{-1}
Chronic Drinker	153.78 mg.kg^{-1}.h^{-1}	3.9 ml.kg^{-1}.h^{-1}	1.95 ml.kg^{-1}.h^{-1}	0.4 ml.kg^{-1}.h^{-1}

- Continue ethanol therapy until ethylene glycol is no longer detected

- Haemodialysis is indicated if:

  Ethylene glycol concentrations >500 mg.l^{-1}

  Unresponsive or severe acidosis

  Renal insufficiency

- Dialysis should be continued until the ethylene glycol concentration <200 mg.l^{-1}; during dialysis, ethanol therapy should be continued; as the ethanol is also readily dialysed, it will be necessary to increase the dosage by 100 mg.kg^{-1}.h^{-1} or more to maintain a blood concentration of 1–1.5 g.l^{-1}; it may be preferable in this situation to add ethanol to the dialysate to achieve a concentration of 1g.l^{-1}

- Although less effective, peritoneal dialysis has been used and could be considered where haemodialysis facilities are not available

- **Fomepizole (Antizol, 4-methylpyrazole):** a new treatment that may be used as an alternative to ethanol therapy.

- Dose: The most effective dosing regime has yet to be determined but the manufacturers recommend:

  a loading dose of 15mg.kg^{-1} followed by 10mg.kg^{-1} every 12 hours for four doses and then 15mg.kg^{-1} every 12 hours. Continue until the blood ethylene glycol concentration is <200mg.l^{-1}

- Fomepizole is not licensed for use in children. Contact a poisons information service for further guidance and paediatric information

- Vitamin B$_1$ and B$_6$ therapy: Thiamine and pyridoxine (100 mg.d^{-1} IV) to all patients may promote biotransformation of glycoxylate to non-toxic metabolites, although clinical evidence is lacking

- Calculation of the osmolar gap: Serum osmolarity may be measured directly by freezing point depression, or by calculation, see below

- The difference between the measured and the calculated osmolarity is known as the osmolar gap, and is usually <10 mOsm.kg^{-1} H$_2$O

- Although a normal osmolar gap does not exclude poisoning from ethylene glycol, an elevated osmolar gap may suggest the possibility of severe poisoning

- Contact a poisons information service for further guidance and paediatric information

---

Calculated osmolarity (mOsm.kg^{-1} H$_2$O) =

$$\frac{(1.86 \times [\text{Na in mmol}.l^{-1}]) + (\text{urea in mmol}.l^{-1}) + (\text{glucose in mmol}.l^{-1})}{0.93}$$

Summary of Environmental Hazards

- In the atmosphere ethylene glycol exists mainly in the vapour-phase; it is degraded in the atmosphere by reaction with photochemically produced hydroxyl radicals with an estimated half-life of about 50 hours

- Ethylene glycol is not expected to concentrate in the food chain

- Drinking Water Standards: no data available

- Soil Guidelines: no data available

- Air Quality Standards: no data available

REFERENCES

1. Budavari S, O'Neil MJ, Smith A, Heckelman PE & Kinneary JF (eds.) *The Merck Index*, 12th edn. Merck & Co., Inc., Whitehouse Station, 1996

2. Hathaway GJ, Proctor NH & Hughes JP. *Proctor and Hughes' Chemical Hazards of the Workplace*, 4th edn. Van Nostrand Reinhold, New York, 1996

3. Walder AD & Tyler CKG, 1994. Ethylene glycol antifreeze poisoning. *Anaesthesia*; 49: 964–7

4. Wetherall DJ, Ledingham JGG & Warrell DA (eds.) *Oxford Textbook of Medicine* (Vol 1) 3rd edn. Oxford University Press, Oxford, 1984

5. Hall AH & Rumack BH (eds.) *TOMES System* ® Micromedex, Englewood, Colorado. CD ROM. vol 41 (exp. 31 July 1999)

6. Brent J, McMartin K, Phillips S, Burkhart KK, Donovan JW, Wells M & Kulig K, 1999. Fomepizole for the treatment of ethylene glycol poisoning. *N Eng J Med* 340: 832–8

7. Grant MW & Schuman JS. *Toxicology of the Eye*, 4th edn. Charles C Thomas, Springfield, 1993

8. Catchings TT, Beamer WC, Lundy L & Prough DS, 1985. Adult respiratory distress syndrome secondary to ethylene glycol ingestion. *Ann Emerg Med*; 14: 6

9. Ellenhorn MJ, Schonwalds S, Ordog G & Wasserberger J. *Ellenhorn's Medical Toxicology – Diagnosis and Treatment of Human Poisoning*, 2nd edn. Williams & Wilkins, London, 1997

10. McCoy HG, Cipolle RJ, Ethlers SM, Sawchuk RJ & Zaske DE, 1979. Severe methanol poisoning – application of a pharmacokinetic model for ethanol therapy and hemodialysis. *Am J Med*; 67: 804–7

Jennifer Butler

ETHYLENE OXIDE

Key Points

- Ethylene oxide is a colourless gas with an ether-like odour at room temperature
- Ethylene oxide is an irritant to the respiratory tract, eyes and skin; at high concentrations it causes CNS depression
- Peripheral neuropathies have been reported after chronic exposure
- *In the event of a large spill, stay upwind and out of low areas. Ventilate closed spaces. Protective clothing, eye protection and breathing apparatus should be worn*

FIRST AID

- Terminate exposure and support vital functions
- The casualty should be moved to an uncontaminated area
- Rescuers should, ideally, be trained personnel and must be careful *not to put themselves at risk* and *so wear appropriate protective clothing and, if available, breathing apparatus*
- If the casualty is unconscious a clear airway should be established and maintained; give 100% oxygen if available
- **Inhalation Exposure:** If the patient stops breathing, expired air resuscitation should be started immediately using a pocket mask with a one way valve, if available. It is important where the face is contaminated that expired air resuscitation is NOT attempted unless an airway with rescuer protection is used
- **Dermal Exposure:** If frostbite is present *do not* remove clothing, flush skin with water
- If frostbite is not present remove contaminated clothing, if possible under a shower and place in double, sealed, clear bags and label; store the bags in a secure area away from patients and staff
- Wash the skin thoroughly with copious amounts of water
- **Eye Exposure:** If the eye tissue is frozen seek medical advice as soon as possible
- If eye tissue is not frozen, irrigate thoroughly with water or saline for 15 minutes
- **Oral Exposure:** Encourage small quantities of oral fluids (no more than 50–100 ml in total)

Detailed Information

- Ethylene oxide is a colourless gas with an ether-like odour at room temperature
- *Common synonyms* dimethylene oxide, 1,2-epoxyethane, oxacyclopropane, oxane, oxidoethane, oxirane
- CAS 75-21-8
- UN 1040
- NIOSH/RTECS KX 2450000
- Chemical formula C_2H_4O
- Molecular weight 44.05
- Ethylene oxide is used in a variety of chemical processes including the production of antifreeze, ethylene glycol,

glycol ethers, polyester fibres, nonionic surfactants and chlorine; used to fumigate and sterilise medical equipment and food due to its antimicrobial and insecticidal properties

- It is commonly mixed with carbon dioxide (e.g. 10–30% ethylene oxide with 70–90% carbon dioxide) or fluorocarbons (e.g. 12% ethylene oxide with 88% fluorocarbons)
- Ethylene oxide is heavier than air; it condenses to a liquid at 10.4°C
- Ethylene oxide is soluble in water but will evaporate;[1] it is slowly hydrolysed to ethylene glycol but the extent is unknown[2]
- It is completely miscible with acetone, benzene, diethyl ether, methanol, carbon tetrachloride and other organic solvents[2]
- Both liquid and gas are highly reactive with many compounds, ethylene oxide is relatively stable in water or when mixed with carbon dioxide or halocarbons[1]
- Violent polymerisation of the liquid may occur on contact with acids, various metal chlorides and hydroxides or heat, this reaction is exothermic
- Ethylene oxide is highly flammable at room temperature and normal atmospheric pressure
- It may react explosively at high temperatures[2]

Summary of Human Toxicity

- Ethylene oxide is absorbed by inhalation and orally; dermal absorption can occur, severity of effect is related to concentration and duration[2]
- Ethylene oxide is water and lipid soluble so it is well absorbed and rapidly metabolised, the half-life in human tissue is approximately 10–30 minutes[2]
- Ethylene oxide alkylates protein and DNA by replacing a hydrogen atom in molecules with a hydroethyl group, which accounts for the mutagenic and carcinogenic effects
- Residues remaining in sterilised medical equipment which has been improperly aerated may cause local effects if ethylene oxide enters the tissues and circulatory system[1]
- Renal failure and anaphylaxis have been reported from exposure to contaminated dialysis equipment[1]
- After food fumigation, ethylene oxide either reacts with constituents or evaporates and so toxicity from this type of exposure is low
- Odour detection is 700 ppm; ethylene oxide concentration may be hazardous prior to sensory detection
- Exposure to 30 mg.kg^{-1} in animals caused nausea, vomiting and diarrhoea for 2 hours[3]
- Maximum Exposure Limit:
  Long term exposure limit: 5 mg.m^{-3} (9.2 ppm)
  Short term exposure limit: no data available

Acute Clinical Effects

Inhalation Effects

- Ethylene oxide is a respiratory tract irritant; effects including coughing, headache, irritation to eyes and nose, and dyspnoea

- High concentrations may lead to delayed pulmonary oedema and chemical pneumonitis

- Cardiovascular collapse has been reported following anoxia

Dermal Effects

- Irritant effects are often delayed for 1–5 hours post exposure[4]

- Effects include erythema, oedema, severe dermatitis, blisters and burns[1,3]

- Concentrations >40% may cause severe burns in less than one minute[3]

- Vaporising liquid ethylene oxide may cause frostbite injury

- Dermal absorption may lead to systemic effects; see below

Eye Effects

- Irritant effects are often delayed

- Contact with liquid or vapour exposure may cause irritation, corneal injury and conjunctivitis

- Severe corneal burns have been reported[5,6]

- Vaporising liquid may cause frostbite injury

Oral Effects

- Not the usual route of exposure

- Exposure may lead to irritation of mouth, throat and gastrointestinal tract

- Systemic effects may occur; see below

Systemic Effects

- Commonly causes headache, nausea and vomiting, which may be delayed unless exposure is serious

- Less frequently reported effects include diarrhoea, abdominal pain, insomnia, excitement, dizziness, twitching, muscle spasms, convulsive movements and in serious exposures, coma and respiratory arrest[3,4]

Chronic Clinical Effects

Repeated dermal exposure can lead to allergic contact dermatitis. Repeated ocular exposure to high concentrations (approximately $900\,mg\,.m^{-3}$) of vapour may cause cataract formation.[1]

Chronic exposure may lead to sensory and motor neuropathies, insomnia, cognitive impairments including emotional lability, impaired concentration and memory, and convulsions.

Exposure to ethylene oxide has been linked with spontaneous abortion. It is considered to be a probable carcinogen; exposure has been associated with leukaemia, stomach and pancreatic cancer and Hodgkin's disease. IARC has determined that there is limited evidence in humans for cancer although there is sufficient evidence in animals.

Management

Inhalation Management

- Maintain a clear airway, give humidified oxygen and ventilate if necessary

- If respiratory irritation occurs assess respiratory function and if necessary perform chest X-rays to check for chemical pneumonitis

- Treat pulmonary oedema with PEEP or CPAP ventilation

- If serious exposure, see systemic management below

Dermal Management

- Remove any remaining contaminated clothing, place in double, sealed, clear bags and label; store in a secure area away from patients and staff

- Allow liquid ethylene oxide to evaporate before washing with water

- Irrigate with copious amounts of water

- An emollient may be required; treat burns symptomatically

- Burns usually heal within 21 days but may leave some brown discoloration

- If serious exposure, see systemic management below

Eye Management

- Irrigate thoroughly with running water or saline for 15 minutes

- Stain with fluorescein and refer to an ophthalmologist if there is any uptake of the stain

Oral Management

- Encourage oral fluids

- DO NOT INDUCE EMESIS due to risk of aspiration

- Activated charcoal may be of use for substantial ingestion; it has been shown to adsorb ethylene oxide *in vitro*

- Contact a poisons information service for further guidance on gut decontamination

- If serious exposure, see systemic effects below

Systemic Management

- Observe for 72 hours in case of delayed effects[3]

- Symptomatic and supportive care

Summary of Environmental Hazards

- In the atmosphere ethylene oxide degrades slowly by reacting with hydroxyl-radicals; the estimated half-life is 211 days

- In water it is removed by volatilisation, hydrolysis and to a lesser extent, biodegradation[3]

- Volatilisation half-lives for a model river and lake are estimated to be 5.9 hours and 38 days, respectively

- It is not strongly adsorbed to soil and is not expected to bioconcentrate in the food chain

- Drinking Water Standards: pesticide: $10\,\mu g\,.l^{-1}$ (UK max)

- Soil Guidelines: no data available

- Air Quality Standards: no data available

REFERENCES

1. International Programme on Chemical Safety. *Health and Safety Guide 16. Ethylene Oxide*: WHO, Geneva 1988

2. German Chemical Society. *BUA Report 141. Ethylene Oxide.* S. Hirzel, WVS 1995

3. Hall AH & Rumack BH (eds.) *TOMES System* ® Micromedex, Englewood, Colorado. CD ROM. vol 41 (exp. 31 July 1999)

4. International Programme on Chemical Safety. *Environmental Health Criteria 55: Ethylene Oxide.* WHO, Geneva, 1985

5. ECETOC. Technical Report No. 5. *Toxicity of ethylene oxide and its relevance to man.* 1982

6. ECETOC. Technical Report No. 11. *Ethylene oxide toxicology and its relevance to man: an up-dating of Technical Report No. 5.* 1984

Elizabeth Schofield

FORMALDEHYDE

Key Points

- Formaldehyde is a colourless gas with a pungent, irritating odour
- It is an irritant to the skin, mucous membranes and respiratory tract
- Ingestion may cause corrosive injury with gastrointestinal irritation, nausea, vomiting and perforation
- Formaldehyde is a mild sensory irritant and potent sensitiser to which some people are more sensitive than others
- Formaldehyde-induced asthma may occur following brief exposure and may provoke late asthma-like reactions in sensitised individuals
- *In the event of a large spill, stay upwind and out of low areas. Ventilate closed spaces. Protective clothing, eye protection and breathing apparatus should be worn*

First Aid

- Terminate exposure and support vital functions
- The casualty should be moved to an uncontaminated area
- Rescuers should, ideally, be trained personnel and must be careful *not to put themselves at risk* and *so wear appropriate protective clothing and, if available, breathing apparatus*
- If the casualty is unconscious a clear airway should be established and maintained; give 100% oxygen if available
- **Inhalation Exposure:** If the patient stops breathing, expired air resuscitation should be started immediately using a pocket mask with a one way valve, if available. It is important where the face is contaminated that expired air resuscitation is NOT attempted unless an airway with rescuer protection is used
- **Dermal Exposure:** Remove contaminated clothing, if possible under a shower and place in double, sealed, clear bags and label; store the bags in a secure area away from patients and staff
- Wash the skin thoroughly with copious amounts of water
- **Eye Exposure:** Irrigate thoroughly with water or saline for 15 minutes
- **Oral Exposure:** Encourage small quantities of oral fluids (no more than 50–100 ml in total) unless perforation is suspected

Detailed Information

- Formaldehyde is a colourless gas with a pungent, irritating odour
- *Common synonyms* formic aldehyde, formalin, methanal, methyl aldehyde, methylene oxide, oxo-methane, oxymethylene
- CAS 50-00-0
- UN 2209
- NIOSH/RTECS LP 8925000
- Molecular formula H_2CO

- Molecular weight 30.03
- Formaldehyde is used as a disinfectant, tissue preservative, antiseptic, deodorant and feedstock for synthetic chemical processes; industrial uses include the manufacture of urea formaldehyde foam, paint pigment, rubber, photographic film, leather, cosmetics, paper, plywood and plastic moulding; the formaldehyde processes produce wrinkle-free, crease-resistant textiles
- Urea-formaldehyde fumes may be released from urea-formaldehyde foam insulation, plywood, particle board, chip board, gas appliances and carpeting; this may be of concern in confined spaces
- Commercially available formalin is 37–50% aqueous solution of formaldehyde that contains up to 15% methanol in order to inhibit polymerisation
- In aqueous solutions a pH of 2.8–4.0 is expected
- Formaldehyde is soluble in water and alcohol
- It decomposes when heated giving off formic acid; it is a strong reducing agent which reacts violently with oxidants and various organic substances
- Formaldehyde reacts with hydrochloric acid giving off bis(chloromethyl)ether

Summary of Human Toxicity

- Absorption of formaldehyde is usually rapid; delayed absorption of methanol might occur following ingestion of formalin if the formaldehyde causes fixation of the stomach
- Most formaldehyde is metabolised by alcohol dehydrogenase in the liver to formic acid; metabolism also occurs in erythrocytes, brain, kidney and muscle to a lesser extent; this conversion is extremely rapid with a half-life of about 1.5 minutes[1]
- Formic acid is then metabolised to carbon dioxide and water via the folate-dependent enzymatic pathway
- Formic acid is excreted in the urine as formates; the approximate half-life of formate is 80–90 minutes
- Odour tolerance to formaldehyde occurs quickly; odour is therefore not a reliable warning of the presence of formaldehyde
- Moisture and heat causes decomposition of formaldehyde-containing products into formaldehyde vapour; release of these vapours indoors has been associated with headaches and respiratory and dermal irritation
- Ingestion of 30 ml of a 37% formaldehyde solution resulted in death
- Dermal exposure to 2–10% solutions resulted in blisters, fissures and urticaria
- In certain individuals, formaldehyde can combine with proteins in the epidermis (e.g. Langerhans' cells) to produce a hapten-protein complex capable of sensitising T lymphocytes; subsequent exposure causes a type IV hypersensitivity reaction

- Formaldehyde-induced asthma may occur following brief exposure to as little as 3 ppm and may provoke late asthma-like reactions in sensitised individuals

- It is a mild sensory irritant causing allergic dermatitis

Toxicity of Inhaled Formaldehyde [2,3]

Concentration	Clinical effects
0.1–0.3 ppm	Irritation to mucous membranes
0.5–1.0 ppm	Odour threshold
1–3 ppm	Irritation of eyes, nose and throat
2–5 ppm	Irritation to mucous membranes, eyes, nose, pharynx and respiratory tract, and lacrimation
10 ppm	Profuse lacrimation, coughing and inability to tolerate after a short time
10–20 ppm	Difficulty in breathing, severe burning nose and throat; lacrimation subsides promptly post exposure but nasal and respiratory irritation may last for about an hour
50–100 ppm	Pulmonary oedema, inflammation and pneumonia
>100 ppm	Death

- Maximum Exposure Limits:
  Long-term exposure limit: 2 ppm (2.5 mg.m^{-3})
  Short-term exposure limit: 2 ppm (2.5 mg.m^{-3})

Acute Clinical Effects

Inhalation Effects

- Exposure may result in mucous membrane irritation, burning sensation and lacrimation

- Headache, rhinitis and dyspnoea may occur

- Severe respiratory tract irritation and dyspnoea at high concentrations

- Cough, chest pain, wheezing, laryngospasm, bronchitis, pulmonary oedema, pneumonitis and death have been reported

- Sensitive individuals may develop acute asthma attacks

- Hepatotoxicity has been reported following inhalation of formaldehyde gas

- Systemic effects may occur; see below

Urea-formaldehyde fumes and home exposure

- Effects reported include mucous membrane irritation, upper respiratory tract irritation, dyspnoea, eye irritation, skin rashes, itching, nausea, rhinitis, headache, dizziness and general fatigue

Dermal Effects

- Irritation dermatitis characterised by redness and thickening of the affected area is common

- In more severe cases there may be blistering, scaling and fissure formation

- Allergic contact dermatitis including drying, erythema, urticaria, eczematous lesions, desquamation, hyperaesthesias and angioneurotic oedema

Eye Effects

- Mild eye irritation with lacrimation following vapour exposure may occur

- Low concentration solutions may produce transient discomfort and irritation

- Concentrated solutions in the eye may result in severe injury including immediate pain, epithelial clouding, corneal opacification and loss of vision[4]

- Note: the eye may appear normal or minimally affected for several hours; delayed severe corneal injury and reduction of vision may develop over 12 or more hours

Oral Effects

- Mucosal irritation and ulceration of the mouth, oesophagus, and stomach results in nausea, vomiting, severe abdominal pain and diarrhoea may occur

- Severe complications include stricture formation, bleeding, perforation and circulatory shock

- Hyperbilirubinaemia has been reported following ingestion of formalin

- Systemic effects may occur; see below

Systemic Effects

- Severe metabolic acidosis and hyperlactataemia results from rapid conversion of formaldehyde to formic acid; tachypnoea secondary to metabolic acidosis is common

- Acute renal failure associated with a high plasma concentration of formic acid

- CNS depression including dizziness, lethargy, ataxia and coma

- Tachycardia, hypotension and cardiovascular collapse

- Hypothermia, renal failure and apnoea following severe ingestion

- Respiratory distress and ARDS have followed ingestion and transdermal absorption of formaldehyde-containing compounds[1]

Chronic Clinical Effects

Formaldehyde is a suspected carcinogen. Occupational exposure has been strongly linked to the development of buccal and nasopharyngeal metaplasia/neoplasia, and to a lesser extent cancers of the nasal cavities.[5,6] The role of formaldehyde in lower respiratory tract cancer aetiology is controversial.

Management

Measurement of blood formaldehyde or formic acid concentrations has no clinical relevance other than confirming exposure when this is in doubt. Formate concentrations may be of use following a large exposure.[7] Monitor methanol concentrations following significant exposure.

Inhalation Management

- Maintain a clear airway, give humidified oxygen and ventilate if necessary

- If respiratory irritation occurs assess respiratory function and if necessary perform chest X-rays to check for chemical pneumonitis

- Consider the use of steroids to reduce the inflammatory response

- Treat pulmonary oedema with PEEP or CPAP ventilation

- See systemic management below

Dermal Management

- Remove any remaining contaminated clothing, place in double, sealed, clear bags and label; store in a secure area away from patients and staff

- Irrigate with copious amounts of water

- An emollient may be required for irritation; treat burns symptomatically

- Symptomatic and supportive care

Eye Management

- Irrigate thoroughly with running water or saline for 15 minutes

- Stain with fluoroscein and refer to an ophthalmologist if there is any uptake of the stain

Oral Management

- NO GASTRIC LAVAGE OR EMETIC

- Encourage oral fluids

- Consider the use of steroids to reduce the inflammatory response

- If facilities are available, early gastro-oesophagoscopy should be undertaken within 12 hours of the event to assess the extent and severity of the injury

- See systemic management below

Systemic Management

- Monitor electrolytes, arterial blood gases, renal and liver function

- Administer sodium bicarbonate if indicated

- Monitor methanol concentration and treat accordingly

- Symptomatic and supportive care

Summary of Environmental Hazards

- Formaldehyde is ubiquitous in the environment as a contaminant of smoke and as photochemical smog

- In the atmosphere, formaldehyde both photolyses and reacts with reactive free radicals (primarily hydroxyl radicals); half-lives in the sunlit troposphere are 1.25 to 6 hours for photolysis, and 7.13 to 71.3 hours for reaction with hydroxyl radicals[1]

- Reaction with nitrate radicals, insignificant during the day, may be an important removal process at night

- Due to its solubility, formaldehyde will efficiently transfer to rain and surface water; one model predicts dry deposition and wet removal half-lives of 19 and 50 hours, respectively[1]

- In water, formaldehyde will biodegrade to low concentrations within a few days; adsorption to sediment and volatilisation are not expected to be significant routes

- In soil, aqueous solutions of formaldehyde will leach through the soil; at high concentrations, the gas may adsorb to clay mineral

- Although biodegradable under both aerobic and anaerobic conditions the fate of formaldehyde in soil is unknown

- It does not bioconcentrate in the food chain

- Drinking Water Standards:
    hydrocarbon total: $10\,\mu g.l^{-1}$ (UK max)
    pesticide: $0.1\,\mu g.l^{-1}$ (UK max)
    formaldehyde: $900\,\mu g.l^{-1}$ (WHO guideline)

- Soil Guidelines: no data available

- Air Quality Standards:
    $< 0.1\,mg.m^{-3}$ as a 30 min average, indoor air, non-industrial buildings (WHO guideline)

REFERENCES

1. Hall AH & Rumack BH (eds.) *TOMES System* ® Micromedex, Englewood, Colorado. CD ROM. vol 41 (exp. 31 July 1999)

2. Hathaway GJ, Proctor NH & Hughes JP. *Proctor and Hughes' Chemical Hazards of the Workplace*, 4th edn. Van Nostrand Reinhold, New York, 1996

3. Ellenhorn MJ, Schonwalds S, Ordog G & Wasserberger J. *Ellenhorn's Medical Toxicology – Diagnosis and Treatment of Human Poisoning*, 2nd edn. Williams & Wilkins, London, 1997

4. Grant MW & Schuman JS. *Toxicology of the Eye*, 4th edn. Charles C Thomas, Springfield, 1993

5. Partanen T, Kauppinen T, Hernber S, Nickels J, Luukkonen R, Hakulinen T & Pukkala E, 1990. Formaldehyde exposure and respiratory cancer among woodworkers – an update. *Scand J Work Environ Health*; 16(6): 394–400

6. Stayner L, Smith A, Reeve G, Blade L, Elliot L, Keenlyside & Halperin W, 1985. Proportionate mortality study of workers in the garment industry exposed to formaldehyde. *Am J Ind Med*; 7(3): 229–40

7. Burkhart KK & Kulig KW, 1990. Formate levels following a formalin ingestion. *Vet Hum Toxicol*; 32(2): 135–7

Henrietta Wheeler

FORMIC ACID

Key Points

- Formic acid is a colourless liquid with a pungent odour
- It is extremely irritating and corrosive to the eyes, mucous membranes, and skin
- Formic acid fumes cause respiratory irritation and possibly pulmonary oedema
- Ingestion can lead to oral irritation, haematemesis, renal failure, shock and perforation
- *In the event of a large spill, stay upwind and out of low areas. Ventilate closed spaces. Protective clothing, eye protection and breathing apparatus should be worn*

FIRST AID

- Terminate exposure and support vital functions
- The casualty should be moved to an uncontaminated area
- Rescuers should, ideally, be trained personnel and must be careful *not to put themselves at risk* and *so wear appropriate protective clothing and, if available, breathing apparatus*
- If the casualty is unconscious a clear airway should be established and maintained; give 100% oxygen if available
- **Inhalation Exposure:** If the patient stops breathing, expired air resuscitation should be started immediately using a pocket mask with a one way valve, if available. It is important where the face is contaminated that expired air resuscitation is NOT attempted unless an airway with rescuer protection is used
- **Dermal Exposure:** Remove contaminated clothing, if possible under a shower and place in double, sealed, clear bags and label; store the bags in a secure area away from patients and staff
- Wash the skin thoroughly with copious amounts of water
- **Eye Exposure:** Irrigate thoroughly with water or saline for 15 minutes
- **Oral Exposure:** Encourage small quantities of oral fluids (no more than 50–100 ml in total) unless perforation is suspected

Detailed Information

- Formic acid is a strong acid in aqueous solution, it is a colourless, fuming liquid with a pungent odour and sour taste
- *Common synonyms* aminic acid, formylic acid, hydrogen carboxylic acid, methanoic acid
- CAS 64-18-6
- UN 1779
- NIOSH/RTECS LQ 4900000
- Molecular formula C_5H_4O
- Molecular weight 46.02
- Formic acid is used as a descaler (available in 44 to 60% concentrations as domestic kettle descalers), wool dye reducer, decalcifying agent for processing animal bones, tanning agent for hides, insecticide, in refrigerants, coagulating latex in rubber production, perfume solvent, silage preservative, and in food as a preservative and flavouring[1,2]
- Formic acid is miscible with water, alcohol, ether and glycerol
- Formic acid is incompatible with strong oxidisers, strong caustics and sulphuric acid
- Formic acid is a synthetic chemical that is highly corrosive; it is flammable when exposed to heat and reacts violently with oxidants and bases[3]
- When heated to decomposition, formic acid emits acrid smoke, irritating fumes, carbon monoxide and hydrogen

Summary of Human Toxicity

- Formic acid can directly damage clotting factors leading to an increase in bleeding and haemorrhage[4]
- It has an elimination half life of about 2.5 hours
- In experiments on excised corneas, formic acid is one of the most rapidly penetrating acids[5]
- Ingestion of under 20 g tends to cause superficial minor oropharyngeal burns only; 30–45 g may produce more serious effects[6]
- Adult fatal dose on ingestion is about 45–60 g or approximately 100 ml of a domestic descaling fluid[7]
- Workers exposed to 15 ppm of a mixture of formic and acetic acids complained of nausea[2]
- Odour threshold range for formic acid is <1 ppm to 38 ppm
- Occupational Exposure Standard:
  Long-term exposure limit: 5 ppm (9.6 mg.m^{-3})
  Short-term exposure limit: no data available

Acute Clinical Effects

Inhalation Effects

- Formic acid is irritating to the respiratory tract leading to coughing, wheezing, tight chest, choking and dypsnoea
- Nausea, vomiting, dizziness, eye irritation, and headache may occur
- Severe exposure causes ulceration of the tongue and buccal irritation; inflammation of the throat, nose and larynx may occur occasionally
- Chemical pneumonitis and respiratory failure are severe complications

Dermal Effects

- Skin contact may cause immediate burns, with vesiculation
- Skin rash, irritation, blisters, inflammation, necrosis and ulceration

Eye Effects

- Formic acid fumes in the eye may cause irritation and pain

- Splashes may cause redness, irritation, oedema, mydriasis and epithelial loss

- Higher concentrations cause corneal erosions, perforation, permanent damage and enucleation[5]

Oral Effects

- Ingestion of small amounts of formic acid may cause gastric irritation with salivation, vomiting, haematemesis, abdominal pain, burning sensation in the mouth and diarrhoea

- Larger quantities or concentrated solutions cause burns to the mouth, oesophagus and stomach; shock, perforation and circulatory collapse

- Aspiration, pulmonary oedema and ARDS may occur

- Metabolic acidosis, haemolysis, renal damage and DIC have been observed

- Late complications may include oesophageal stricture and pyloric stenosis which may follow several weeks later

Chronic Clinical Effects

Chronic occupational exposures may cause nausea and albumin or blood in urine. It is a sensitiser and exposure may lead to allergic dermatitis or occupational asthma.

Management

Measurement of blood or urine formic acid concentrations has no clinical relevance other than confirming exposure when this is in doubt.

Inhalation Management

- Maintain a clear airway, give humidified oxygen and ventilate if necessary

- If respiratory irritation occurs assess respiratory function and if necessary perform chest X-rays to check for chemical pneumonitis

- Consider the use of steroids to reduce the inflammatory response

- Treat pulmonary oedema with PEEP or CPAP ventilation

- Symptomatic and supportive care

Dermal Management

- Remove any remaining contaminated clothing, place in double, sealed, clear bags and label; store in a secure area away from patients and staff

- Irrigate with copious amounts of water

- Treat burns symptomatically

Eye Management

- Irrigate thoroughly with running water or saline for 15 minutes

- Stain with fluorescein and refer to an ophthalmologist if there is any uptake of the stain

Oral Management

- NO GASTRIC LAVAGE OR EMETIC

- Encourage oral fluids, unless perforation is suspected

- Give plasma expanders/blood or IV fluids for shock and analgesics for pain

- Consider the use of steroids to reduce the inflammatory response

- Take abdominal X-ray to check for perforation

- Folinic acid (1 mg.kg^{-1} IV 4–6 hourly until clinical improvement) may enhance formate metabolism by the liver and should be considered in severely poisoned patients [6,7]

- Correct acidosis with sodium bicarbonate; urinary alkalinisation may enhance the elimination of formate and decrease the risk of acute renal necrosis (see regimen below)

- Monitor renal function and arterial pH; clotting should be monitored carefully to detect early DIC[7]

- Frusemide IV, 20 mg 4 hourly, may be useful to block the formate chloride exchanger, preventing renal tubular re-absorption of formate[6]

- Renal failure may require haemodialysis

- Symptomatic and supportive care

- If facilities are available, early gastro–oesophagoscopy should be undertaken within 12 hours of the event to assess the extent and severity of the injury

- *Urinary alkalinisation*: 1 litre of 1.26 % sodium bicarbonate (isotonic) IV over 4 hours and/or 50 ml boluses of 8.4 % sodium bicarbonate IV (ideally via a central line)

- Contact a poisons information service for further guidance and paediatric doses

Summary of Environmental Hazards

- In the atmosphere formic acid reacts with photochemically produced hydroxyl radicals with a half-life of 34 days[8]

- Formic acid is highly soluble in water; it is non–persistent with a half-life of 2 to 20 days

- Formic acid should leach into some soils where it would probably biodegrade

- It does not bioconcentrate in the food chain

- Drinking water standards: no data available

- Soil Guidelines: no data available

- Air Quality Standards: no data available

REFERENCES

1. Clayton GD & Clayton FE (eds.) *Patty's Industrial hygiene and toxicology*, 4th edn. John Wiley & Sons, Inc., New York, 1994

2. Hathaway GJ, Proctor NH & Hughes JP. *Proctor and Hughes' Chemical Hazards of the Workplace*, 4th edn. Van Nostrand Reinhold, New York, 1996

3. Dutch Institute for the Working Environment and the Dutch Chemical Industry Association (eds.) *Chemical Safety Sheets – Working Safely with Hazardous Chemicals.* Samson Chemical Publishers, The Netherlands, 1991

4. Ellenhorn MJ, Schonwalds S, Ordog G & Wasserberger J. *Ellenhorn's Medical Toxicology – Diagnosis and Treatment of Human Poisoning*, 2nd edn. Williams & Wilkins, London, 1997

5. Grant MW & Schuman JS. *Toxicology of the Eye*, 4th edn. Charles C Thomas, Springfield, 1993

6. Jefferys DB & Wiseman HM, 1980. Formic acid poisoning. *Postgrad Med J*; 56(661): 761–2

7. Moore DF, Bentley AM, Dawling S, Hoare AM & Henry JA, 1994. Folinic acid and enhanced renal elimination in formic acid intoxication. *J Toxicol-Clin Toxicol*; 32(2): 199–204

8. Hall AH & Rumack BH (eds.) *TOMES System* ® Micromedex, Englewood, Colorado. CD ROM. vol 41 (exp. 31 July 1999)

Henrietta Wheeler

GLUTARALDEHYDE

Key Points

- Glutaraldehyde is an oily, colourless liquid with a pungent odour
- It is an irritant to the skin and mucous membranes
- Absorption can occur via inhalation, orally and to some extent dermally
- Glutaraldehyde is a strong sensitiser and can cause an allergic contact dermatitis
- It may be capable of inducing asthma
- *In the event of a large spill, stay upwind and out of low areas. Ventilate closed spaces. Protective clothing, eye protection and breathing apparatus should be worn*

FIRST AID

- Terminate exposure and support vital functions
- The casualty should be moved to an uncontaminated area
- Rescuers should, ideally, be trained personnel and must be careful *not to put themselves at risk* and *so wear appropriate protective clothing and, if available, breathing apparatus*
- If the casualty is unconscious a clear airway should be established and maintained; give 100% oxygen if available
- **Inhalation Exposure:** If the patient stops breathing, expired air resuscitation should be started immediately using a pocket mask with a one way valve, if available. It is important where the face is contaminated that expired air resuscitation is NOT attempted unless an airway with rescuer protection is used
- **Dermal Exposure:** Remove contaminated clothing, if possible under a shower and place in double, sealed, clear bags and label; store the bags in a secure area away from patients and staff
- Wash the skin thoroughly with copious amounts of water
- **Eye Exposure:** Irrigate thoroughly with water or saline for 15 minutes
- **Oral Exposure:** Encourage small quantities of oral fluids (no more than 50–100 ml in total) unless perforation is suspected

Detailed Information

- Glutaraldehyde is an oily, colourless liquid with a pungent odour
- *Common synonyms* glutaral, glutaric dialdehyde, 1,5-pentanedial
- CAS 111-30-8
- UN specific number not available
- NIOSH/RTECS MA 2450000
- Molecular formula $C_5H_8O_2$
- Molecular weight 100.13
- Glutaraldehyde is used for sterilising rubber or plastic clinical equipment which cannot be heat sterilised; as a tanning agent for leather; as an embalming fluid; a preservative in cosmetics, skin care products, fabric softeners; for medical treatment of warts, hyperhidrosis, and herpes simplex[1, 2]
- It is soluble in water and organic solvents; glutaraldehyde solutions in water are slightly acidic
- It is a strong reducing agent which reacts violently with oxidants
- When heated glutaraldehyde gives off acid vapours

Summary of Human Toxicity

- Glutaraldehyde is well absorbed both orally and by inhalation, although at normal temperatures its low vapour pressure reduces the potential for inhalation exposures
- It is less well absorbed dermally
- Glutaraldehyde is a potent irritant to the eyes, nasal passages, respiratory tract, and skin[3]
- These irritant effects are enhanced by alkalinisation[3]
- Glutaraldehyde is a sensitiser and can cause contact dermatitis and asthma
- A toxic dose has not been established in humans; a 10% solution caused dermatitis when used therapeutically[4]
- Maximum Exposure Limit: not yet set

Acute Clinical Effects

Inhalation Effects

- Glutaraldehyde is irritating to the respiratory tract causing nose and throat irritation, epistaxis, chest tightness, headache and nausea[3]
- In sensitive individuals bronchospasm has been reported
- In severe or prolonged exposure pulmonary oedema may occur
- It is a strong sensitiser and may cause asthma
- Palpitations and tachycardia may occur, these are reversible on cessation of exposure[2,4]

Dermal Effects

- Glutaraldehyde is an irritant to the skin causing dryness and erythema
- Brown staining to skin and nails can occur
- It is a strong sensitiser resulting in allergic contact dermatitis; highly keratinised skin is less likely to be affected as it binds to glutaraldehyde reducing local absorption[3]

Eye Effects

- Glutaraldehyde is an irritant to the eyes causing inflammation and oedema
- Strong solutions may cause severe eye injury[5]

Oral Effects

- Glutaraldehyde is irritating to the mucous membranes and gastrointestinal tract

- Exposure may cause nausea, vomiting and diarrhoea

- High doses and strong solutions may cause gastrointestinal bleeding

Chronic Clinical Effects

Occupational contact dermatitis is frequently observed in health care workers after exposure to glutaraldehyde. Tachycardia has been reported with occupational exposure.[4] Hand eczema is prevalent. Rubber gloves do not appear to be protective.[2]

Occupational asthma may occur, often delayed for several hours post exposure. The severity of attacks may increase in frequency and severity, and are often related to the duration of the exposure.

Management

Inhalation Management

- Maintain a clear airway, give humidified oxygen and ventilate if necessary

- If respiratory irritation occurs assess respiratory function and if necessary perform chest X-rays to check for chemical pneumonitis

- Consider the use of steroids to reduce the inflammatory response

- Treat pulmonary oedema with PEEP or CPAP ventilation

- Symptomatic and supportive care

Dermal Management

- Remove any remaining contaminated clothing, place in double, sealed, clear bags and label; store in a secure area away from patients and staff

- Irrigate with copious amounts of water

- An emollient may be required

- Symptomatic and supportive care

Eye Management

- Irrigate with water or saline for 15 minutes

- Stain with fluorescein and refer to an ophthalmologist if there is any uptake of the stain

Oral Management

- NO GASTRIC LAVAGE OR EMETIC

- Encourage oral fluids

- Consider the use of steroids to reduce the inflammatory response

- Symptomatic and supportive care

- If facilities are available, early gastro-oesophagoscopy should be undertaken within 12 hours of the event to assess the extent and severity of the injury

Summary of Environmental Hazards

- Glutaraldehyde is not expected to bioconcentrate in the food chain

- In water glutaraldehyde is expected to undergo moderate biodegradation probably producing glutamic acid[4]

- Drinking Water Standards: no data available

- Soil Guidelines: no data available

- Air Quality Standards: no data available

REFERENCES

1. Budavari S, O'Neil MJ, Smith A, Heckelman PE & Kinneary JF (eds.) *The Merck Index*, 12th edn. Merck & Co., Inc., Whitehouse Station, 1996

2. Ellenhorn MJ, Schonwalds S, Ordog G & Wasserberger J. *Ellenhorn's Medical Toxicology – Diagnosis and Treatment of Human Poisoning*, 2nd edn. Williams & Wilkins, London, 1997

3. Wiggins P, McCurdy SA & Zeidenberg W, 1989. Epistaxis due to glutaraldehyde exposure. *J Occp Med*; 31(10): 854–6

4. Hall AH & Rumack BH (eds.) *TOMES System ®* Micromedex, Englewood, Colorado. CD ROM. vol 41 (exp. 31 July 1999)

5. Grant MW & Schuman JS. *Toxicology of the Eye*, 4th edn. Charles C Thomas, Springfield, 1993

Jennifer Butler

HYDRAZINE

Key Points

- Hydrazine is a colourless, oily liquid that fumes in air; it has a pungent ammonia-like odour
- It is absorbed by inhalation, ingestion and dermal exposure
- Hydrazine is a CNS stimulant and depressant causing increased excitability, convulsions and coma from all routes of exposure
- Renal and liver damage may occur
- *In the event of a large spill, stay upwind and out of low areas. Ventilate closed spaces. Protective clothing, eye protection and breathing apparatus should be worn*

FIRST AID

- Terminate exposure and support vital functions
- The casualty should be moved to an uncontaminated area
- Rescuers should, ideally, be trained personnel and must be careful *not to put themselves at risk* and *so wear appropriate protective clothing and, if available, breathing apparatus*
- If the casualty is unconscious a clear airway should be established and maintained; give 100% oxygen if available
- **Inhalation Exposure:** If the patient stops breathing, expired air resuscitation should be started immediately using a pocket mask with a one way valve, if available. It is important where the face is contaminated that expired air resuscitation is NOT attempted unless an airway with rescuer protection is used
- **Dermal Exposure:** Remove contaminated clothing, if possible under a shower and place in double, sealed, clear bags and label; store the bags in a secure area away from patients and staff
- Wash the skin thoroughly with copious amounts of water
- **Eye Exposure:** Irrigate thoroughly with water or saline for 15 minutes
- **Oral Exposure:** Encourage small quantities of oral fluids (no more than 50–100 ml in total)

Detailed Information

- Hydrazine is a colourless, oily liquid that fumes in air; it has a pungent ammonia-like odour
- *Common synonyms* diamide, diamine, hydrazine aqueous, hydrazine base
- CAS 302-01-2
- UN 2029
- UN 2030
- UN 3293
- NIOSH/RTECS MU 7175000
- Molecular formula N_2H_4
- Molecular weight 32.06

- Hydrazine is used as a rocket propellant; as an intermediate in the synthesis of agricultural chemicals, blowing agents for plastic, drugs and solder flux; aqueous hydrazine is used as a corrosion inhibitor in boiler water
- It is miscible with water and alcohol, insoluble in chloroform and ether; it is a strong reducing agent
- Decomposes on contact with many metals and metal oxides, with risk of fire or explosion
- It is a strong base which reacts violently with acids and corrodes metals
- Can ignite spontaneously in air and when in contact with porous materials
- Pure hydrazine decomposes on heating or when exposed to ultraviolet radiation to form ammonia, hydrogen and nitrogen; this may be an explosive reaction[1]
- It burns with a violet flame

Summary of Human Toxicity

- Hydrazine can be absorbed by inhalation, ingestion and through the skin
- Hydrazine appears to exert damage to the kidneys and liver through free radical mechanism after oxidative metabolism by cytochrome P4502E1
- Hydrazine exposure causes liver and CNS dysfunction by producing vitamin B_6 deficiency, thus the use of pyridoxine (vitamin B_6) may improve the clinical condition[2]
- The half-life of hydrazine in dogs is approximately 48 hours; probably similar in man[3]
- The odour perception 3–9 $mg.m^{-3}$ [1]
- Occupational exposure to 0.7 ppm once a week for 6 months lead to fatal liver, renal and cardiac damage
- Ingestion of a mouthful of hydrazine resulted in confusion, lethargy, restlessness, and hepatotoxicity; 10 g of pyridoxine IV was given resulting in improved mental state and liver function[2]
- Ingestion of 20–30 ml of a 6% aqueous solution led to vomiting, weakness, and irregular breathing with recovery in 5 days
- Hydrazine burns involving 22% of the body surface lead to coma followed by liver damage, recovery was seen in 5 weeks[3]
- Maximum Exposure Limits:
  Long-term exposure limit: 0.02 ppm (0.03 $mg.m^{-3}$)
  Short-term exposure limit: 0.1 ppm (0.13 $mg.m^{-3}$)

Acute Clinical Effects

Inhalation Effects

- Headache, dizziness, nausea, vomiting, sore throat, tight chest and flushed skin may be observed[4]

- Respiratory irritation may occur with irritation to the mucous membranes, rhinitis, salivation, coughing, choking and dyspnoea

- Acute low dose exposure may cause bronchial mucous destruction and pulmonary oedema which may be delayed for several days[4]

- Systemic effects may occur; see below

Dermal Effects

- Irritation may develop over several hours

- Rash, blisters, vesicles and crusts may develop[5]

- Inflammation, dermatitis and oedema

- Dilated and constricted pupils have been reported following dermal exposure[6]

- Systemic effects may be delayed for 14 hours or more;[3] see systemic effects below

Eye Effects

- Hydrazine vapours may cause irritation to the eyes, which may be delayed

- Itching, burning and swelling of the eyes may occur

- High vapour concentration can cause temporary blindness[7]

- Liquid hydrazine can lead to conjunctivitis, burns or permanent damage to the eyes

Oral Effects

- Ingestion may lead to nausea, vomiting, irritation to mucous membranes and swelling

- Oesophageal or gastrointestinal tract irritation or burns may occur[8]

- Weakness, lethargy, coma and delayed arrhythmia have been reported

- Systemic effects may occur; see below

Systemic Effects

- CNS stimulation with twitching of extremities, clonic movements, hyperreflexia, convulsions and pyrexia commonly occur,[4] and may progress to lethargy, ataxia, confusion, coma and hypotension

- Oliguria, haematuria, hyperglycaemia and/or hypoglycaemia and elevated LFTs are common[2]

- Leucocytosis, paraesthesia, peripheral neuropathy may be delayed over a few days

- Sequelae of deficits in concentration, comprehension, memory, task performance and mood status have been reported following respiratory and dermal exposure[8]

Chronic Clinical Effects

Allergic contact dermatitis and sensitisation may develop from short or prolonged dermal contact.[1,5]

Chronic exposure can cause lethargy, conjunctivitis, tremor, pyrexia, vomiting, diarrhoea, pneumonia, hepatocellular damage, nephritis and acute tubular necrosis.[9]

Hydrazine is an animal carcinogen and a suspected human carcinogen, but limited epidemiological studies have not indicated that occupational exposure to hydrazine is a risk for cancer.[8]

Management

Measurement of blood concentration is of no clinical relevance other than to confirm exposure when this is in doubt.

Inhalation Management

- Maintain a clear airway, give humidified oxygen and ventilate if necessary

- If respiratory irritation occurs assess respiratory function and if necessary perform chest X-rays to check for chemical pneumonitis

- Consider the use of steroids to reduce the inflammatory response

- Treat pulmonary oedema with PEEP or CPAP ventilation

- See systemic management below

Dermal Management

- Remove any remaining contaminated clothing, place in double, sealed, clear bags and label; store in a secure area away from patients and staff

- Irrigate with copious amounts of water

- An emollient may be required

- Treat irritation and burns symptomatically

Eye Management

- Irrigate thoroughly with running water or saline for 15 minutes

- Stain with fluorescein and refer to an ophthalmologist if there is any uptake of the stain

Oral Management

- NO GASTRIC LAVAGE OR EMETIC

- Encourage oral fluids

- Symptomatic and supportive care

- For large exposure, see systemic management below

Systemic Management

- Monitor electrolytes, cardiac, renal and liver function

- Pyridoxine may be of use to relieve CNS complications.[4] Contact a poisons information service for further guidance and paediatric doses:

    Suggested Dose: 25 mg.kg^{-1} [3] repeated if coma or convulsions persist to a **maximum** of 150 mg.kg^{-1}.d^{-1} **or** 10g (whichever is the lower)

    Side effects: large doses have been used (10 g IV) and sensory polyneuropathy developed after a week which resolved over 6 months[2]

- Symptomatic patients should be observed in hospital for at least 48 hours

- Symptomatic and supportive care

Summary of Environmental Hazards

- In the atmosphere hydrazine will exist solely in the vapour phase in the ambient atmosphere, it degrades by reacting with photochemically-produced hydroxyl radicals and ozone with estimated half-lives of about 6 and 9 hours, respectively[8]

- It is expected to degrade in soils containing a high percentage of organic carbon and strongly adsorbed in soils containing high clay content; in other soils, especially sandy soils, hydrazine may have high mobility

- Volatilisation from moist soil surfaces is not expected; the potential for volatilisation of hydrazine from dry soil surfaces may exist

- In water hydrazine undergoes rapid degradation, especially in water containing high concentrations of organic matter and dissolved oxygen; the estimated half-life of hydrazine in pond water is 8.3 days[8]

- It may bind to clay and organic matter found in sediments and particulate material in water; it should not strongly adsorb to other types of particulates

- Biodegradation is not expected to be an important environmental fate process in the presence of a large amount of hydrazine due to its toxicity to micro-organisms; it may be important at low hydrazine concentrations

- Low bioconcentration in fish is expected

- Drinking Water Standards: no data available

- Soil Guidelines: no data available

- Air Quality Standards: no data available

REFERENCES

1. International Programme on Chemical Safety. *Environmental Health Criteria 68: Hydrazine*. WHO, Geneva 1987

2. Harati Y & Niakan E, 1986. Hydrazine toxicity, pyridoxine therapy, and peripheral neuropathy. *Ann Intern Med*; 104(5): 728–9

3. Kirklin JK, Watson M, Bondoc CC & Burke JF, 1976. Treatment of hydrazine-induced coma with pyridoxine. *N Engl J Med*; 294(17): 938–9

4. Frierson WB, 1965. Use of pyridoxine HCL in acute hydrazine and UDMH intoxication. *Ind Med Surg*; 34: 650–1

5. Evan DM, 1959. Two cases of hydrazine hydrate dermatitis without systemic intoxication. *Brit J Indust Med*; 16: 12–7

6. Dhannin C, Vesint L & Feauveaux J, 1988. Burns and the toxic effects of a derivative of hydrazine. *Burns*; 14(2): 130–4

7. Hathaway GJ, Proctor NH & Hughes JP. *Proctor and Hughes' Chemical Hazards of the Workplace*, 4th edn. Van Nostrand Reinhold, New York, 1996

8. Hall AH & Rumack BH (eds.) *TOMES System* ® Micromedex, Englewood, Colorado. CD ROM. vol 41 (exp. 31 July 1999)

9. Sotaniemi E, Hirvonen J, Isomaki H, Takkunen J & Kaila J, 1971. Hydrazine toxicity in the human. *Ann Clin Res*; 3(1): 30–3

Henrietta Wheeler

HYDROCHLORIC ACID

Key Points

- Hydrochloric acid is a colourless or slightly yellow fuming pungent liquid
- Hydrochloric acid is a solution of hydrogen chloride gas in water
- It is extremely irritating to the eyes, mucous membranes, and skin
- Ingestion of hydrochloric acid is extremely corrosive to the gastrointestinal tract
- *Hydrochloric acid vapours are heavier than air and may accumulate in low or confined areas; in the event of a large spill, stay upwind and out of low areas. Ventilate closed spaces. Protective clothing, eye protection and breathing apparatus should be worn*

FIRST AID

- Terminate exposure and support vital functions
- The casualty should be moved to an uncontaminated area
- Rescuers should, ideally, be trained personnel and must be careful *not to put themselves at risk* and *so wear appropriate protective clothing and, if available, breathing apparatus*
- If the casualty is unconscious a clear airway should be established and maintained; give 100% oxygen if available
- **Inhalation Exposure:** If the patient stops breathing, expired air resuscitation should be started immediately using a pocket mask with a one way valve, if available. It is important where the face is contaminated that expired air resuscitation is NOT attempted unless an airway with rescuer protection is used
- **Dermal Exposure:** Remove contaminated clothing, if possible under a shower and place in double, sealed, clear bags and label; store the bags in a secure area away from patients and staff
- Wash the skin thoroughly with copious amounts of water
- **Eye Exposure:** Irrigate thoroughly with water or saline for 15 minutes
- **Oral Exposure:** Encourage small quantities of oral fluids (no more than 50–100 ml in total) unless perforation is suspected

Detailed Information

- Hydrochloric acid is the colourless, fuming aqueous solution of hydrogen chloride; it has a pungent odour and is non-flammable[1]
- *Common synonyms* anhydrous hydrochloric acid, aqueous hydrogen chloride, HCl, spirits of salt
- CAS 7647-01-0
- UN 1050 (anhydrous)
- UN 1789 (solution)
- UN 2186 (liquid)
- NIOSH/RTECS MW 4025000

- Molecular formula HCl
- Molecular weight 36.46
- Hydrochloric acid is used in the production of chlorides; production of dyes and dye intermediates; steel pickling, in the production of pharmaceutical hydrochlorides and chlorine, in certain household cleaners such as some drain and toilet cleaners
- The commercial 'concentrated' or fuming acid contains 38% hydrochloric acid[2]
- Hydrochloric acid reacts with air to form corrosive acid fumes which are heavier than air and spread at ground level
- Hydrochloric acid is a strong acid that reacts violently with bases and is corrosive; it attacks metals, giving off hydrogen which is flammable
- Reacts violently with strong oxidants to produce chlorine

Summary of Human Toxicity

- Hydrochloric acid exerts its effects by virtue of its strong acidity
- Hydrochloric acid is corrosive by inhalation, ingestion, to the eyes, skin and respiratory tract
- On ingestion it is more corrosive to the stomach (especially the pyloric region) and intestinal tract than the oesophagus (as opposed to alkalis)
- Concentrated hydrochloric acid burns the tissue by removing water
- Concentrations of 1500 to 2000 ppm are fatal to humans within a few minutes[3]
- Minimum lethal exposures in humans ranged from 3000 ppm for 5 minutes to 1300 ppm for 30 and 81 minutes[3]

Toxicity of Hydrochloric Acid[4,5]

Concentration	Clinical effects
0.067–0.134 ppm	Odour threshold and change in respiratory pattern
5 ppm	No organic damage
10 ppm	Irritation; work undisturbed
10–50 ppm	Work difficult but possible
35 ppm	Short exposure irritation of throat
50–100 ppm	Exposure for 1 h barely tolerable
1000–2000 ppm	Brief exposure dangerous; laryngospasm
1300–2000 ppm	Lethal after a few min

- Occupational Exposure Standards:
  Long-term exposure limit: no data available
  Short-term exposure limit: 5 ppm (7.6 mg.m^{-3})

Acute Clinical Effects

Inhalation Effects

- Hydrochloric acid is irritating to the eyes and respiratory tract

- Initially there may be coughing with tight chest and choking

- Headache, nausea and dyspnoea may follow

- Inflammation, and occasionally ulceration, of the throat, nose and larynx

- Laryngeal spasm and pulmonary oedema may occur in acute exposure

Dermal Effects

- Mild exposures to a dilute form may cause irritation and erythema to the skin[4]

- The liquid or concentrated vapours may cause immediate, severe and penetrating burns

- Concentrated solutions on the skin may lead to chemical burns and deep ulcers

- Severe skin burns can occur with necrosis resulting in scarring

- Sudden circulatory collapse can occur with shock if large areas of the skin have been burnt

Eye Effects

- Hydrochloric acid is very irritating to the eyes and may cause conjunctivitis

- High vapour concentrations may cause corneal necrosis and impaired vision

- Splashes to the eye can cause redness and irritation of the conjunctiva, white coagulation of the corneal and conjunctival epithelium and total corneal opacification and eye loss

- Damage to the conjunctival epithelium usually returns to normal after a few days[6]

Oral Effects

- Ingestion of small amounts of hydrochloric acid may cause epigastric pain, local irritation, nausea and vomiting and possible haematemesis

- With larger amounts severe burns of the mouth, oesophagus and stomach, especially the pyloric region can develop, stricture may also occur

- Acidosis, shock and circulatory collapse

- Pyloric stenosis often follows several weeks to years later in patients who survive an acute episode of ingestion[3]

Chronic Clinical Effects

Repeated or prolonged exposure to diluted solutions may cause dermatitis and erosion of teeth,[3] it may also be associated with changes in pulmonary function, chronic bronchitis and overt upper respiratory tract abnormalities.

Management

Inhalation Management

- Maintain a clear airway, give humidified oxygen and ventilate if necessary

- If respiratory irritation occurs assess respiratory function and if necessay perform chest X-rays to check for chemical pneumonitis

- Consider the use of steroids to reduce the inflammatory response

- Treat pulmonary oedema with PEEP or CPAP ventilation

- Symptomatic and supportive care

Dermal Management

- Remove any remaining contaminated clothing, place in double, sealed, clear bags and label; store in a secure area away from patients and staff

- Irrigate with copious amounts of water

- Treat burns symptomatically

Eye Management

- Irrigate thoroughly with running water or saline for 15 minutes

- Stain with fluorescein and refer to an ophthalmologist if there is any uptake of the stain

Oral Management

- NO GASTRIC LAVAGE OR EMETIC

- Encourage oral fluids, unless perforation is suspected

- Give plasma expanders/blood or IV fluids for shock and analgesics for pain

- Consider the use of steroids to reduce the inflammatory response

- Take abdominal X-ray to check for perforation

- Symptomatic and supportive care

- If facilities are available, early gastro-oesophagoscopy should be undertaken within 12 hours of the event to assess the extent and severity of the injury

Summary of Environmental Hazards

- Hydrogen chloride in water dissociates almost completely releasing hydrogen and chloride ions; the hydrogen ions are captured by water molecules to form hydronium ions[3]

- In soil hydrochloric acid will begin to infiltrate, the presence of water influencing the movement rate; during transport through the soil, hydrochloric acid will dissolve soil material and neutralise to some degree[3]

- Drinking Water Standards:
  chloride: 400 mg.l^{-1} (UK max)
  250 mg.l^{-1} (WHO guideline)

- Soil Guidelines: no data available

- Air Quality Standards: not data available

REFERENCES

1. Dutch Institute for the Working Environment and the Dutch
   Chemical Industry Association (eds.) *Chemical Safety Sheets –
   Working Safely with Hazardous Chemicals*. Samson Chemical
   Publishers, The Netherlands, 1991

2. Sax NI & Lewis RJ. *Hawley's Condensed Chemical Dictionary*,
   11th edn. Van Nostrand Reinhold Company, New York,
   1987

3. Hall AH & Rumack BH (eds.) *TOMES System®*
   Micromedex, Englewood, Colorado. CD ROM. vol 41.
   (exp. 31 July 1999)

4. Hathaway GJ, Proctor NH & Hughes JP. *Proctor and Hughes'
   Chemical Hazards of the Workplace*, 4th edn. Van Nostrand
   Reinhold, New York, 1996

5. Harbison RD (ed.) *Hamilton and Hardy's Industrial Toxicology*,
   5th edn. Mosby-Year Book Inc., St Louis, 1998

6. Grant MW & Schuman JS. *Toxicology of the Eye*, 4th edn.
   Charles C Thomas, Springfield, 1993

Henrietta Wheeler

HYDROFLUORIC ACID

Key Points

- Hydrofluoric acid is a colourless to green fuming liquid which is extremely corrosive and has a strong irritating odour
- All routes of exposure to hydrofluoric acid may be fatal; all exposures must be taken seriously
- It can cause severe burns to any tissue with which it comes into direct contact
- Inhalation exposures may lead to severe respiratory irritation and pulmonary oedema which may be immediate or delayed
- Systemic effects are possible from all routes of exposure, but are most likely from oral and dermal routes
- They include electrolyte imbalances particularly hypomagnesaemia and hypocalcaemia which can lead to cardiac arrhythmias and cardiac arrest
- *Hydrofluoric acid vapours are heavier than air and may accumulate in low or confined areas; in the event of a large spill, stay upwind and out of low areas. Ventilate closed spaces. Protective clothing, eye protection and breathing apparatus should be worn*

FIRST AID

- Terminate exposure and support vital functions
- The casualty should be moved to an uncontaminated area
- Rescuers should, ideally, be trained personnel and must be careful *not to put themselves at risk* and *so wear appropriate protective clothing and, if available, breathing apparatus*
- If the casualty is unconscious a clear airway should be established and maintained; give 100% oxygen if available
- **Inhalation Exposure:** If the patient stops breathing, expired air resuscitation should be started immediately using a pocket mask with a one way valve, if available. It is important where the face is contaminated that expired air resuscitation is NOT attempted unless an airway with rescuer protection is used
- **Dermal Exposure:** Remove contaminated clothing, if possible under a shower and place in double, sealed, clear bags and label; store the bags in a secure area away from patients and staff
- Wash the skin thoroughly with copious amounts of water and then massage *calcium gluconate gel* into the affected areas. If calcium gluconate gel is not available then either 10% calcium gluconate solution or 25% magnesium sulphate solution should be applied
- NB – In cases of dermal exposure the use of topical and systemic analgesics is best avoided initially since pain is a valuable indicator of the sites and extent of the injury
- **Eye Exposure:** Irrigate thoroughly with water or saline for 15 minutes
- **Oral Exposure:** Encourage small quantities of oral fluids (no more than 50–100 ml in total) unless perforation is suspected

Detailed Information

- Hydrofluoric acid is the colourless to green, fuming, corrosive aqueous solution of hydrogen fluoride; it has a strong irritating odour
- *Common synonyms* anhydrous hydrofluoric acid, fluohydric acid, fluoric acid, hydrofluoride
- CAS 7664-39-3
- UN 1052 (anhydrous)
- UN 1790 (solution)
- NIOSH/RTECS MW 7875000
- Molecular formula HF
- Molecular weight 20.01
- Hydrofluoric acid is used for cleaning cast metal, for frosting, etching and polishing glass and for removing sand from metal castings
- Hydrofluoric acid is the aqueous form of hydrogen fluoride; anhydrous hydrogen fluoride and hydrofluoric acid are the most acidic substances known[1,2]
- The vapour is heavier than air and reacts with water to form fumes which spread at ground level[3]
- In aqueous solution it gives off large quantities of heat; it reacts with alcohol and unsaturated compounds; it attacks glass and other silicatious materials[4]
- When heated to decomposition hydrofluoric acid emits highly toxic fumes of fluorides

Summary of Human Toxicity

- Hydrofluoric acid is extremely toxic and all exposures should be regarded seriously
- Systemic toxicity can occur from all routes of exposure, especially oral and dermal
- It readily penetrates intact skin, getting under nails and into deep tissue layers causing liquefactive necrosis of soft tissues and decalcification and corrosion of bone; this may be extremely painful and prolonged for days
- The severe pain associated with the tissue destruction is thought to be due to the calcium precipitating property of the fluoride ion which results in the immobilisation of tissue calcium and a relative excess of potassium in the tissues
- The fluoride ion also inhibits glycolytic enzymes and may produce a severe hypocalcaemia resulting in ventricular dysrhythmias and depressed contractility
- A fatality has occurred from systemic poisoning following exposure of 2.5% of the body surface to concentrated hydrofluoric acid[5]
- Severe ocular damage may occur with concentrations greater than 0.5%[6]
- Fatalities from oral exposures have been reported from 1.5 g (concentration unknown) 6.5 h post ingestion; 3–4 oz of a

rust remover (hydrofluoric acid concentration unknown) 45–60 min post ingestion; 15 ml of a 9% solution[6]

- Treatment involves the inactivation of the fluoride ion by complexing it with magnesium and calcium salts

Toxicity of Dermal Hydrofluoric Acid[6]

Concentration	Clinical effects
<20%	Symptoms can be delayed for up to 24 h, pain and erythema
20–50%	Symptoms usually occur within 30 to 60 min, but may be delayed for up to 8 h
>50%	Symptoms likely to occur within 5 to 10 min, immediate pain and tissue destruction is rapid

Toxicity of Inhaled Hydrofluoric Acid[1,7,8]

Concentration	Clinical effects
5 ppm	Fumes irritating to eyes and nose
2.6–4.8 ppm	For periods up to 50 d: slight irritation of nose, eyes and skin
30 ppm	Several min: mild irritation of eyes, nose and respiratory tract
50 ppm	Severe respiratory irritation
120 ppm	1 min caused conjunctival and respiratory irritation with stinging of skin
250 ppm	Exposure to fumes for 5 min fatal

- Occupational Exposure Standards, as fluoride:
     Long-term exposure limit: not available
     Short-term exposure limit: 3 ppm (2.5 mg . m^{-3})

Acute Clinical Effects

Inhalation Effects

- Initially there may be hoarseness, coughing fits and choking, and nosebleeds, although pulmonary function may be normal[1]

- Severe respiratory irritation, dyspnoea, fever and cyanosis may be delayed for up to 24 hours

- Laryngeal oedema, chemical pneumonitis, tracheobronchitis and haemorrhagic pulmonary oedema may be delayed

- Renal failure has been reported (cause unknown) in an ultimately fatal inhalation exposure

Dermal Effects

- Skin irritation, rash and burning sensation may occur

- Erythema, central blanching with peripheral erythema, pain which may be severe, swelling and vesiculation can occur from minimal exposure

- In severe cases there may also be ulceration, a blue-grey discoloration of the skin and necrosis and bone decalcification

- Tendinitis and tenosynovitis may result

- Systemic effects may occur; see below

Eye Effects

- Conjunctivitis, corneal necrosis and opacification with conjunctival ischaemia

- Loss of the epithelium and extensive oedema of the lids, conjunctivae and cornea is possible[7]

- Severe exposures may destroy the eye and require enucleation

Oral Effects

- Ingestion of hydrofluoric acid causes severe epigastric pain, nausea, vomiting, abdominal pain, oral burns, necrotic lesions and haemorrhagic gastritis

- Severe acidosis and shock may occur

- Pancreatitis has been reported (cause unknown) in an ultimately fatal oral exposure[6]

- Late complications may include oesophageal stricture and pyloric stenosis which may follow several weeks later

- Systemic effects may occur; see below

Systemic Effects

- Hypocalcaemia and hypomagnesaemia which may lead to cardiac arrhythmias, and metabolic acidosis[1]

- Pulmonary oedema may occur immediately or be delayed

Chronic Clinical Effects

Chronic low level exposures may result in nasal congestion and bronchitis. Chronic low level dermal contact may lead to redness and irritation. Effects of chronic exposure may lead to systemic fluoride toxicity (fluorosis), osteosclerosis and teeth discoloration. See fluoride salts (page 242).

Management

Inhalation Management

- Maintain a clear airway, give humidified oxygen and ventilate if necessary

- If respiratory irritation occurs assess respiratory function and if necessary perform chest X-rays to check for chemical pneumonitis

- Consider administration of 2.5 to 3% calcium gluconate solution via nebulizer[9] and if bronchial lavage is needed, the use of calcium gluconate solution instead of bicarbonate solution will be of benefit

- Consider the use of steroids to reduce the inflammatory response

- Treat pulmonary oedema with PEEP or CPAP ventilation

- Symptomatic and supportive care

Dermal Management

- Remove any remaining contaminated clothing, place in double, sealed, clear bags and label; store in a secure area away from patients and staff

- Irrigate immediately with copious amounts of water

- Massage calcium gluconate gel into the affected area for a minimum of 30 minutes; if the gel is not available apply calcium gluconate (10%) solution or magnesium sulphate (25%) solution

- If the pain persists or the hydrofluoric acid solution was greater than 20% in strength, then calcium gluconate (10%) solution (*not* calcium chloride because it is irritant) should be infiltrated under the site of injury; multiple small volume (0.5 ml) injections are best

- If the pain recurs injections may be repeated and extended in area

- For severe exposures to the extremities consider a Bier's block technique (using calcium gluconate together with a local anaesthetic); seek advice from a local poisons information centre

- For burns >5% in area or for suspected hypocalcaemia IV calcium should be given as follows: calcium gluconate (10%) 0.1 to 0.2 ml.kg^{-1} (0.02–0.04 mmol.kg^{-1})

- See systemic management below

- Contact a poisons information service for further guidance

Eye Management

- Irrigate thoroughly with running water or saline for 15 minutes

- Stain with fluorescein

- **Immediately** refer to an ophthalmologist for all cases

Oral Management

- NO GASTRIC LAVAGE OR EMETIC

- Neutralising chemical should never be given because heat is produced during neutralisation and this could exacerbate any injury

- Give plasma expanders/blood or IV fluids for shock and analgesics for pain

- Consider the use of steroids to reduce the inflammatory response

- For hypocalcaemia IV calcium should be given as follows: calcium gluconate (10%) 0.1 to 2 ml.kg^{-1} (0.02 – 0.04 mmol.kg^{-1})

- Symptomatic and supportive care

- If facilities are available, early gastro-oesophagoscopy should be undertaken within 12 to 24 hours of the event to assess the extent and severity of the injury

- See systemic management below

- For further guidance, contact a poisons information service

Systemic Management

- Monitor ECG; check electrolytes, particularly calcium and magnesium, hourly initially and thereafter as determined by clinical condition of the patient

- Monitor renal function

- Contact a poisons information service for further guidance

Summary of Environmental Hazards

- Hydrofluoric acid is harmful to aquatic life at low concentrations, however natural alkalinity will slowly dissipate the acidity[6]

- All plant species may be threatened by hydrofluoric exposure although 0.1 ppm for 3–4 hours had no effect on plants[6]

- It is not expected to bioconcentrate in the food chain

- Drinking Water Standards: fluoride:
        1.5 mg.l^{-1} (UK max)
        1.5 mg.l^{-1} (WHO guideline)

- Soil Guidelines: no data available

- Air Quality Standards: no data available

REFERENCES

1. Hathaway GJ, Proctor NH & Hughes JP. *Proctor and Hughes' Chemical Hazards of the Workplace*, 4th edn. Van Nostrand Reinhold, New York, 1996

2. Budavari S, O'Neil MJ, Smith A, Heckelman PE & Kinneary JF (eds.) *The Merck Index*, 12th edn. Merck & Co., Inc., Whitehouse Station, 1996

3. Dutch Institute for the Working Environment and the Dutch Chemical Industry Association (eds.) *Chemical Safety Sheets – Working Safely with Hazardous Chemicals.* Samson Chemical Publishers, The Netherlands, 1991

4. Sax NI. *Dangerous Properties of Industrial Materials*, 6th edn. Van Nostrand Reinhold, New York, 1984

5. Tepperman PB, 1990. Fatality due to systemic fluoride poisoning following a hydrofluoric acid skin burn. *J Occup Med*; 22(10): 691–2

6. Hall AH & Rumack BH (eds.) *TOMES System®* Micromedex, Englewood, Colorado. CD ROM. vol 41. (exp. 31 July 1999)

7. Grant MW & Schuman JS. *Toxicology of the Eye*, 4th edn. Charles C Thomas, Springfield, 1993

8. Harbison RD (ed.) *Hamilton and Hardy's Industrial Toxicology*, 5th edn. Mosby-Year Book Inc., St Louis, 1998

9. Lee DC, Wiley JF & Synder JW, 1993. Treatment of inhalational exposure to hydrofluoric acid with nebulised calcium gluconate (letter). *J Occup Med*; 35(5): 470

Henrietta Wheeler

HYDROGEN CYANIDE AND CYANIDE SALTS

Key Points

- **N.B.** Unless otherwise stated all figures, values and reference numbers relate to hydrogen cyanide (HCN) only; cyanide salts are toxic except for metal cyanides which rarely produce cyanide poisoning
- Clinical effects appear within seconds or minutes after ingestion or inhalation
- Initially giddiness, headache, anxiety, confusion and dyspnoea may occur
- Rapidly followed by coma, convulsions, bradycardia, hypotension and metabolic acidosis
- Exposure to HCN may be fatal within minutes
- *In the event of a large spill, stay upwind and out of low areas. Ventilate closed spaces. Protective clothing, eye protection and breathing apparatus should be worn*
- *Cyanide vapours may exude from heavily contaminated casualties and so there is a risk that emergency personnel may become contaminated*
- *Casualties should be transported in such a way that there is no risk of the drivers of the emergency vehicles becoming contaminated by the fumes*

FIRST AID

- *Rescuers must not enter a contaminated area without full personal protective equipment and self-contained breathing apparatus*
- Terminate exposure and support vital functions
- The casualty should be moved to an uncontaminated area
- If the casualty is unconscious a clear airway should be established and maintained; give 100% oxygen if available
- **Inhalation, Oral and Symptomatic Exposure:** If the patient stops breathing, expired air resuscitation should be started immediately using a pocket mask with a one way valve, if available. It is important where the face is contaminated that expired air resuscitation is NOT attempted unless an airway with rescuer protection is used
- Amyl nitrite by inhalation may be given, if available; preferably 0.2 to 0.4 ml via an Ambu bag or one ampoule on to a cloth and held to the casualty's nose for 30 seconds per minute with a new ampoule every 3 minutes to a maximum of 6 ampoules
- If medical help is available, intravenous antidotes should be administered to **unconscious** adult patients:
- *Note: before an intravenous antidote is given a blood sample should be taken for estimation of cyanide concentrations later; cyanide concentrations measured in blood taken after the administration of an antidote cannot be interpreted accurately*
- Give 50ml of 25% sodium thiosulphate solution (12.5 g) IV over 10 minutes followed by:

  Either

  - 20ml of 1.5% dicobalt edetate solution, i.e. 300 mg, IV over one minute followed immediately by 50 ml of 50% dextrose IV

(**N.B.** This drug should only be given if the diagnosis is certain, because serious cobalt toxicity may occur)

Or

- 10ml of 3% sodium nitrite solution (300 mg) IV over 5 to 20 minutes
- Once intravenous therapy has been started amyl nitrite is no longer necessary
- **Details of all antidotal therapy administered should accompany the patient to hospital**
- **Dermal Exposure:** Remove contaminated clothing, if possible under a shower and place in double, sealed, clear bags and label; store the bags in a secure area away from patients and staff
- Wash the skin thoroughly with copious amounts of water
- **Eye Exposure:** Irrigate thoroughly with water or saline for 15 minutes

Detailed Information

- HCN is a colourless gas with a with a distinct odour of bitter almonds; in the workplace it may exist as either a water-white or pale blue liquid or a gas
- Potassium, sodium and calcium cyanides are white, deliquescent, non-combustible solids with faint bitter almond odour
- Cyanogen is colourless, flammable gas with a pungent almond odour; cyanogen bromide is a colourless or white crystalline solid
- Cyanogen chloride is either a colourless irritant gas or liquid; cyanogen azide is a clear, colourless, oily liquid and cyanogen iodine a colourless solid

Synonyms of Cyanide and Cyanide Compounds

Compound	Synonyms
Calcium cyanide	calcid, calcyanide, cyanogas, calcyan
Cyanide	carbon nitride ion, cyanide anion, cyanide ion, isocyanide
Cyanogen	carbon nitride, cyanogen gas, oxalic nitrile, oxalonitrile, oxalyl cyanide, nitriloacetonitrile, ethanedinitrile
Cyanogen bromide	bromine cyanide, bromocyanogen, cyanobromide, bromocyanide, bromocyan
Cyanogen chloride	chlorine cyanide, chlorocyanide, chlorcyan, chlorocyanogen
Cyanogen iodide	iodine cyanide
Hydrogen cyanide	formonitrile, HCN, hydrocyanic acid, prussic acid
Potassium cyanide	cyanide of potassium
Sodium cyanide	cyanide of sodium

Identification Numbers of Cyanide and Cyanide Compounds

Chemical / Compound	CAS	UN	NIOSH / RTECS
Calcium cyanide	592-01-8	1575	EW 0700000
Cyanide	57-12-5	1935	GS 7175000
Cyanogen	460-19-5	1026	GT 1925000
Cyanogen azide	764-05-6	n.a.	n.a.
Cyanogen bromide	506-68-3	1889	GT 2100000
Cyanogen chloride	506-77-4	1589	GT 2275000
Cyanogen fluoride	1495-50-7	n.a.	n.a.
Cyanogen iodide	506-78-5	n.a.	NN 1750000
Hydrogen cyanide	74-90-8	1051 (stabilised, anhydrous) 1613 (solution) 1614 (stabilised, anhydrous, absorbed)	MW 6825000
Potassium cyanide	151-50-8	1680	n.a.
Sodium cyanide	143-33-9 13998-03-3	1689	VZ 7525000
Zinc cyanide	557-21-1	1713	ZH 1575000

n.a. = number not available

Molecular Formula and Weight of Cyanide Compounds

Chemical/ Compound	Molecular formula	Molecular weight
Calcium cyanide	$Ca(CN)_2$	92.12
Cyanide	CN	26.02
Cyanogen	$(CN)_2$	52.04
Cyanogen bromide	$BrCN$	105.93
Cyanogen chloride	$ClCN$	61.47
Cyanogen iodide	ICN	152.92
Hydrogen cyanide	HCN	27.03
Potassium cyanide	KCN	65.12
Sodium cyanide	$NaCN$	49.01
Zinc cyanide	$Zn(CN)_2$	117.41

- Cyanide salts are used mainly in industry and for pest control; hydrogen cyanide is used in fumigation of ships, large buildings or aeroplanes

- Cyanide salts are used in metal cleaning, gardening, in ore-extracting processes, dyeing, printing and photography; used in resin monomer production (e.g. acrylates)

- HCN may become unstable and explosive on prolonged storage (>90 days), if exposed to high temperatures (>184°C) and pressure

- HCN is miscible with water, soluble in alcohol and ether

- HCN reacts spontaneously with alkalis and acetaldehyde

Summary of Human Toxicity

- Cyanide is absorbed by inhalation, ingestion, and through the eye and intact skin

- Cyanides inhibit cellular respiration by binding reversibly with cytochrome oxidase thus preventing oxidation of reduced cytochrome c; clinical effects are therefore due to tissue hypoxia

- In massive cyanide poisonings other mechanisms may contribute to clinical effects such as pulmonary and coronary arterial vasoconstriction causing heart pump failure and decreasing cardiac output

- Cyanides act extremely quickly once absorbed

- Hydrogen cyanide inhalation can be lethal within minutes; ingestion of inorganic cyanide salts such as potassium or sodium cyanide may produce clinical effects and sometimes fatalities within minutes

- Dermal exposures to cyanides have resulted in fatalities

- The odour detection of cyanide is 2–5 ppm; olfactory fatigue is rapid and some 20% or more of the population may not detect HCN since this ability is genetically determined

- There is significant variation in reported toxic doses for HCN:

Fatal Inhalation Concentrations for Hydrogen Cyanide[1,2,3]

Concentration	Fatality time
110–135 ppm	30 min to 1 h
270 ppm	Immediate
180–550 ppm	30 min to 1 h
1301 ppm	3 min
3630 ppm	30 sec

- Oral fatal dose for HCN has been estimated as approximately 50–100 mg; 5 ml of a 20% HCN solution has been fatal on ingestion; 150–300 mg of inorganic cyanide salts (e.g. KCN and NaCN) would be expected to cause fatalities in adults[1,2,3]

- Dermal exposure (area 18500 cm² of skin) to 10000 ppm of HCN vapour for 10–19 minute may be fatal; exposure of 1000 cm² skin to HCN vapour at 55000 ppm may cause fatality after 45–85 minutes[1,2]

- Maximum Exposure Limits:
  Long-term exposure limit: no data available
  Short-term exposure limit: 10 ppm (11 mg . m⁻³)

Acute Clinical Effects

All routes: Systemic Effects

- Mild effects from low level exposure usually are headache, anxiety, nausea and vomiting

- Clinical effects or fatalities may occur within minutes of exposure

- Principal signs and symptoms are headache, dizziness, vomiting, anxiety, confusion, weakness, ataxia, hyperventilation, dyspnoea, hypotension, bradycardia and collapse

- Loss of vision and hearing may occur

- Coma and convulsions may occur, as may other features such as cardiac arrhythmias and pulmonary oedema

- The patient may have a lactic acidosis

- Arterial oxygen tension is likely to be normal but venous oxygen tension will be high and similar to that of arterial blood

- Some sources describe a bright-red colouration of the skin, and cyanosis as a late finding following cyanide exposures, although these effects are rarely noted in case reports of exposures

- Most cases of acute poisoning either die or recover, but rare cases of long-term sequelae include development of parkinsonian syndrome, memory deficits, personality changes and extrapyramidal syndromes have been reported

Eye Effects

- Irritant effects; conjunctivitis and eye lid oedema has been reported following exposure to cyanogen chloride[4]

- Theoretically, systemic toxicity could arise from this route of exposure

- Transient blindness has occasionally been reported from exposure to sublethal cyanide[4]

- Blindness is possible with cyanide-induced damage to optic nerves and retina[4]

Chronic Clinical Effects

Respiratory tract irritation, chest discomfort, exertional dyspnoea and varying degrees of rhinitis, nasal obstruction and bleeding have been seen in workers chronically exposed to cyanide.[5] Headaches, vertigo, fatigue, nausea, poor appetite, sleeping disturbances and convulsions may occur with chronic occupational exposure.[3,6] Hoarseness, skin and mucous membrane irritation may be found.[3]

Functional hearing changes have been reported in chronic exposure to cyanide.[5] Enlarged thyroid glands and decreased iodine uptake have been reported.[5,6]

Hospital Management for all Exposures[7,8,9]

All medical staff should wear full personal protective equipment when decontaminating patients; if they also become symptomatic they should be treated as below

Cyanide concentrations may be determined in blood and correlate well with toxicity

- **Mild Poisoning** – patients with nausea, dizziness, drowsiness only (blood cyanide concentration <2 mg.l^{-1})

  - *Inhalation* – patient unlikely to deteriorate further following arrival at hospital

  - Oxygen should be given if required, together with reassurance, and the patient should be observed

  - *Ingestion* – consider gastric lavage within 1 hour of ingestion

  - Contact a poisons information service for further guidance on gut decontamination

  - As toxicity may be delayed following the ingestion of cyanide salts, the patient should be observed, ideally in ITU, for several hours and treatment given as required

- **Moderate poisoning** – patients who have suffered a short-lived period of unconsciousness, convulsions, vomiting and/or cyanosis (blood cyanide concentrations 2–3 mg.l^{-1})

  - Give 100% oxygen for no more than 4 hours and observe in ITU

  - Correct acidosis with intravenous bicarbonate

  - Give patient 50 ml of 25% sodium thiosulphate solution (12.5 g) IV over 10 minutes

  - If patient deteriorates treat as for severe poisoning

  - If cyanide salts were ingested then once the patient is stabilised gastric lavage may be considered

- **Severe poisoning** – (deep coma with dilated, non-reactive pupils and deteriorating cardiorespiratory function (blood cyanide concentrations 3–4 mg.l^{-1})

  - Give 100% oxygen and cardiorespiratory support

  - Correct acidosis with intravenous bicarbonate

  - Before administering parenteral antidotal therapy ascertain whether antidote(s) have already been given prior to hospital admission

  - Some antidotes are themselves toxic if used to excess (see adverse effects of antidotes below)

  - If patient has not already received parenteral therapy then the following drugs may be given:

Either

20 ml of 1.5% dicobalt edetate solution, i.e. 300 mg, IV over one minute followed immediately by 50 ml of 50% dextrose IV

N.B. This drug should only be given if the diagnosis is certain, because serious toxicity is otherwise likely to occur (see adverse effects of antidotes below)

Or

10 ml of 3% sodium nitrite solution (300 mg) IV over 5 to 20 minutes

Then

50 ml of 25% sodium thiosulphate solution (12.5 g) IV over 10 minutes

- If patient does not respond or responds only partially then further doses of sodium thiosulphate may be given as required; a second dose of dicobalt edetate may also be given but the patient should not receive more than this because of the danger of cobalt toxicity

- It is inadvisable to repeat the dose of sodium nitrite because of the risk of excess methaemoglobinaemia and impaired oxygen transport (see adverse effects of antidotes below)

- For further guidance on the use of antidotes or for paediatric doses, the physician should consult a poisons information service

- If cyanide salts were ingested then once the patient is stabilised a gastric lavage may be considered; discuss with a poisons information service

Notes

- Administration of 100% oxygen is probably the single most important aspect of treatment and should be continued while patient is receiving antidotal therapy

- Oxygen should be given using a mask and bag with a non-return valve to prevent inspiration of exhaled gases

- Patients have survived severe cyanide poisoning with supportive therapy alone (because the diagnosis was made retrospectively)

Adverse Effects of Antidotes

Dicobalt Edetate

- Nausea, vomiting, urticarial rash, bronchospasm, chest pain, tacycardia, hypotension, periorbital oedema and/or convulsions may be precipitated if the drug is injected too rapidly or if given in the absence of cyanide

- If severe reaction develops injection should be stopped and appropriate supportive measures instituted

Sodium Nitrite

- Nausea, vomiting, headache, hypotension, syncope and methaemoglobinaemia

- Methaemoglobinaemia is necessary for the antidotal action of the drug because methaemoglobin competes for cyanide with cytochrome oxidase, forming an inert cyanmethaemoglobin complex

- If methaemoglobinaemia becomes excessive (i.e. greater than 40%), oxygen transport to the tissues will become significantly impaired, effectively negating the antidotal effect of the sodium nitrite

- Severe methaemoglobinaemia, in the order of 60 to 70% will need to be corrected by exchange transfusion

- Methylene blue should not be given to reverse excess methaemoglobin because it will result in the release of cyanide as cyanmethaemoglobin is converted back to haemoglobin

Summary of Environmental Hazards

- Most cyanide in the atmosphere is expected to exist almost entirely as hydrogen cyanide gas, although small amounts of metal cyanides may be present as particulate matter in air (for nitriles see acetonitrile page 12)

- Hydrogen cyanide vapour reacts with photochemically generated hydroxyl radicals fairly slowly, with an estimated half-life of 334 days[3]

- Hydrogen cyanide is expected to be resistant to direct photolysis

- The relatively slow rate of degradation of hydrogen cyanide suggests that this compound has the potential to be transported over long distances before being removed by physical or chemical processes

- Wet deposition may be an important fate for hydrogen cyanide; metal cyanide particles are expected to be removed from air by both wet and dry deposition

- The fate of cyanides in soil are expected to be pH dependent; at soil surfaces with pH < 9.2 hydrogen cyanide is expected to volatilise[3]

- In soil with pH < 9.2 hydrogen cyanide is expected to be highly mobile, and in cases where cyanide levels are toxic to micro-organisms (i.e. landfills, spills), it may leach into groundwater[3]

- In subsurface soil, cyanide present at low concentrations would probably biodegrade

- Hydrogen cyanide is harmful to aquatic life in very low concentrations

- Drinking water standards:
  cyanide: $50\,\mu g.l^{-1}$ (UK max)
                    $70\,\mu g.l^{-1}$ (WHO guideline)
  chloride: $400\,mg.l^{-1}$ (UK max)
                    $250\,mg.l^{-1}$ (WHO guideline)
  fluoride: $1.5\,mg.l^{-1}$ (UK max)
                    $1.5\,mg.l^{-1}$ (WHO guideline)
  sodium: $150\,mg.l^{-1}$ (UK max)
  zinc: $5000\,\mu g.l^{-1}$ (UK max)

- Soil Guidelines:
  free cyanide: $1\,mg.kg^{-1}$ (target)
                             $20\,mg.kg^{-1}$ (intervention)
  complex cyanide (pH < 5): $5\,mg.kg^{-1}$ (target)
                             $650\,mg.kg^{-1}$ (intervention)
  complex cyanide (pH > 5): $5\,mg.kg^{-1}$ (target)
                             $50\,mg.kg^{-1}$ (intervention)
  zinc: $140\,mg.kg^{-1}$ (target)
                             $720\,mg.kg^{-1}$ (intervention)

- Air Quality Standards: no data available

REFERENCES

1. International Programme on Chemical Safety Poisons Information Monograph. *Cyanides.* IPCS/INTOX/PIM.159, September 1992

2. Clayton GD & Clayton FE (eds.) *Patty's Industrial Hygiene and Toxicology*, 4th edn. John Wiley & Sons, Inc., New York, 1994

3. Hall AH & Rumack BH (eds.) *TOMES System* ® Micromedex, Englewood, Colorado. CD ROM. vol.41 (exp. 31 July 1999)

4. Grant MW & Schuman JS. *Toxicology of the Eye*, 4th edn. Charles C Thomas, Springfield, 1993

5. Hathaway GJ, Proctor NH & Hughes JP. *Proctor and Hughes' Chemical Hazards of the Workplace*, 4th edn. Van Nostrand Reinhold, New York, 1996

6. Harbison RD (ed.) *Hamilton and Hardy's Industrial Toxicology*, 5th edn. Mosby-Year Book Inc., St Louis, 1998

7. Wurtzburg H, 1996. Treatment of cyanide poisoning in an industrial setting. *Vet Hum Toxicol*; 38(1): 44–7

8. Meredith TJ, Jacobsen D, Haines JA, Berger J-C & van Heijst ANP (eds.) *Antidotes for poisoning by Cyanide*. IPCS/CEC Evaluation of Antidotes Series 2, University Press, Cambridge, 1993

9. Ballantyne B & Marrs TC (eds.) *Clinical and Experimental Toxicology of Cyanides*. Wright, Bristol, 1987

Alexander Campbell

HYDROGEN PEROXIDE

Key Points

- Hydrogen peroxide is a colourless, odourless liquid with a bitter taste
- It is a strong oxidiser
- By ingestion it causes gastric irritation and occasionally gas embolism
- It is irritating to the skin and respiratory tract
- *In the event of a large spill, stay upwind and out of low areas. Ventilate closed spaces. Protective clothing, eye protection and breathing apparatus should be worn*

FIRST AID

- Terminate exposure and support vital functions
- The casualty should be moved to an uncontaminated area
- Rescuers should, ideally, be trained personnel and must be careful *not to put themselves at risk* and *so wear appropriate protective clothing and, if available, breathing apparatus*
- If the casualty is unconscious a clear airway should be established and maintained; give 100% oxygen if available
- **Inhalation Exposure:** If the patient stops breathing, expired air resuscitation should be started immediately using a pocket mask with a one way valve, if available. It is important where the face is contaminated that expired air resuscitation is NOT attempted unless an airway with rescuer protection is used
- **Dermal Exposure:** Remove contaminated clothing, if possible under a shower and place in double, sealed, clear bags and label; store the bags in a secure area away from patients and staff
- Wash the skin thoroughly with copious amounts of water
- **Eye Exposure:** Irrigate thoroughly with water or saline for 15 minutes
- **Oral Exposure:** Encourage small quantities of oral fluids (no more than 50–100 ml in total)

Detailed Information

- Hydrogen peroxide is a colourless, odourless liquid with a bitter taste
- *Common synonyms*: hydrogen dioxide, hydroperoxide
- CAS 7722-84-1

Identification Numbers of Hydrogen Peroxide

UN	NIOSH/RTECS
2015 (stabilised)	MX 0887000 (3% solution)
2014 (solution 20–52%)	MX 0890000 (8–20% solution)
2984 (solution 8–20%)	MX 0899000 (30% solution)
	MX 0899500 (20–60% solution)
	MX 0900000 (90% solution)

- Molecular formula H_2O_2

- Molecular weight 34.02
- Hydrogen peroxide is used as a 6% solution for bleaching hair and some disinfectant solutions for contact lenses contain 3% hydrogen peroxide; chlorine free (environmentally-friendly) bleaches contain 6% hydrogen peroxide; industrial strengths are up to 90% and are used mainly as bleaching and oxidising agents; 90% solution is used as rocket fuel
- The strength of a solution may be described as a percentage or volume, where 1% hydrogen peroxide releases 3.3 volumes of oxygen during decomposition; a 3% solution is equivalent to 10 volume and a 6% solution to 20 volume, etc.
- Hydrogen peroxide is miscible with water, soluble in ether and decomposed by many organic solvents
- Hydrogen peroxide is an oxidising agent which in the presence of organic matter or if permitted to become alkaline, vigorously and exothermically decomposes to oxygen and water

Summary of Human Toxicity

- Hydrogen peroxide acts on exposed tissues by releasing oxygen; for each volume of 3% solution, 10 volumes of oxygen may be produced[1]
- Exposure to 3%, household strengths, of hydrogen peroxide by ingestion, dermally or in the eye does not normally give rise to toxic effects
- Inhalation of 90% hydrogen peroxide causes nasal irritation, increased salivation, a scratchy feeling of the throat and respiratory tract inflammation[2]
- Solutions of >10% on the skin may cause burns
- Ingestion is the main route of exposure and causes gastrointestinal irritation
- There is also a risk of gas embolism which although not common from ingestion has occurred after the use of hydrogen peroxide for irrigation of body cavities
- Several deaths from ingestion are reported in the literature; in most cases the exposures were to concentrated solutions of 30–40%. One case reports a 49-year-old female ingesting 240 ml of a 35% solution, from which she died 78 hours later[3]
- Cerebral infarction, believed to have resulted from gas embolisation of the cerebral vasculature, has been reported in an 84-year-old man who took 30 ml of 35% hydrogen peroxide diluted in 100–300 ml water[4]
- Multiple brain emboli occurred in a 63-year-old who ingested 120 ml of 36% solution; he recovered [5]

Toxicity of Ingested Hydrogen Peroxide [3,4,5]

Concentration	Clinical effects
30 ml (35% sol.) diluted 100–300 ml H_2O	Cerebral infarction, probably from a gas embolisation of the cerebral vasculature
120 ml (35% sol.)	Multiple brain emboli
240 ml (35% sol.)	Fatal 78 h later

- Occupational Exposure Standards:
  Long-term exposure limit: 1 ppm (1.4 mg.m⁻³)
  Short-term exposure limit: 2 ppm (2.8 mg.m⁻³)

Acute Clinical Effects

Inhalation Effects

- Transient dyspnoea and cough; with concentrated solutions there may be more severe irritation and inflammation of the respiratory tract

Dermal Effects

- Skin irritation with paraesthesia, blistering and whitening; solutions >10% may cause burns

Eye Effects

- Irritation with a burning sensation, conjunctival hyperaemia, lacrimation and severe pain which should resolve in a few hours, but with more concentrated solutions resolution may by up to 24 hours

- There are rare cases of temporary corneal injury resulting from application of 3% solution to the eye[6]

Oral Effects

- Irritation of the gastrointestinal tract, the severity of which depends on the concentration of the solution

- Vomiting is common, and the vomitus may be frothy due to the liberation of oxygen (risk of aspiration)

- Haematemesis and gastric distension, due to the release of oxygen, may occur

- Lethargy, coma, convulsions and respiratory arrest have been reported[7]

- Gas embolism has been reported in several cases of ingestion of hydrogen peroxide

- In severe cases, ischaemic ECG changes and EMD (electromechanical dissociation) may be seen because of embolisation of the heart, restricting blood flow[8]

- Cerebral infarction and multiple brain emboli may occur in severe cases

Chronic Clinical Effects

Chronic exposure to an aerosol of hydrogen peroxide in a concentration of 41 mEq.m⁻³ (1 ppm = 1.41 mg.m⁻³) resulted in chronic diffuse interstitial lung disease.[9] Although no cases have been reported, chronic ingestion could cause gastritis.

Management

Inhalation Management

- Maintain a clear airway, give humidified oxygen and ventilate if necessary

- If respiratory irritation occurs assess respiratory function and if necessary perform chest X-rays to check for chemical pneumonitis

- Patients should be kept at rest

- Treat pulmonary oedema with PEEP or CPAP ventilation

- Symptomatic and supportive care

Dermal Management

- Remove any remaining contaminated clothing, place in double, sealed, clear bags and label; store in a secure area away from patients and staff

- Irrigate with copious amounts of water

- Treat irritation symptomatically

- Bleaching of the skin usually resolves within a few hours

Eye Management

- Irrigate thoroughly with running water or saline for 15 minutes

- Stain with fluorescein and refer to an ophthalmologist if there is any uptake of the stain

Oral Management

- Gastric decontamination is not worthwhile for ingestion of hydrogen peroxide due to its rapid dissociation

- Encourage oral fluids unless there is evidence of severe injury

- If gastric distension is severe a fine bore gastric tube may be passed to aid the release of gas

- In cases with severe clinical effects patients should be X-rayed to check for the presence of gas emboli

- If facilities are available, early endoscopic evaluation should be undertaken within 12 hours of the event in any patient with haematemesis, persistent vomiting or other evidence of gastric burns to assess the extent and severity of the injury

- Monitor ECG in severe cases

- Hyperbaric oxygen therapy has been suggested for patients with evidence of cerebral embolism due to hydrogen peroxide[4]

- Contact a poisons information service for further guidance

Summary of Environmental Hazards

- Gaseous hydrogen peroxide is a common air contaminant; it may concentrate in cloud water

- Persistence is unlikely because of the strong oxidising activity of this chemical

- It decomposes to water and hydrogen

- In water, hydrogen peroxide decomposes to water and oxygen

- Hydrogen peroxide does not concentrate in the food chain

- Drinking Water Standards: no data available

- Soil Guidelines: no data available

- Air Quality Standards: no data available

REFERENCES

1. Gosselin RE, Smith RP & Hodge HC. *Clinical Toxicology of Commercial Products*, 5th edn. Williams & Wilkins, Baltimore, 1984

2. Oberst FW, Comstock CC & Hackley EB, 1954. Inhalation toxicity of ninety percent hydrogen peroxide vapor. *Indust Hyg Occup Med*; 10: 319–27

3. Litovitz TL, Felberg L, Soloway RA, Ford M & Geller R, 1995. 1994 Annual Report of the AAPCC toxic exposure surveillance system. *Am J Emerg Med*; 13(5): 551–97

4. Sherman SJ, Boyer LV & Sibley WA, 1994. Cerebral infarction immediately after ingestion of hydrogen peroxide. *Stroke*; 25: 1065–7

5. Ijichi I, Itoh T, Sakai R, Nakaji K, Miyauchi T, Takahashi R, Kadosaka S, Hirata M, Yoneda S, Kajita Y & Fujita Y, 1997. Multiple brain embolism after ingestion of concentrated hydrogen peroxide. *Neurology*; 48(1): 277–9

6. Grant MW & Schuman JS. *Toxicology of the Eye*, 4th edn. Charles C Thomas, Springfield, 1993

7. Giberson TP, Kern JD, Pettigrew DW, Eaves CC & Haynes JF, 1989. Near-fatal hydrogen peroxide ingestion. *Ann Emerg Med*; 18: 778–9

8. Christensen DW, Faught WE, Black RE, Woodward GA & Timmons OD, 1992. Fatal oxygen embolization after hydrogen peroxide ingestion. *Crit Care Med*; 20(4): 543–4

9. Ellenhorn MJ, Schonwalds S, Ordog G & Wasserberger J. *Ellenhorn's Medical Toxicology – Diagnosis and Treatment of Human Poisoning*, 2nd edn. Williams & Wilkins, London, 1997

Nicola Bates

HYDROGEN SULPHIDE

Key Points

- Hydrogen sulphide has a characteristic pungent 'rotten egg' odour
- It is rapidly absorbed almost exclusively by inhalation
- It is irritant to the respiratory tract, eyes and skin
- At high concentrations it causes rapid collapse, respiratory paralysis, imminent coma, followed by death within minutes
- *In the event of a large spill, stay upwind and out of low areas. Ventilate closed spaces. Protective clothing, eye protection and breathing apparatus should be worn*

FIRST AID

- Terminate exposure and support vital functions
- The casualty should be moved to an uncontaminated area
- Rescuers should, ideally, be trained personnel and must be careful *not to put themselves at risk* and *so wear appropriate protective clothing and, if available, breathing apparatus*
- If the casualty is unconscious a clear airway should be established and maintained; give 100% oxygen if available
- **Inhalation Exposure:** If the patient stops breathing, expired air resuscitation should be started immediately using a pocket mask with a one way valve, if available. It is important where the face is contaminated that expired air resuscitation is NOT attempted unless an airway with rescuer protection is used
- **Dermal Exposure:** Remove contaminated clothing, if possible under a shower and place in double, sealed, clear bags and label; store the bags in a secure area away from patients and staff
- Wash the skin thoroughly with copious amounts of water
- **Eye Exposure:** If the eye tissue is frozen, seek medical advice as soon as possible
- If tissue is not frozen, irrigate thoroughly with water or saline for 15 minutes

Detailed Information

- Hydrogen sulphide is a colourless gas, it is heavier than air with a characteristic pungent 'rotten egg' odour
- *Common synonyms* hydrosulphuric acid, stink damp, sulphureted hydrogen, sulphur hydride
- CAS 7783-06-4
- UN 1053
- NIOSH/RTECS MX 1225000
- Molecular formula H_2S
- Molecular weight 34.08
- Hydrogen sulphide is used in farming, brewing, tanning, glue making, metal recovery processes and gas exploration; in processing artificial silk, fur-dressing, abattoirs and dye production

- Hydrogen sulphide is produced by decaying organic matter and is a by-product of the petroleum industry, tanning, rubber vulcanising, and heavy water production
- Hydrogen sulphide is produced by reacting dilute sulphuric acid with iron sulphide, by reacting hydrogen with sulphur in the vapour phase, or by heating sulphur with paraffin
- Soluble in water and alcohol
- It reacts violently with strong oxidants with risk of fire and explosion and is incompatible with most metals (forming sulphides)
- It is extremely flammable; it burns with a blue flame decomposing to toxic sulphur dioxide, water and elemental sulphur

Summary of Human Toxicity

- Hydrogen sulphide is rapidly absorbed almost exclusively by inhalation and it is a respiratory toxin[1]
- It reacts with surface moisture of mucous membranes to form sodium sulphide, which produces the irritant effect
- Metabolism of hydrogen sulphide involves three different pathways: oxidation to sulphate (thought to be the major metabolic pathway) and methylation, both of which represent detoxification with the major conversion site being the liver, and reaction with proteins which accounts for the toxic action of hydrogen sulphide[2,3]
- Hydrogen sulphide toxicity is considered to be due primarily to the reversible inhibition of cytochrome oxidase which impairs cell respiration[4]
- Concentrations of 0.1 to 0.2% in the atmosphere may be fatal in a few minutes[5]

Toxicity of Hydrogen Sulphide[6]

Concentration	Clinical Effects
0.02–0.025 ppm	Odour threshold
0.3 ppm	Distinct odour
3–5 ppm	Offensive, moderately intense odour
10 ppm	Obvious, unpleasant odour; sore eyes
20–30 ppm	Strong, intense odour but not intolerable
50 ppm	Conjunctival irritation first noticeable
50–100 ppm	Mild irritation to respiratory tract and eyes after 1 h
100 ppm	Loss of smell in 3–15 min, may sting eyes and throat
250 ppm	Prolonged exposure may cause pulmonary oedema
1000 ppm	Rapid collapse, respiratory paralysis, imminent coma, followed by death within mins

- Maximum Exposure Limit: not yet set

Acute Clinical Effects

The tissues most sensitive to hydrogen sulphide are those with exposed mucous membranes and those with high oxygen demands.

Inhalation Effects

- Symptoms of respiratory tract irritation develop on prolonged exposure: rhinitis, pharyngitis, pneumonia, bronchitis and pulmonary oedema may develop[4]

- Systemic effects may include vomiting, diarrhoea, headache, lethargy, horizontal or vertical nystagmus, vertigo, olfactory paralysis, drowsiness, tremors, numbness and weakness of the extremities, dyspnoea, tachypnoea, tachycardia and hypotension[7]

- Albuminuria, casts, haematuria and acidosis have been reported[5]

- At high doses the central nervous system is the target organ with rapid collapse, respiratory paralysis resulting in asphyxia, convulsions, imminent coma and cardiac arrhythmias, followed by death within minutes

Dermal Effects

- Cutaneous absorption is negligible and is not a significant route for systemic poisoning

- Discoloration of the skin has been reported

- Direct contact with the liquefied material or escaping compressed gas can cause frostbite injury, burning and blistering

Eye Effects

- Symptoms of eye irritation develop on prolonged exposure

- Effects are notable at concentrations lower than those which cause systemic toxicity, usually starting after several hours of exposure, and may not appear until after the patient has finished work for the day[8]

- Gradual onset of irritation and inflammation, with lacrimation, burning, conjunctival hyperaemia, redness and swelling of the lids

- Symptoms progress to greater burning discomfort and photophobia, painful conjunctivitis, injection of the conjunctivae, ocular pain, blepharospasm and seeing coloured haloes around lights and clouding of vision

- 'Gas eyes': fine grey stippling (keratoconjunctivitis) and vesiculation of the corneal epithelium and optical irregularity of the corneal epithelium may become evident[8]

- In the most severe cases the corneal surface may become lustreless and eroded from loss of epithelial cells

- Recovery is usually spontaneous and complete, but there may also be permanent damage or loss of vision

- Exposure to the liquefied material or escaping compressed gas can cause frostbite injury, burning and blistering

Oral Effects

- Not applicable

Chronic Clinical Effects

Chronic CNS effects may be cumulative over several 'knockdowns'. Prolonged exposure may lead to chronic headache, fatigue, dizziness, irritability, conjunctivitis, digestive disturbances, weight loss.[5,9] Neurological effects may persist in survivors of high-concentration exposures with evidence of cognitive function abnormalities, labile effects, personality changes and anosmia, and a suggestion of increasing severity associated with length of time of unconsciousness being reported.[9]

Management

Blood sulphide measurements can be used in fatal cases to establish the cause of death, but sulphide can be lost from biological materials very quickly and therefore analyses must be carried out without delay.

Inhalation Management

- Maintain a clear airway, give humidified oxygen and ventilate if necessary

- Hyperbaric oxygen may be of benefit but remains unproven[5]

- If respiratory irritation occurs assess respiratory function and if necessary perform chest X-rays to check for chemical pneumonitis

- Consider the use of steroids to reduce the inflammatory response

- Treat pulmonary oedema with PEEP or CPAP ventilation

- As the mechanism of hydrogen sulphide toxicity is similar to cyanide, amyl nitrite and sodium nitrite may be of use (see cyanide, page 124). However, the efficacy of nitrite therapy and its use is controversial; excessive sulphmethaemoglobin formation may result in hypoxia adding to that already caused by the hydrogen sulphide

- Sodium thiosulphate injection should NOT be used

- Contact a poisons information service for further guidance

Dermal Management

- Remove any remaining contaminated clothing, place in double, sealed, clear bags and label; store in a secure area away from patients and staff

- Irrigate with copious amounts of water

- Skin burns should be treated as a thermal injury

Eye Management

- Irrigate thoroughly with running water or saline for 15 minutes

- Stain with fluorescein and refer to an ophthalmologist if there is any uptake of the stain

Oral Management

- Not applicable

Summary of Environmental Hazards

- In the atmosphere hydrogen sulphide will disperse with residence times ranging from about one day to more than 40 days depending upon season, latitude and atmospheric conditions

- Micro-organisms in soil and water are involved in oxidation-reduction reactions which oxidise hydrogen sulphide to elemental sulphur

- Hydrogen sulphide is harmful to aquatic life in low concentrations; it does not bioconcentrate in the food chain[5]

- Drinking Water Standards:
  0.05 mg.l^{-1} (WHO level where customers may complain)

- Soil Guidelines: no data available

- Air Quality Standards:
  0.15 mg.m^{-3} averaging time 24 hours (WHO guideline)

REFERENCES

1. Clayton GD & Clayton FE (eds.) *Patty's Industrial hygiene and toxicology*, 4th edn. John Wiley & Sons, Inc., New York, 1994

2. Haddad LM, Shannon MW & Winchester JF (eds.) *Clinical Management of Poisoning and Drug Overdose*, 3rd edn. WB Saunders Co., Philadelphia, 1998

3. Beauchamp RO, Bus JS, Popp JA, Boreiko CJ & Andjelkovich DA, 1984. A critical review of the literature on hydrogen sulphide toxicity. *CRC Crit Rev Toxicol*; 13: 25–97

4. Stine RJ, Slosberg B & Beaucham BE, 1976. Hydrogen sulphide intoxication – A case report and discussion of treatment. *Ann Intern Med*; 85(6): 756–8

5. Hall AH & Rumack BH (eds.) *TOMES System* ® Micromedex, Englewood, Colorado. CD ROM. vol 41 (exp. 31 July 1999)

6. Ellenhorn MJ, Schonwalds S, Ordog G & Wasserberger J. *Ellenhorn's Medical Toxicology – Diagnosis and Treatment of Human Poisoning*, 2nd edn. Williams & Wilkins, London, 1997

7. Gosselin RE, Smith RP & Hodge HC. *Clinical Toxicology of Commercial Products*, 5th edn. Williams & Wilkins, Baltimore, 1984

8. Grant MW & Schuman JS. *Toxicology of the Eye*, 4th edn. Charles C Thomas, Springfield, 1993

9. Guidotti TL, 1996. Hydrogen Sulphide. *Occup Med*; 46(5): 367–71

Catherine Farrow

ISOPROPANOL

Key Points

- Isopropanol is a colourless, flammable, bitter tasting liquid, with a slight odour of alcohol
- It is toxic by inhalation and ingestion
- Isopropanol is absorbed dermally although systemic effects are rarely seen on acute exposure
- It is an irritant to skin, eyes, and respiratory tract
- Exposure may cause CNS and respiratory depression, cardiac arrythmias and renal damage
- *In the event of a large spill, stay upwind and out of low areas. Ventilate closed spaces. Protective clothing, eye protection and breathing apparatus should be worn*

FIRST AID

- Terminate exposure and support vital functions
- The casualty should be moved to an uncontaminated area
- Rescuers should, ideally, be trained personnel and must be careful *not to put themselves at risk* and *so wear appropriate protective clothing and, if available, breathing apparatus*
- If the casualty is unconscious a clear airway should be established and maintained; give 100% oxygen if available
- **Inhalation Exposure:** If the patient stops breathing, expired air resuscitation should be started immediately using a pocket mask with a one way valve, if available. It is important where the face is contaminated that expired air resuscitation is NOT attempted unless an airway with rescuer protection is used
- **Dermal Exposure:** Remove contaminated clothing, if possible under a shower and place in double, sealed, clear bags and label; store the bags in a secure area away from patients and staff
- Wash the skin thoroughly with copious amounts of water
- **Eye Exposure:** Irrigate thoroughly with water or saline for 15 minutes
- **Oral Exposure:** Encourage small quantities of oral fluids (no more than 50–100 ml in total)

Detailed Information

- Isopropanol is a clear, colourless, flammable, bitter tasting liquid, with a slight odour of alcohol
- *Common synonyms* dimethyl carbinol, isopropyl alcohol, 2-propanol, Propan-2-ol
- CAS 67-63-0
- UN 1219
- NIOSH/RTECS NT 8050000
- Molecular formula C_3H_8O
- Molecular weight 60.10
- Isopropanol is used as a rubbing alcohol, solvent, cleaning agent, disinfectant, preservative for pathology specimens; as a component of some window cleaners, liquid soaps, cosmetics, pharmaceuticals and antifreeze solutions; in the manufacture of acetone

- Soluble in water, alcohol and ether
- Reacts violently with strong oxidisers, inorganic acids, aldehydes, and isocyanates
- Stable under normal conditions but ignites on contact with many metallic oxides or ignition sources

Summary of Human Toxicity

- Isopropanol is toxic by inhalation and ingestion; it is absorbed dermally although systemic effects are rarely seen on acute exposure
- Isopropanol is approximately twice as potent a CNS depressant as ethanol and since it is metabolised at a slower rate it has a longer duration of action[1]
- It is partly oxidised to acetone in the liver[1] and 25–50% excreted unchanged via the kidneys
- The minimum lethal oral dose is reported to be 100–250 ml[2]
- Aspiration may occur with solutions of >70% isopropanol[3,4]
- Occupational Exposure Standards:
  Long-term exposure limit: 400 ppm (999 mg.m^{-3})
  Short-term exposure limit: 500 ppm (1250 mg.m^{-3})

Acute Clinical Effects

Inhalation Effects

- Irritation of eyes, nose, throat, and respiratory tract
- Systemic effects are possible; see below

Dermal Effects

- Exposure may cause irritation at site of contact
- Drying and cracking of the skin may occur
- Dermatitis may occur in sensitive individuals
- Isopropanol is absorbed dermally although systemic effects are unlikely from acute exposure

Eye Effects

- Isopropanol is irritant to the eyes but injury is transient
- Prolonged contact may cause patchy epithelial loss with rapid recovery[5]

Oral Effects

- Symptoms occur within 30–60 minutes after ingestion[6]
- Initially a burning sensation in the mouth and throat, gastrointestinal irritation with abdominal pain, vomiting and haematemesis
- Aspiration may occur
- For systemic effects, see below

Systemic Effects

- CNS effects include dizziness, disorientation, headache and confusion

- Drowsiness progressing to coma

- Tachycardia is commonly seen, although there may be bradycardia and arrythmias

- Hypo- or hyperglycaemia may occur

- Renal effects include, oliguria, ketonuria, and anuria

- Severe intoxication can result in coma, hypothermia, hyporeflexia, respiratory depression and arrest

- Marked hypotension is a poor prognostic sign

- Acute tubular necrosis, rhabdomyolysis, hepatic dysfunction, haemolytic anaemia, and myoglobinuria have also been reported, but may be due to hypotension and prolonged coma rather than a direct toxic effect

Chronic Clinical Effects

Repeated skin contact with isopropanol can cause defatting dermatitis with drying and cracking. Rare cases of allergic contact dermatitis have been reported. Early studies suggested an association between the manufacture of isopropanol and paranasal sinus cancer. There may also be an increased risk of laryngeal cancer among workers involved in the manufacturing process. It is unclear whether the increased incidence of cancer is due to the isopropanol itself or some intermediary in the process.[7]

Management

Serum acetone and isopropanol concentrations may be determined

Inhalation Management

- Maintain a clear airway, give humidified oxygen and ventilate if necessary

- If respiratory irritation occurs assess respiratory function and if necessary perform chest X-rays to check for chemical pneumonitis

- Treat pulmonary oedema with PEEP or CPAP ventilation

- Consider the use of steroids to reduce inflammatory response

- Symptomatic and supportive care

- Systemic toxicity is likely with large or prolonged exposure; see systemic management below

Dermal Management

- Remove any remaining contaminated clothing and place in double, sealed, clear bags and label; store in a secure area away from patients and staff

- Wash area with copious amounts of water

- An emollient may be required

- Treat irritation symptomatically

- Systemic effects possible with prolonged exposure, see systemic management below

Eye management

- Irrigate thoroughly with running water or saline for 15 minutes

- Stain with fluoroscein and refer to an ophthalmologist if there is any uptake of the stain

Oral Management

- Due to the rapid absorption of isopropanol, gastric decontamination is of little use

- Isopropanol does not bind to activated charcoal

- Close monitoring of the patient's respiration is required in the first few hours of ingestion because of the risk of respiratory depression

Systemic management

- In severe cases electrolytes, creatinine, glucose, full blood count, arterial blood gases, serum acetone and isopropanol, renal and liver function should be monitored

- Hypotension should be treated with IV fluids and ionotropes if necessary

- Deep coma and refractory hypotension are indications for haemodialysis or peritoneal dialysis, it should also be considered in patients with an isopropanol level greater than 4 g.l^{-1}. However, optimal supportive care will generally be effective if dialysis is not available [6]

Summary of Environmental Hazards

- In the atmosphere isopropanol will photodegrade with an estimated half-life ranging from one to several days; due to its solubility in water, rainout may be significant [2]

- When released into water, isopropanol will volatilise (half-life estimated to be 5.4 days) and may biodegrade

- It would not be expected to adsorb to sediment or concentrate in the food chain

- When spilled on the soil, isopropanol will both evaporate quickly and leach into the ground due to its high vapour pressure and low adsorption to soil

- Drinking Water Standards: no data available

- Soil Guidelines: no data available

- Air Quality Standards: no data available

REFERENCES

1. Smith MS, 1983. Solvent toxicity: isopropanol, methanol, and ethylene glycol. *Ear Nose Throat J*; 62: 11–26

2. Hall AH & Rumack BH (eds.) *TOMES System®* Micromedex, Englewood, Colorado. CD ROM. vol 41 (exp. 31 July 1999)

3. Adelson L, 1962. Fatal intoxications with isopropyl alcohol (rubbing alcohol). *Am J Clin Pathol*; 38(2): 144–51

4. Mecikalski MB & Depner TA, 1982. Peritoneal dialysis for isopropanol poisoning. *West J Med*; 137: 322–5

5. Grant MW & Schuman JS. *Toxicology of the Eye*, 4th edn. Charles C Thomas, Springfield, 1993

6. Lacoutre PG, Wason S, Abrams A & Lovejoy FH, 1983. Acute isopropyl alcohol intoxication, diagnosis and management. *Am J Med*; 75: 680–6

7. Hathaway GJ, Proctor NH & Hughes JP. *Proctor and Hughes' Chemical Hazards of the Workplace*, 4th edn. Van Nostrand Reinhold, New York, 1996

Jennifer Butler

KEROSENE

Key Points

- Kerosene is a mixture of petroleum hydrocarbons, chiefly of the methane series
- It is a pale yellow or water-white, mobile, oily liquid; it has also been described as colourless to light brown
- It is irritating to both the respiratory system and gastrointestinal tract
- Kerosene causes CNS depression and can sensitise the myocardium to endogenous catecholamines in large exposures
- Kerosene vapours are heavier than air, can cause an explosive mixture with air and may travel great distances to an ignition source to flash back and cause fire or explosion
- *Kerosene vapours are heavier than air and may accumulate in low or confined areas; in the event of a large spill, stay upwind and out of low areas. Ventilate closed spaces. Protective clothing, eye protection and breathing apparatus should be worn*

First Aid

- Terminate exposure and support vital functions
- The casualty should be moved to an uncontaminated area
- Rescuers should, ideally, be trained personnel and must be careful *not to put themselves at risk* and *so wear appropriate protective clothing and, if available, breathing apparatus*
- If the casualty is unconscious a clear airway should be established and maintained; give 100% oxygen if available
- **Inhalation Exposure:** If the patient stops breathing, expired air resuscitation should be started immediately using a pocket mask with a one way valve, if available. It is important where the face is contaminated that expired air resuscitation is NOT attempted unless an airway with rescuer protection is used
- **Dermal Exposure:** Remove contaminated clothing, if possible under a shower and place in double, sealed, clear bags and label; store the bags in a secure area away from patients and staff
- Wash the skin thoroughly with copious amounts of water
- **Eye Exposure:** Irrigate thoroughly with water or saline for 15 minutes
- **Oral Exposure:** Encourage small quantities of oral fluids (no more than 50–100 ml in total)

Detailed Information

- Kerosene is a mixture of petroleum hydrocarbons, chiefly of the methane series having the form 10–16 carbon atoms per molecule: C_nH_{2n+2}
- It is a pale yellow or water-white, mobile, oily liquid; it has also been described as colourless to light brown
- *Common synonyms* coal oil, Deobase, kerosine, kerosine (petroleum)
- CAS 8008-20-6

- UN 1223
- NIOSH/RTECS OA 5500000
- Used as fuel and solvent
- Kerosene possesses a petroleum odour; it is soluble in water and is miscible with other petroleum solvents
- It can react with oxidising materials and in a fire may produce irritating, corrosive and /or toxic gases

Summary of Human Toxicity

- Note: the toxic response to kerosene will vary according to the compound's origin and utilisation; the deodorised and refined kerosenes are the least toxic, while others may contain benzene or alkylbenzenes, which are potentially more toxic
- Inhalation or contact with material may irritate or burn skin and eyes
- Vapours may cause dizziness or suffocation
- Occupational Exposure Standard: no data available

Acute Clinical Effects

Inhalation Effects

- Low to moderate concentrations can cause transient euphoria similar to alcohol intoxication[1]
- Burning sensation in the chest, weakness, inco-ordination and confusion and disorientation may occur
- High concentrations can cause headache, drowsiness, increased respiration, convulsions and coma
- Death may occur from ventricular arrhythmias as at high concentrations the solvent sensitises the myocardium to adrenaline and other catecholamines

Dermal Effects

- Defatting action on the skin can lead to irritation and possible secondary infection

Eye Effects

- Lacrimation, irritation and pain may occur

Oral Effects

- Ingestion is hazardous due to the risk of aspiration which may occur during ingestion or subsequent vomiting
- Nausea, vomiting, abdominal pain and diarrhoea
- Gagging, coughing and wheezing leading to pulmonary oedema and chemical pneumonitis

Chronic Clinical Effects

Chronic exposure to kerosene may cause headache, neuralgia, loss of memory, respiratory irritation and polyneuritis. One case of hypoplastic anaemia has been reported from chronic kerosene exposure.[2]

Kerosene can cause defatting, erythema and eczema-like skin lesions with chronic exposure.

Management

Measurement of blood concentration is of no clinical relevance other than to confirm exposure when this is in doubt.

Inhalation Management

- Maintain a clear airway, give humidified oxygen and ventilate if necessary

- If respiratory irritation occurs assess respiratory function and if necessary perform chest X-rays to check for chemical pneumonitis

- See systemic effects below

Dermal Management

- Remove any remaining contaminated clothing, place in double, sealed, clear bags and label; store in a secure area away from patients and staff

- Irrigate with copious amounts of water with a few drops of detergent added

- An emollient may be required

- Treat irritation and burns symptomatically

- For severe or prolonged exposure, see systemic effects below

Oral Management

- Encourage oral fluids for small ingestion

- DO NOT INDUCE EMESIS due to risk of aspiration

- Consider gastric lavage within 1 hour of a substantial ingestion, ensuring airway is protected

- Contact a poisons information service for further guidance on gut decontamination

- Symptomatic and supportive care; see systemic effects below

Eye Management

- Irrigate thoroughly with running water or saline for 15 minutes

- Stain with fluorescein and refer to an ophthalmologist if there is any uptake of the stain

Systemic Management

- Patients should be kept at rest

- With severe exposures the patient should be kept on a cardiac monitor for 6 hours, avoiding the use of all stimulants, except for resuscitation

- Treat pulmonary oedema with PEEP or CPAP ventilation

- Symptomatic and supportive care

Summary of Environmental Hazards

- If released to soil kerosene is expected to biodegrade under both aerobic and anaerobic conditions

- Some components of kerosene may adsorb very strongly to soil

- Kerosene may rapidly volatilise from both moist and dry soil, although its expected strong adsorption may significantly attenuate the rate of this process

- If released to water, kerosene is expected to biodegrade under both aerobic and anaerobic conditions

- Some components of kerosene may significantly bioconcentrate in fish and aquatic organisms and strongly adsorb to sediment and suspended organic matter

- The estimated half-life for volatilisation of kerosene from a model river is 3–6 hours while that from a model lake is approximately 130 days (the first does not account for the attenuating effect of adsorption)[1]

- In the atmosphere kerosene may undergo oxidation by a gas-phase reaction with photochemically produced hydroxyl radicals with an estimated half-life of 2–3.4 days

- Drinking Water Standards:
  hydrocarbon total: $10 \, \mu g \cdot l^{-1}$ (UK max)

- Soil Guidelines: Dutch Criteria:
  mineral oil: $50 \, mg \cdot kg^{-1}$ (target)
  $5000 \, mg \cdot kg^{-1}$ (intervention)

Air Quality Standards: no data available

REFERENCES

1. Hall AH & Rumack BH (eds.) *TOMES System* ® Micromedex, Englewood, Colorado. CD ROM. vol. 41 (exp. 31 July 1999)

2. Harbison RD (ed.) *Hamilton and Hardy's Industrial Toxicology*, 5th edn. Mosby-Year Book Inc., St Louis, 1998

Henrietta Wheeler

LEAD AND LEAD COMPOUNDS

Key Points

- The following information refers only to adults; infants and children are more sensitive to lead. Contact a poisons information centre for paediatric cases
- Lead is present in both inorganic and organic forms; it is important to establish the exposure type
- Organic lead is absorbed by ingestion, inhalation and via the skin
- Dermal exposure may cause local irritation
- On inhalation organic lead can cause severe toxicity with symptoms occurring a few hours to 10 days post exposure
- Chronic exposure may result in various symptoms including gastrointestinal disturbances, hypermineralisation, neuromuscular dysfunction, personality changes, cerebal oedema, renal failure and gout
- Chelation therapy is available for severely poisoned individuals
- *In the event of a large spill, stay upwind and out of low areas. Ventilate closed spaces. Protective clothing, eye protection and breathing apparatus should be worn*

FIRST AID

- Terminate exposure and support vital functions
- The casualty should be moved to an uncontaminated area
- Rescuers should, ideally, be trained personnel and must be careful *not to put themselves at risk* and *so wear appropriate protective clothing and, if available, breathing apparatus*
- If the casualty is unconscious a clear airway should be established and maintained; give 100% oxygen if available
- **Dermal Exposure:** Remove contaminated clothing, if possible under a shower and place in double, sealed, clear bags and label; store the bags in a secure area away from patients and staff
- Wash the skin thoroughly with copious amounts of water
- **Eye Exposure:** Irrigate thoroughly with water or saline for 15 minutes
- **Oral Exposure:** Encourage small quantities of oral fluids (no more than 50–100 ml in total)

Detailed Information

Synonyms of Lead and Lead Compounds:

Chemical	Synonyms
Metallic lead	lead flake, lead inorganic, lead metal, plumbum
Lead acetate	dibasic lead acetate, lead (II) acetate, plumbous acetate, salt of saturn, sugar of lead, lead acetate trihydrate
Lead chloride	lead dichloride, lead (II) chloride, plumbous chloride
Lead chromate	chrome yellow, Cologne yellow, King's yellow, Leipzig yellow, Paris yellow

Chemical	Synonyms
Lead dioxide	brown lead, lead peroxide, lead superoxide, lead (IV) oxide
Lead fluoride	lead difluoride, plumbous fluoride
Lead hydroxide	lead oxide hydrate, basic lead hydroxide
Lead monoxide	lead oxide yellow, plumbous oxide, litharge, massicot, lead protoxide
Lead tetraoxide	lead oxide, mineral orange, mineral lead, Paris red, orange lead, red lead, red lead oxide, trilead tetraoxide
Tetraethyl lead	tetraethylplumbane, lead tetraethide
Tetramethyl lead	tetramethylplumbane

Identification Numbers of Lead and Lead Compounds

Chemical/compound	CAS	UN	NIOSH/RTECS
Lead	7439–92–1	not available	OF 7525000
Lead acetate	301–04–2	1616	AI 5250000
Lead azide	13424–46–9	0129	OF 8650000
Lead bromide	10031–22–8	not available	not available
Lead chloride	7758–95–4	not available	OF 9450000
Lead chromate	7758–97–6	not available	GB 2975000
Lead dioxide	1309–60–0	1872	OG 0700000
Lead fluoride	7783–46–2	not available	OG 1225000
Lead hydroxide	19783–14–3	not available	not available
Lead iodide	10101–63–0	not available	not available
Lead monoxide	1317–36–8	not available	OG 1750000
Lead nitrate	10099–74–8	1469	OG 2100000
Lead phosphate	7446–27–7	not available	OG 2675000
Lead sulphate	7446–14–2	not available	OG 4375000
Lead sulphide	1314–87–0	not available	OG 4550000
Lead tetraoxide	1314–41–6	not available	OG 5425000
Tetraethyl lead	78–00–2	not available	TP 4550000
Tetramethyl lead	75–74–1	1649	TP 4725000

Molecular Formula and Weight of Lead Compounds

Compound	Molecular formula	Molecular weight
Lead acetate	$C_4H_6O_4Pb$	325.28
Lead azide	$Pb(N_3)_2$	291.26
Lead borate	$Pb(BO_2)_2.H_2O$	309.79 (approx)
Lead bromate	Br_2O_6Pb	463.01
Lead bromide	$PbBr_2$	367.04
Lead butyrate	$Pb(C_4H_7O_2)_2$	381.40
Lead chlorate	$Pb(ClO_3)_2$	372.12
Lead chloride	$PbCl_2$	278.12
Lead chromate	$PbCrO_4$	323.22
Lead dioxide	PbO_2	239.21
Lead fluoride	PbF_2	245.21
Lead hydroxide	$3PbO.H_2O$	687.59
Lead iodide	PbI_2	461.05

Compound	Molecular formula	Molecular weight
Lead monoxide	PbO	223.21
Lead nitrate	$Pb(NO_3)_2$	331.23
Lead phosphate	$Pb_3(PO_4)_2$	811.54
Lead sulphate	$PbSO_4$	303.28
Lead sulphide	PbS	239.28
Lead tetraoxide	Pb_3O_4	685.63
Tetraethyl lead	$Pb(C_2H_5)_4$	323.45
Tetramethyl lead	PbC_4H_{12}	267.33

- Lead and its compounds are used as construction material for tank linings, piping and other equipment handling corrosive gases and liquids; in the manufacture of sulphuric acid, petroleum refining, halogenation, sulphonation, extraction, condensation; for X-ray and atomic radiation protection; manufacture of tetraethyl lead, pigments for paints and other organic and inorganic lead compounds; bearing metal and alloys; storage batteries; in ceramics, plastics and other lead alloys; in the metallurgy of steel and other metals; in Asian cosmetics and as a constituent or contaminant of traditional medicines

- Many inorganic lead compounds give off toxic lead fumes on combustion (lead chloride, lead chromate, lead sulphate, lead sulphide)

- Tetraethyl lead and tetramethyl lead are used as anti-knocking additives in petrol

- Organic lead compounds (tetraethyl lead and tetramethyl lead) have a characteristic fruity odour, are flammable and volatile

Summary of Human Toxicity

- Approximately 5% to 10% of an ingested lead dose is absorbed in the gut, the remainder appears in the faeces[1]

- Gastric acid solubilises lead salts and lead absorption occurs in the small bowel[1]

- Lead may also be absorbed via lungs; absorption of inorganic lead may depend on particulate size and physical state of the compound, 50% to 70% of an inhaled dose is absorbed if the particle size (< 1 μm) allows the material to reach the alveoli, this is the primary route of absorption of organic lead[1]

- There is limited evidence to suggest that inorganic lead may be absorbed via skin[2]

- Organic lead however is absorbed via skin in sufficient quantities to produce toxicity[3]

- Absorbed lead is initially distributed in various tissues and then gradually redistributed to form an exchangeable compartment (blood and soft tissues) and a storage compartment (bone)[1]

- In continued exposure, the lead which is at first only loosely deposited in bone, gradually becomes fixed to bone, probably as inert and insoluble lead phosphate

- Total body burden of lead may be divided into at least two kinetic pools, which have different rates of turnover:

  - the larger, and kinetically slower pool is the skeleton with a half-life of more than 20 years

  - a much more labile soft tissue and a rapidly exchangeable bone fraction pool with a half-life of 30–60 days

- Blood lead concentrations (prior to any treatment or other action to abate exposure) of:

  a) greater than 400 μg.l⁻¹ (40 μg.dl⁻¹) may require treatment with chelating agents if the patient is symptomatic, although this is unlikely

  b) greater than 800 μg.l⁻¹ (80 μg.dl⁻¹) will require treatment with chelating agents if the patient is symptomatic which is quite likely

  c) greater than 1200 μg.l⁻¹ (120 μg.dl⁻¹) will require treatment with chelating agents as patients are very likely to have symptoms

- Occupational Exposure Standards:

  Long-term exposure limit:
  Lead and lead compounds (except tetraethyl lead) 0.15 mg.m⁻³ (as Pb)
  Tetraethyl lead: 0.10 mg.m⁻³ (as Pb)

  Short-term exposure limit: no data available

Acute Clinical Effects

Inhalation Effects

Inorganic lead

- Lead is not an irritant but may cause the same effects as for ingestion if enough is inhaled

Organic lead

- Organic lead compounds may cause severe toxicity by inhalation, the latent period before onset of symptoms varies from a few hours in the more severe cases to as much as 10 days

- Initial effects are anorexia, vomiting, insomnia, tremor, weakness, fatigue, nausea, headache, aggression, depression, irritability, restlessness, hyperactivity, confusion and memory impairment

- From the onset of the initial effects there may be a delay of hours or even days before acute mania, convulsions, delirium, fever and coma

- Apparent complete recovery can take 2–6 months to occur and some effects may persist for longer[4]

Dermal Effects

Inorganic lead

- Mild local irritation

Organic lead

- Acute poisoning occurs from dermal exposure to small amounts of organic lead compounds; inhalation is the more common route; see inhalation effects above

Eye Effects

- Lead metal foreign bodies in the eye may cause mechanical injury

Oral Effects

- Commonly gastrointestinal colic with nausea, vomiting, anorexia and abdominal pain

- Malaise, convulsions, coma, encephalopathy, hepatic and renal damage, anaemia, hypertension and bradycardia may occur[5,6,7]

Chronic Clinical Effects

Inorganic lead

Clinical effects from chronic exposures include severe gastrointestinal disturbances with constipation, abdominal pain and tenderness. Other effects include anaemia, weakness, pallor, anorexia, insomnia, renal hypertension and mental fatigue. Rarely there may be a bluish 'lead line' on the gums. Lead may also be drawn to areas of the skeleton that grow most rapidly and in some cases hypermineralisation of the radius, tibia and femur can be seen on X-ray with the development of metaphyseal lines.[8] Neuromuscular dysfunction may result in signs of motor weakness and paralysis of the extensor muscles of the wrist and ankles. Encephalopathy can occur in patients with previously mild symptoms. Effects include vomiting, confusion, ataxia, apathy, bizarre behaviour and coma and convulsions due to cerebral oedema. Nephropathy may occur and is characterised by albuminuria, glycosuria and renal tubular acidosis. Gout has also been reported.

Although there is no evidence that lead is carcinogenic in humans, there are several reports in which benign and malignant tumours were produced in experimental animals.

Lead chromate is a suspected human carcinogen of the lung but a study of 1152 workers at a pigment factory did not find any relationship between exposure to lead chromate and lung cancer.[9,10]

Organic lead

Initial symptoms of poisoning are nonspecific and include asthenia, weakness, fatigue, pallor, headache, nausea, vomiting, diarrhoea, insomnia, anorexia and weight loss. Ataxia, tremor, hypotonia, bradycardia and hypothermia may also develop. In more severe poisoning, disorientation, hallucinations, facial contortions and episodes of intense hyperactivity may occur. In severe cases maniacal behaviour and convulsions may develop which may lead to coma and death. Recovery may take weeks or months and may be incomplete.[9]

Management

Blood lead concentrations are important indicators of toxicity

Inhalation Management

- Remove from exposure

- Symptomatic and supportive care

- See systemic management below

Dermal Management

- Remove any remaining contaminated clothing, place in double, sealed, clear bags and label; store in a secure area away from patients and staff

- Irrigate with copious amounts of water

- Treat irritation symptomatically

Eye Management

- Irrigate thoroughly with running water or saline for 15 minutes

- Stain with fluorescein and refer to an ophthalmologist if there is any uptake of the stain

Oral Management

Inorganic lead:

- An X-ray should be performed to confirm the acute ingestion of a lead weight or similar foreign body; gastric decontamination is not indicated, although a mild laxative may be given

- A further X-ray should be done if the foreign body has not been passed in the stools after 48 hours and surgical removal may be indicated if it has become lodged in the gastrointestinal tract or the patient develops marked symptoms

- Consider gastric lavage within 1 hour of an acute ingestion of lead salts and an X-ray will confirm effective removal

- Activated charcoal does not adsorb lead

- Contact a poisons information service for further guidance on gut decontamination

Organic lead

- NO EMETIC, encourage oral fluids

- Consider gastric lavage with a cuffed ET tube for large exposure

- Symptomatic and supportive care

- See systemic management below

Systemic Management

- Gastrointestinal disturbances and renal impairment should be treated supportively

- Most standard tests of renal function do not appear to detect early effects of lead exposure;[11] however renal function should be assessed as renal tubular damage, characterised by aminoaciduria, hypophosphataemia and glycosuria, sometimes followed by hyperphosphataemia, has been reported

- Increased urinary uric acid and hyperuricaemia are seen as a consequence of chronic renal damage

- Since several nutritional and dietary deficiencies will increase the absorption of lead they must be excluded; assess plasma concentrations of calcium, phosphorous, iron and zinc as well as total intake of calories, fat, ascorbic acid, vitamin D and protein

- Anaemia should be treated with iron supplements

- Full blood count and iron studies should be performed as microcytic hypochromic anaemia is common; reticulocytosis and punctate basophilia can occur

- Convulsions as a result of acute encephalopathy may be treated with diazepam; mannitol and corticosteroids should be used for cerebral oedema

- Bone X-rays are used in the diagnosis of lead poisoning as increased density is an indication of exposure to lead of several months duration

Chronic Management

- An abdominal X-ray should be performed to determine whether lead is present in the gut; specks or flecks of highly radio-opaque material will be seen, almost as if the film has faults

- Chelation therapy should not begin until the X-ray is clear so a mild laxative should be given or whole bowel irrigation performed, depending on the clinical condition of the patient and the urgency to start chelation therapy

Chelation Therapy

- Blood lead concentration should be determined to confirm the diagnosis, urgency of this analysis will depend on the clinical condition of the patient

- Chelation therapy is usually indicated if the blood lead concentration is $> 600\,\mu g . l^{-1}$

- The aim of chelation therapy is to reduce the lead concentration to $< 150\,\mu g . l^{-1}$ and ideally to $< 100\,\mu g . l^{-1}$

- After therapy there is a period of re-equilibriation of 10–14 days after which the lead concentration should be measured in order to determine the need for further courses of chelation

- Chelating agents can be nephrotoxic (sodium calcium edetate), hepatotoxic (succimer) and deplete systemic copper and zinc; if these drugs are being used, regular monitoring must take place

- Chelating agents are of limited benefit in organic lead poisoning

- Contact a poisons information service for further guidance and paediatric doses

Dosages of chelation agents

- **SODIUM CALCIUM EDETATE (Ledclair)**

  - 30–$40\,mg . kg^{-1}$ by IV infusion in 5% dextrose or 0.9% saline twice daily for up to 5 days, repeated if necessary after 48 hours for a maximum of another 5 days

- **SUCCIMER (MESO-2,3-DIMERCAPTOSUCCINIC ACID, DMSA) (Chemet – 100 mg only)**

  - Orally $30\,mg . kg^{-1}$ body weight daily, in three divided doses, for 5 days, then $20\,mg . kg^{-1}$, in two divided doses, for 14 days

- **PENICILLAMINE (Distamine, Pendramine)**

- Orally, 1–$2\,g$ daily and children $20\,mg . kg^{-1}$ body weight per day in divided doses until blood lead concentration falls below toxic level

- **DIMERCAPROL**

  - By IM injection 2.5–$3\,mg . kg^{-1}$ every 4 hours for 2 days, 2–4 times on the third day, then 1–2 times daily for 10 days or until recovery

Identifying the Source of the Lead

- An important aspect in the successful treatment of lead poisoning is the identification and elimination of the source of the lead

- In adults this is usually occupational, although it may result from contaminated food or water or through the use of traditional remedies

- A comparison of the isotopes of lead in samples of suspected sources (e.g. water, air, soil, food, cooking utensils, paint or jewellery) to the lead isotopes found in the blood can be used to confirm the source of lead

- The blood lead concentrations of the patient's immediate family should also be determined to identify others at risk

Summary of Environmental Hazards

- Metallic lead is insoluble in water; tetraethyl and tetramethyl lead are poorly soluble

- Trialkyl lead compounds are more soluble in water

- Lead is strongly adsorbed onto sediment and soil particles; if released or deposited onto soil, lead will be retained in the upper 2–5 cm of soil, especially in soils with at least 5% organic matter of pH 5 or above

- Leaching is not important under normal conditions although there is some evidence to suggest that lead is taken up by some plants. It is expected to slowly undergo speciation to the more insoluble sulphate, sulphide, oxide and phosphate salts

- Lead released into the atmosphere partitions to surface water, soil and sediment. Lead is transported in the atmosphere and in surface water

- Organic lead compounds are subject to photolysis

- Drinking Water Standards:
      lead: $50\,\mu g . l^{-1}$ (UK max)
          $10\,\mu g . l^{-1}$ (WHO guideline)

- Soil Guidelines: Dutch Criteria:
      lead: $85\,mg . kg^{-1}$ (target)
          $530\,mg . kg^{-1}$ (intervention)

- Air Quality Standards:
      0.5–$1.0\,\mu g . m^{-3}$ long-term average (e.g. annual mean) (WHO guideline)

REFERENCES

1. Friberg L, Nordberg GF & Vouk VB. *Handbook on the Toxicology of Metals*. Elsevier, Amsterdam, 1989

2. Ellenhorn MJ, Schonwalds S, Ordog G & Wasserberger J. *Ellenhorn's Medical Toxicology – Diagnosis and Treatment of Human Poisoning*, 2nd edn. Williams & Wilkins, London, 1997

3. Florence TM, Lilley SG & Stauber JL, 1988. Skin absorption of lead. *Lancet*; 2: 157

4. Grandjean P & Nielsen T, 1979. Organolead compounds: environmental health aspects. *Residue Rev*; 72: 97–148

5. Carton JA, Maradona JA & Arribas JM, 1987. Acute-subacute lead poisoning. *Arch Intern Med*; 147: 697–703

6. Khan AJ, Patel U, Rafecq M, Myerson A, Kumar K & Evans HE, 1983. Reversible acute renal failure in lead poisoning. *Clin Lab Obs*; 102(1): 147–9

7. Parras F, Patier JL & Ezpeleta C, 1989. Lead contaminated heroin as a source of inorganic lead intoxication. *N Engl J Med*; 316(12): 755

8. Goldfrank LR, Flomenbaum NE, Lewin NA, Weisman RS, Howland MA & Hoffman RS. *Goldfrank's Toxicological Emergencies*, 5th edn. Appleton & Lange, Norwalk, 1994

9. Hathaway GJ, Proctor NH & Hughes JP. *Proctor and Hughes' Chemical Hazards of the Workplace*, 4th edn. Van Nostrand Reinhold, New York, 1996

10. Davies JM, 1984. Lung cancer mortality among workers making lead chromate and zinc chromate pigments at three English factories. *Br J Ind Med*; 41: 158–69

11. Winship KA, 1989. Toxicity of lead: a review. *Adverse Drug React Acute Pois Rev*; 8(3): 117–52

Nicola Bates

LINDANE

Key Points

- Lindane is a white, crystalline, non-combustible powder used as an organochlorine insecticide
- It is defined as not less than 99% of pure gamma hexachlorocyclohexane
- The primary target of action is the CNS, leading to excitation and convulsions
- Exposure to lindane fumes may cause CNS depression and can sensitise the myocardium to endogenous catecholamines
- Ingestion may cause sudden convulsions, especially in children

FIRST AID

- Terminate exposure and support vital functions
- The casualty should be moved to an uncontaminated area
- Rescuers should, ideally, be trained personnel and must be careful *not to put themselves at risk* and *so wear appropriate protective clothing and, if available, breathing apparatus*
- If the casualty is unconscious a clear airway should be established and maintained; give 100% oxygen if available
- **Inhalation Exposure:** If the patient stops breathing, expired air resuscitation should be started immediately using a pocket mask with a one way valve, if available. It is important where the face is contaminated that expired air resuscitation is NOT attempted unless an airway with rescuer protection is used
- **Dermal Exposure:** Remove contaminated clothing, if possible under a shower and place in double, sealed, clear bags and label; store the bags in a secure area away from patients and staff
- Wash the skin thoroughly with copious amounts of water
- **Eye Exposure:** Irrigate thoroughly with water or saline for 15 minutes
- **Oral Exposure:** Encourage small quantities of oral fluids (no more than 50–100 ml in total)

Detailed Information

- Lindane is a white, crystalline, volatile powder
- *Common synonyms* γ-benzene hexachloride, γ-hexachlorane, γ-hexachlorocyclohexane, HCH
- CAS 58-89-9
- UN 2761
- NIOSH/RTECS GV 4900000
- Molecular formula $C_6H_6Cl_6$
- Molecular weight 209.85
- Lindane is a widely used organochlorine insecticide; it is used in ant killers, wood preservatives, garden and agricultural insecticides (up to 60%) and formerly in head lice shampoos (usually 1%) and flea powders

- It is used in many forms, including powders, pellets, sprays, vaporising fluids, emulsions, gels and solutions which are often in organic solvents
- Lindane is a non-combustible powder, although it is remarkably volatile even in solid form; usually used in dissolved form in a flammable solvent, thus making an explosion hazard
- Decomposes when heated or burned, giving off toxic and corrosive fumes of phosgene and hydrochloric acid

Summary of Human Toxicity

- Lindane can be absorbed by inhalation, ingestion and from skin and eye contact
- It acts as a CNS stimulant and also may cause cardiac excitability
- Plasma half-life is 20 hours in children and possibly up to 20 days in repeated occupational exposure
- Lindane is highly lipophilic; it is stored in the adipose tissues[1]
- A single dose of 840 mg.kg^{-1} of body weight in adults and 180 mg.kg^{-1} of body weight in children has been reported as being lethal
- Ingestion of 45 g by an adult caused convulsions and 392 g led to apnoea and status epilepticus[2]
- A single application of lindane to the skin of an adult has led to convulsions[2]
- An 18 h whole body dermal application of 1% lindane lotion to a 2-month-old baby for the treatment of scabies resulted in death
- Odour threshold of 12 mg.l^{-1}
- Occupational Exposure Standards:
  Long-term exposure limit: 0.1 mg.m^{-3}
  Short-term exposure limit: no data available

Acute Clinical Effects

Inhalation Effects

- Lindane is highly irritant to the upper respiratory tract
- Vapours may irritate the eyes, nose, throat and lead to severe headache and vomiting
- May sensitise the myocardium to catecholamines and other systemic effects; see systemic effects below

Dermal Effects

- Prolonged exposure to lindane may result in dermatitis
- Topical use may cause local sensitivity reactions
- Absorption may lead to convulsions and other systemic effects; see systemic effects below

Eye Effects

- Lindane is irritating to the eye

- Prolonged exposure may cause conjunctivitis and systemic effects; see below

Oral Effects

- Nausea, vomiting, diarrhoea and abdominal pain

- Systemic effects are likely; see below

Systemic Effects

- Convulsions may occur in the absence of other symptoms following lindane exposure

- Other effects include dizziness, faintness, irritability, headache, hyperreflexia, muscle twitching and ataxia

- In severe cases metabolic acidosis, hypotension, coma, cyanosis and respiratory depression

- Death may occur from ventricular arrhythmias as at high concentrations lindane or the solvent carrier may sensitise the myocardium to adrenaline and other catecholamines

- Liver failure and rhabdomyolysis leading to renal failure and myoglobinuria have also been reported

- Symptoms may be prolonged for several days; in severe cases, weeks

Chronic Clinical Effects

Chronic exposure to lindane can lead to anorexia and fatigue. Aplastic anaemia, peripheral neuropathy, hepatic necrosis, renal damage and acute myeloid leukaemia have been reported following lindane exposure. However, much of this evidence is retrospective and failed to establish any causal link with lindane.[3]

Cases of allergic and toxic dermatitis, rhinitis, conjunctivitis and eczema have been reported in workers exposed during the manufacturing of lindane.[3]

Management

Lindane concentrations should be measured in potentially severe cases; if the patient is symptomatic serial samples should be taken. Concentrations in blood do not appear to increase with increased duration of exposure but primarily reflect recent lindane absorption.[4]

Inhalation Management

- Maintain a clear airway, give humidified oxygen and ventilate if necessary

- If respiratory irritation occurs assess respiratory function and if necessary perform chest X-rays to check for chemical pneumonitis

- For large or prolonged exposure, see systemic management below

Dermal Management

- Remove any remaining contaminated clothing, place in double, sealed, clear bags and label; store in a secure area away from patients and staff

- Irrigate with copious amounts of water

- An emollient may be required

- For large or prolonged exposure, see systemic management below

Eye Management

- Irrigate thoroughly with running water or saline for 15 minutes

- Stain with fluorescein and refer to an ophthalmologist if there is any uptake of the stain

- For large or prolonged exposure, see systemic management below

Oral Management

- Encourage oral fluids

- Consider gastric lavage within 1 hour of a substantial ingestion, followed by 50 g of activated charcoal, ensuring the airway is protected

- Contact a poisons information service for further guidance on gut decontamination

- As lindane is lipophilic avoid fats and milk as they may enhance absorption

- Symptomatic and supportive care, see systemic management below

Systemic Management

- Control convulsions with diazepam; maintain blood pressure with IV fluids

- Monitor pH and correct acidosis if necessary

- Ventilate for respiratory depression if required

- With severe exposure the patient should be kept on a cardiac monitor for at least 12 hours, avoiding the use of all stimulants except for resuscitation

- In severe cases monitor CPK and renal function

- Symptomatic and supportive care

Summary of Environmental Hazards

- In the atmosphere lindane is likely to be subject to rain-out and dry deposition; the estimated half-life for the reaction of vapour phase lindane with atmospheric hydroxyl radicals is 1.63 days[3]

- In water lindane is not expected to volatilise significantly

- Lindane has a half-life of 88 to 1146 days in soil[5]

- It bioconcentrates slightly in the food chain

- Drinking Water Standards:
    hydrocarbon total: $10 \, \mu g.l^{-1}$ (UK max)
    pesticide: $0.1 \, \mu g.l^{-1}$ (UK max)
    lindane: $2 \, \mu g.l^{-1}$ (WHO guideline)

- Soil Guidelines: Dutch Criteria:
    HCH total: $2 \, mg.kg^{-1}$ (intervention)

- Air Quality Standards: no data available

REFERENCES

1. Clayton GD & Clayton FE (eds.) *Patty's Industrial hygiene and toxicology*, 4th edn. John Wiley & Sons, Inc., New York, 1994

2. Ellenhorn MJ, Schonwalds S, Ordog G & Wasserberger J. *Ellenhorn's Medical Toxicology – Diagnosis and Treatment of Human Poisoning*, 2nd edn. Williams & Wilkins, London, 1997

3. Hall AH & Rumack BH (eds.) *TOMES System ®* Micromedex, Englewood, Colorado. CD ROM. vol. 41 (exp. 31 July 1999)

4. Hathaway GJ, Proctor NH & Hughes JP. *Proctor and Hughes' Chemical Hazards of the Workplace*, 4th edn. Van Nostrand Reinhold, New York, 1996

5. WHO. *Guidelines for drinking water quality*, 2nd edn. WHO, Geneva, 1996

Henrietta Wheeler

MAGNESIUM

Key Points

- Magnesium is an odourless silver-white shining metal or metal powder
- Magnesium is an essential electrolyte in the body
- There are few data about industrial or environmental exposure to magnesium, the majority of information is about oral exposure
- Hypermagnesaemia may lead to nausea, vomiting, hypotension, bradycardia, conduction depression, convulsions and cardiac arrest
- Progressive renal failure has been reported
- *In the event of a large spill, stay upwind and out of low areas. Ventilate closed spaces. Protective clothing, eye protection and breathing apparatus should be worn*

First Aid

- Terminate exposure and support vital functions
- The casualty should be moved to an uncontaminated area
- Rescuers should, ideally, be trained personnel and must be careful *not to put themselves at risk* and *so wear appropriate protective clothing and, if available, breathing apparatus*
- If the casualty is unconscious a clear airway should be established and maintained; give 100% oxygen if available
- **Dermal Exposure:** Remove contaminated clothing, if possible under a shower and place in double, sealed, clear bags and label; store the bags in a secure area away from patients and staff
- The skin should be washed thoroughly with copious amounts of water
- **Eye Exposure:** Irrigate thoroughly with water or saline for 15 minutes
- **Oral Exposure:** Encourage small quantities of oral fluids (no more than 50–100 ml in total)

Detailed Information

- Magnesium is an odourless silver-white shining metal or metal powder
- *Common synonyms* magnesium powder, magnesium sheet
- CAS 7439-95-4
- UN 1869
- UN 1418 (powder)
- UN 2950 (granules)
- NIOSH/RTECS OM2100000
- Atomic formula Mg
- Atomic weight 24.31
- Magnesium alloys are used for structural parts, die-cast auto parts, missiles, space vehicles; powders are used for pyrotechnics and flash photography, production of iron, nickel, zinc, titanium, zirconium, dry and wet batteries

- Magnesium sulphate is one of the most commonly employed household purgatives; many 'non-absorbed' antacids contain magnesium[1]
- Magnesium particles can be charged electrostatically and can ignite due to friction
- It burns with an intense flame and is a strong reducing agent which reacts violently with oxidants
- Magnesium reacts violently with many substances, with risk of fire and explosion; it reacts with acids, giving off hydrogen gas

Summary of Human Toxicity

- Magnesium is an essential electrolyte in the body, it is a cofactor in enzyme systems; it is also involved in muscle contractions, neuronal transmission and phosphate transfer[1]
- There are few data about industrial or environmental exposure to magnesium, the majority of information is about oral exposure
- Approximately 30–40% of dietary magnesium is absorbed; absorption is influenced by the load of magnesium presented to the gastrointestinal tract: the smaller the load, the greater the percentage of absorption
- Magnesium is excreted mainly renally
- Magnesium excess occurs if the gastrointestinal tract is unusually permeable or renal failure does not permit excretion
- Calcium salts may antagonise respiratory depression induced by hypermagnesaemia, presumably by displacing magnesium from cell membranes
- Death secondary to bowel perforation has been reported following administration of 120 g of magnesium sulphate and 6 doses of sorbitol for treatment of an aspirin overdose[2]
- Hypermagnesaemia with hypercalcaemia occurred following ingestion of Dead Sea water while bathing; Dead sea water contains high concentrations of bromide, calcium, magnesium and sodium[1]
- Hypermagnesaemia has occurred following irrigation of the renal pelvis with magnesium-containing urological solutions for dissolution of struvite urinary stones[1]
- Hypermagnesaemia occurred following ingestion of 200g of magnesium sulphate in a previously healthy female[3]
- Individuals with metabolic derangements (i.e. anorexia and IV drug users) and those suffering renal failure, may develop magnesium toxicity at lower doses
- Normal magnesium concentrations are 0.8–1.3 mmol.l^{-1} in serum
- Occupational Exposure Standards: Magnesium Oxide (as Mg) Total inhalable dust:
  Long-term exposure limit: 10 mg.m^{-3}
  Short-term exposure limit: no data available

Fume and respirable dust:
  Long-term exposure limit: 4 mg.m^{-3}
  Short-term exposure limit: 10 mg.m^{-3}

Acute Clinical Effects

Inhalation Effects

- Exposure to magnesium fumes may cause mild irritation to the eyes and nose[4]

- Metal fume fever may develop following exposure to fresh magnesium oxide fumes, characterised by fever, chills, headache, muscle pain and vomiting which may be delayed in onset for several hours

Dermal Effects

- Magnesium particles embedded in the skin may cause local swelling, vesiculation, necrosis and ulceration

- Slow-healing skin blebs characterised by gas development (chemical gas gangrene) and lymphangitis may develop[1]

Eye Effects

- Magnesium metal foreign bodies in the eye may cause mechanical injury

Oral Effects

- Extreme thirst or subjective feeling of heat may be observed

- Nausea, vomiting and blurred vision may occur with development of hypermagnesaemia

- Hypotension, bradycardia, conduction depression and cardiac arrest may occur

- Lethargy, hyporeflexia, weakness and cutaneous flushing

- CNS depression, coma, loss of deep tendon reflexes and respiratory depression with cyanosis and apnoea have been reported

- Convulsions, hypophosphataemia and severe hyperosmolality have been reported

- Progressive renal failure has been reported

Chronic Clinical Effects

No data available.

Management

Magnesium concentrations, electrolytes and renal function are indicators of toxicity

Inhalation Management

- Maintain a clear airway, give humidified oxygen and ventilate if necessary

- If respiratory irritation occurs assess respiratory function and if necessary perform chest X-rays to check for chemical pneumonitis

- Symptomatic and supportive care

Dermal Management

- Remove any remaining contaminated clothing, place in double, sealed, clear bags and label; store in a secure area away from patients and staff

- Irrigate with copious amounts of water

- An emollient may be required

Eye Management

- Irrigate thoroughly with running water or saline for 15 minutes

- Stain with fluorescein and refer to an ophthalmologist if there is any uptake of the stain

Oral Management

- Encourage oral fluids

- Monitor respiratory function, in severe cases calcium salts may temporarily improve symptoms: calcium gluconate 10% 0.1–0.2 ml.kg^{-1} (0.02–0.04 mmol.kg^{-1}) IV (monitor ECG and stop if heart rate decreases)[1]

- Respiratory depression, coma and hypotension may all be reversed with haemodialysis

- If patient unconscious, monitor renal function

- Symptomatic and supportive care

Summary of Environmental Hazards

- Metallic magnesium reacts slowly with water as it is protected by an oxide layer

- Bioaccumulation is estimated to be low

- Drinking Water Standards: 50 mg.l^{-1} (UK max)

- Soil Guidelines: no data available

- Air Quality Standards: no data available

REFERENCES

1. Hall AH & Rumack BH (eds.) *TOMES System* ® Micromedex, Englewood, Colorado. CD ROM. vol. 41 (exp. 31 July 1999)

2. Brent J, Kulig K & Rumack BH, 1989. Iatrogenic death from sorbitol and magnesium sulfate during treatment for salicylism (Abstract). *Vet Human Toxicol*; 31:334

3. Garcia-Webb P, Bhagat C, Oh T, Hart G & Thompson W, 1984. Hypermagnesaemia and hypophosphataemia after ingestion of magnesium sulphate. *Br Med J*; 288(6419):759

4. Hathaway GJ, Proctor NH & Hughes JP. *Proctor and Hughes' Chemical Hazards of the Workplace*, 4th edn. Van Nostrand Reinhold, New York, 1996

Henrietta Wheeler

MANGANESE AND MANGANESE COMPOUNDS

Key Points

- Manganese is a grey-white metal similar to iron but harder and more brittle
- Acute manganese poisoning is rare but may occur following ingestion of very large amounts of manganese compounds or from inhalation; inorganic manganese is not absorbed through the skin
- Exposure to manganese oxide fumes at high concentrations may cause metal fume fever
- Chronic manganese poisoning, or manganism, occurs after variable periods of heavy exposure, ranging from 6 months to 3 years; manganism resembles Parkinson's disease
- *In the event of a large spill, stay upwind and out of low areas. Ventilate closed spaces. Protective clothing, eye protection and breathing apparatus should be worn*

FIRST AID

- Terminate exposure and support vital functions
- The casualty should be moved to an uncontaminated area
- Rescuers should, ideally, be trained personnel and must be careful *not to put themselves at risk* and *so wear appropriate protective clothing and, if available, breathing apparatus*
- If the casualty is unconscious a clear airway should be established and maintained; give 100% oxygen if available
- **Dermal Exposure:** Remove contaminated clothing, if possible under a shower and place in double, sealed, clear bags and label; store the bags in a secure area away from patients and staff
- Wash the skin thoroughly with copious amounts of water
- **Eye Exposure:** Irrigate thoroughly with water or saline for 15 minutes
- **Oral Exposure:** Encourage small quantities of oral fluids (no more than 50–100 ml in total)

Detailed Information

- Manganese is a grey-white metal similar to iron but harder and more brittle
- CAS 7439-96-5
- UN specific number not available
- NIOSH/RTECS OO 9275000
- Atomic symbol Mn
- Atomic weight 54.94

Identification numbers of Manganese Compounds

Compound	CAS	UN	NIOSH/RTECS
Mancozeb	8018-01-7	n.a.	ZB 3200000
Maneb	12427-38-2	2968	OP 0700000
Manganese acetate	638-38-0	n.a.	AI 5770000
Manganese borate	1303-95-3	n.a.	n.a.
Manganese chloride	7773-01-5	n.a.	OO 9625000
Manganese dioxide	1313-13-9	n.a.	OP 0350000
Manganese fluoride	7783-40-6	n.a.	OP 0875000
Manganese monoxide	1317-35-7	n.a.	OP 0900000
Manganese nitrate	10377-66-9	2724	n.a.
Manganese oxalate	640-67-5	n.a.	n.a.
Manganese (II) oxide	1344-48-5	n.a.	OP 0900000
Manganese (III) oxide	1317-35-7	n.a.	OP 0915000
Manganese pyrophosphate	53731-35-4	n.a.	n.a.
Manganese resinate	9008-34-8	1330	OP 0957000
Manganese sesquioxide	1317-34-6	n.a.	OP 0915000
Manganese silicate	7759-00-4	n.a.	n.a.
Manganese sulphate	7785-87-7	n.a.	OP 1050000
Manganese sulphide	18820-29-6	n.a.	n.a.
Manganese tetroxide	1317-35-7	n.a.	OP 0895000

n.a. = number not available

Molecular Formula and Weight of Manganese compounds

Compound	Molecular formula	Molecular weight
Mancozeb	$C_8H_{12}MnN_4S_8Zn$	541.03
Maneb	$C_4H_6MnN_2S_4$	265.29
Manganese acetate	$Mn(C_2H_3O_2)_2$	173.03
Manganese borate	MnB_4O_7	210.14
Manganese bromide	$MnBr_2$	214.75
Manganese chloride	$MnCl_2$	125.84
Manganese dioxide	MnO_2	86.94
Manganese fluoride	MnF_2	92.93
Manganese iodide	MnI_2	308.75
Manganese monoxide	MnO	70.93
Manganese nitrate	$Mn(NO_3)_2$ (approx.)	178.95
Manganese oxalate	MnC_2O_4	142.96
Manganese (II) oxide	MnO	70.93
Manganese (III) oxide	Mn_2O_3	157.88
Manganese pyrophosphate	$Mn_2O_7P_2$	228.88
Manganese sesquioxide	Mn_2O_3	157.87
Manganese sulphate	$MnSO_4$	151.00
Manganese sulphide	MnS	87.00
Manganese tetroxide	Mn_3O_4	228.81

- Manganese is used in the manufacture of alloys, dry-cell batteries, glass, inks, ceramics, paints, welding rods, rubber and wood preservatives; in fungicides, mining and processing manganese ores

- It is not found in the free state but in several minerals

- Manganese decomposes in water, readily dissolves in dilute mineral acids; dust or powder is flammable

- Manganese and compounds are relatively insoluble in water; many manganese compounds are flammable

- Manganese dioxide decomposes when heated above 535°C giving off manganese oxide and hydrogen, with increased risk of fire

- Manganese dioxide is a strong oxidant which reacts violently with combustible substances and reducing agents, it can explode when mixed with organic solids; reacts violently with hydrochloric acid giving off chlorine

Summary of Human Toxicity

- Acute manganese poisoning is rare but may occur following ingestion of large amounts of manganese compounds or from inhalation[1]

- Inorganic manganese is not absorbed through the skin[2]

- Approximately 3 to 7.5% of ingested manganese is absorbed from the gastrointestinal tract; absorption may increase with iron deficiency[2,3]

- Excretion occurs mainly in the gastrointestinal tract via bile; manganese excretion may be regulated by a homeostatic system which maintains constant tissue concentrations[2]

- Approximately 60 to 70% of inhaled manganese is removed by mucociliary clearance and eventually swallowed[4]

- Manganese half-life is 2 to 5 weeks depending on body stores; it is longer in the brain[4]

- Freshly formed manganese oxide fumes at high concentrations may cause metal fume fever

- The major concern of manganese exposure is its effects on the CNS following chronic exposure

- Chronic manganese poisoning (manganism) occurs after variable periods of heavy exposure, ranging from 6 months to 3 years[1]

- Manganism is extremely similar to Parkinson's disease; in both conditions neuropathological changes occur in the basal ganglia, with selective destruction of dopaminergic neurons[1]

- Parkinson effects may continue to progress after exposure ceases;[2] Parkinonism has been reported following a 5 year exposure to a manganese-containing fungicide[5]

- Manganese is an essential trace element; daily dietary intake is 3–7 mg [3]; exposure to high concentrations of manganese in drinking water (2,000 µg.l⁻¹) has been linked with abnormal neurological scores[6]

- Manganese concentrations do not necessarily correlate with the severity of clinical signs[7]

- Occupational Exposure Standards: Manganese fume (as Mn):
  Long-term exposure limit: 1 mg.m⁻³
  Short-term exposure limit: 3 mg.m⁻³

  Manganese and compounds (as Mn):
  Long-term exposure limit: 5 mg.m⁻³
  Short-term exposure limit: no data available

Acute Clinical Effects

Inhalation Effects

- High concentrations of freshly formed manganese oxide fumes may cause metal fume fever, characterised by an influenza like illness with chills, sweating, pyrexia, nausea and cough; symptoms occur 4–12 hours post exposure and last for 24 hours

- Manganese may penetrate the lower respiratory tract toward the alveolar membrane, leading to pneumonitits, pneumonia and bronchitis

Dermal Effects

- No data available

Eye Effects

- Local contact with manganese compounds does not appear to be a problem; manganese metal foreign bodies in the eye may cause mechanical injury[8]

Oral Effects

- Large oral doses may cause gastrointestinal irritation

Chronic Clinical Effects

Manganism, or 'manganese madness' is characterised by headache, asthenia, irritability; transitory psychological disturbances such as hallucinations, apathy, confusion, insomnia, compulsive behaviour, decreased libido, impotence and emotional instability are often found early in the disease.[2,3] As exposure continues, symptoms include generalised muscle weakness, speech impairment, nystagmus, inco-ordination, memory impairment, tremor, incontinence, paraesthesia and muscle cramps. Advanced stages include excess salivation, inappropriate emotional reactions and Parkinson-like symptoms, such as mask-like face, severe muscle rigidity, gait disorders and other extrapyramidal symptoms.[2,3] Established neurological symptoms tend to persist or even progress in the absence of additional exposure.

Manganese penetrates the lower respiratory tract toward the alveolar membranes, leading to dyspnoea, pneumonitis, pneumonia and bronchitis.[9]

Management

Whole blood manganese concentrations of > 30 µ.l⁻¹ (540 nmol.l⁻¹) are associated with a risk of neurological toxicity.

Normal whole blood manganese concentrations:
7–18 µg.l⁻¹ (120–325 nmol.l⁻¹)

Normal plasma/serum manganese concentrations:
0.5–1.3 µg.l⁻¹ (9–24 nmol.l⁻¹)

Inhalation Management

- If respiratory irritation occurs assess respiratory function and if necessary perform chest X-rays to check for chemical pneumonitis

- Consider the use of steroids to reduce the inflammatory response

- Symptomatic and supportive care

Dermal Management

- Remove any remaining contaminated clothing, place in double, sealed, clear bags and label; store in a secure area away from patients and staff

- Irrigate with copious amounts of water

- Symptomatic and supportive care

Eye Management

- Irrigate thoroughly with running water or saline for 15 minutes

- Stain with fluorescein and refer to an ophthalmologist if there is any uptake of the stain

- Surgical removal of a foreign body if appropriate

Oral Management

- Encourage oral fluids

- Symptomatic and supportive care

Chelation Therapy

- The effectiveness of chelation is questionable as the body burden does not appear to be related to manganese toxicity[2]

- Chelation may be of use early in the psychiatric stage of intoxication, prior to permanent neurological damage; poor response has been seen in patients removed from exposure for years[2]

Doses of chelation agents

- Contact a poisons information service for further guidance and paediatric doses

- **SODIUM CALCIUM EDETATE (Ledclair)**

  - Use may be of benefit early in intoxication, although efficacy may be reduced if no exposure for years

  - 30–40 mg.kg^{-1} by IV infusion in 5% dextrose or 0.9% saline twice daily for up to 5 days, repeat if necessary after 48 hours for a maximum of 5 days

- **LEVODOPA**

  - Response to levodopa is generally good in treatment of manganism[3]

  - 3.5–12 g.d^{-1} has been given without side effects

- **SODIUM PARA-AMINOSALICYLIC ACID**

  - Two patients with chronic manganese poisoning were treated with sodium para-aminosalycylic acid (PAS)

  - 6 g.d^{-1} in 500 ml of 10% glucose IV for 4 days followed by a 3 day rest, repeated for 3.5 months

  - Both patients showed marked improvement in neurological function which remained up to 6 and 19 months later[10]

Summary of Environmental Hazards

- Manganese is one of the most abundant metals in the earth's crust and usually occurs together with iron

- Dissolved manganese concentrations in ground water and surface waters that are poor in oxygen can reach concentrations of several milligrams per litre

- On exposure to oxygen, manganese can form insoluble oxides that may result in undesirable deposits and colour problems in distribution systems

- Drinking Water Standards:
  manganese: 50 µg.l^{-1} (UK max)
  500 µg.l^{-1} (WHO provisional guideline)
  100 µg.l^{-1} (WHO level where customers may complain)

  chloride: 400 mg.l^{-1} (UK max)
  250 mg.l^{-1} (WHO guideline)

  fluoride: 1.5 mg.l^{-1} (UK max)
  1.5 mg.l^{-1} (WHO guideline)

  nitrate: 50 mg.l^{-1} (UK max)
  50 mg.l^{-1} (WHO guideline)

  sulphate: 250 mg.l^{-1} (UK max)

- Soil Guidelines: Dutch Criteria: no data available

- Air Quality Standards: 1 µg.m^{-3} (WHO guideline)

REFERENCES

1. Hathaway GJ, Proctor NH & Hughes JP. *Proctor and Hughes' Chemical Hazards of the Workplace*, 4th edn. Van Nostrand Reinhold, New York, 1996

2. Hall AH & Rumack BH (eds.) *TOMES System* ® Micromedex, Englewood, Colorado. CD ROM. vol. 41 (exp. 31 July 1999)

3. Baselt RC & Cravey RH. *Disposition of Toxic Drugs and Chemicals in Man*, 4th edn. Chemical Toxicology Institute, 1995

4. Friberg L, Nordberg GF & Vouk VB. *Handbook on the Toxicology of Metals*. Elsevier, Amsterdam, 1989

5. Meco G, Bonifati V, Vanacore N & Fabrizo E, 1994. Parkinsonism after chronic exposure to the fungicide meneb (manganese ethylen-bis-dithiocarbamate). *Scan J Work Environ Health*; 20(4):301–5

6. Kondakis X, Makris N, Leotsinidis M, Prinou M & Papapetropoulos T, 1989. Possible health effects of high manganese concentration in drinking water. *Arch Environ Health*; 44(3):175–8

7. Harbison RD (ed.) *Hamilton and Hardy's Industrial Toxicology*, 5th edn. Mosby-Year Book Inc., St Louis, 1998

8. Grant MW & Schuman JS. *Toxicology of the Eye*, 4th edn. Charles C Thomas, Springfield, 1993

9. Ellenhorn MJ, Schonwalds S, Ordog G & Wasserberger J. *Ellenhorn's Medical Toxicology – Diagnosis and Treatment of Human Poisoning*, 2nd edn. Williams & Wilkins, London., 1997

10. Ky SQ, Deng HS, Xie PY & Hu W, 1992. A report of two cases of chronic serious manganese poisoning treated with sodium para-aminosalicylic acid. *Br J Ind Med*; 49(1):66–9

Henrietta Wheeler

MERCAPTANS

Key Points

- Mercaptans generally have an unpleasant garlic-like odour
- They are irritating to the eyes and respiratory tract
- Dermal contact may cause frostbite
- Large exposures may lead to drowsiness, bronchitis and chemical pneumonitis
- *In the event of a large spill, stay upwind and out of low areas. Ventilate closed spaces. Protective clothing, eye protection and breathing apparatus should be worn*

Detailed Information

- Mercaptans are generally colourless liquids which have an unpleasant odour described as rotten cabbage, leeks, garlicky or skunk-like

- Mercaptans are generally used as gas odorants; also as intermediate and starting material in the manufacture of insecticides, antioxidants, jet fuels, and plastics; synthesis of methionine; as a catalyst and as synthetic flavouring agents

- Some mercaptans occur naturally in a wide variety of vegetables such as garlic and onion, in coal tar and petroleum distillates; also in human breath

FIRST AID

- Terminate exposure and support vital functions
- The casualty should be moved to an uncontaminated area
- Rescuers should, ideally, be trained personnel and must be careful *not to put themselves at risk* and *so wear appropriate protective clothing and, if available, breathing apparatus*
- If the casualty is unconscious a clear airway should be established and maintained; give 100% oxygen if available
- **Inhalation Exposure:** If the patient stops breathing, expired air resuscitation should be started immediately using a pocket mask with a one way valve, if available. It is important where the face is contaminated that expired air resuscitation is NOT attempted unless an airway with rescuer protection is used
- **Dermal Exposure:** If frostbite occurs *do not* remove clothing, flush skin with water
- If frostbite is not present remove contaminated clothing, if possible under a shower and place in double, sealed, clear bags and label; store the bags in a secure area away from patients and staff
- Wash the skin thoroughly with copious amounts of water
- **Eye Exposure:** Irrigate thoroughly with water or saline for 15 minutes
- **Oral Exposure:** Encourage small quantities of oral fluids (no more than 50–100 ml in total)

Identification of Mercaptans[1,2]

Mercaptan Compound	CAS number	UN number	NIOSH/RTECS number	Molecular formula	Molecular weight
Methyl–	74–93–1	1064	PB 4375000	CH_4S	48.11
Ethyl–	75–08–1	2363	KI 9625000	C_2H_6S	62.13
Propyl–	107–03–9	2402	TZ 7300000	C_3H_8S	76.17
Isopropyl–	75–33–2	2402	NT 4280000	C_3H_8S	76.17
n–Butyl–	109–79–5	2347	EK 6300000	$C_4H_{10}S$	90.19
Isobutyl–	513–44–0	n.a.	TZ 7630000	$C_4H_{10}S$	90.19
s–Butyl–	513–53–1	n.a.	n.a.	$C_4H_{10}S$	90.19
t–Butyl–	75–66–1	n.a.	TZ 2766000	$C_4H_{10}S$	90.19
n–Amyl–	110–66–7	1111	SA 3150000	$C_5H_{12}S$	104.22
s–Amyl–	2084–19–7	n.a.	n.a.	$C_5H_{12}S$	104.22
Isoamyl–	541–31–1	n.a.	n.a.	$C_5H_{12}S$	104.22
t–Amyl–	1679–09–0	n.a.	n.a.	$C_5H_{12}S$	104.22
n–Hexyl–	111–31–9	n.a.	MO 4550000	$C_6H_{14}S$	118.24
n–Heptyl–	1639–09–4	n.a.	MJ 1400000	$C_7H_{16}S$	132.27
n–Octyl–	111–88–6	n.a.	n.a.	$C_8H_{18}S$	146.30
s–Octyl–	25265–79–6	n.a.	n.a.	$C_8H_{18}S$	146.30
t–Octyl–	141–59–3	3023	SA 3260000	$C_8H_{18}S$	146.30
n–Nonyl–	1455–21–6	n.a.	n.a.	$C_9H_{20}S$	160.32
t–Nonyl–	25360–10–5	n.a.	n.a.	$C_9H_{20}S$	161.21
n–Decyl–	143–10–2	n.a.	n.a.	$C_{10}H_{22}S$	174.35
Undecyl–	5332–52–5	n.a.	n.a.	$C_{11}H_{24}S$	188.37
n–Dodecyl–	112–55–0	n.a.	n.a.	$C_{12}H_{26}S$	202.41
t–Dodecyl–	25103–58–6	n.a.	n.a.	$C_{12}H_{26}S$	202.41
Triisobutyl–	25103–58–6	n.a.	n.a.		200.10
n–Tetradecyl–	2079–95–0	n.a.	n.a.	$C_{14}H_{30}S$	230

Mercaptan Compound	CAS number	UN number	NIOSH/RTECS number	Molecular formula	Molecular weight
n-Hexadecyl-	2917-26-2	n.a.	n.a.	$C_{16}H_{34}S$	258.22
n-Octadecyl-	2885-00-9	n.a.	n.a.	$C_{18}H_{38}S$	286.57
Cyclohexyl-	1596-69-3	n.a.	n.a.	$C_6H_{11}SH$	116.24
Phenyl-	108-98-5	2337	DC 0525000	C_6H_6S	110.18
Benzyl-	100-53-8	n.a.	n.a.	C_7H_8S	124.20
1,2-Ethylenedithiol-	540-63-6	n.a.	n.a.	$C_6H_6S_2$	94.20
Allyl-	870-23-5	n.a.	n.a.	C_3H_6S	74
3-Chloropropyl-	17481-19-5	n.a.	n.a.	C_3H_7ClS	110
Perchloromethyl-	594-42-3	1670	PB 0370000	CCl_4	185.88

n.a. = number not available

Synonyms for Mercaptans[1,3]

Mercaptan Compound	Common Synonyms
Methyl-	methanethiol, mercaptomethane, methylmercaptan, methyl sulfhydrate, thiomethanol, thiomethyl alcohol
Ethyl-	ethanethiol, ethyl hydrosulfide, ethyl sulfhydrate, ethyl thioalcohol, thioethyl alcohol
Propyl-	1-propanethiol, mercaptopropanol, n-propyl mercaptan, n-thiopropyl alcohol
Isopropyl-	1-methylethanethiol, 2-mercaptopropane, isopropanethiol, isopropylmercaptan, isopropylthiol
n-Butyl-	1-butanethiol, n-butyl thioalcohol, thiobutyl alcohol, butyl sulfhydrate
Isobutyl-	1-propanethiol, isobutanethiol
s-Butyl-	sec-butyl mercaptan, 2-butyl mercaptan, 2-mercaptobutane, sec-butanethiol, sec-butyl thioalcohol,
t-Butyl-	2-propanethiol, tert-butanethiol, tert-butyl mercaptan, 2-methyl-2-propanethiol
n-Amyl-	1-pentanethiol, amyl hydrosulfide, amyl mercaptan, n-amyl mercaptan, amyl thioalcohol
n-Hexyl-	hexanethiol
n-Heptyl-	1-heptanethiol, n-heptylmercaptan
n-Octyl-	1-octanethiol, 1-mercaptooctane, 1-octyl thiol, octyl mercaptan, octylthiol
t-Octyl-	2-pentanethiol , tert-octanethiol, 2,4,4-trimethyl-2-pentanethiol
n-Nonyl-	1-nonanethi, 1-mercaptononane, nonylthiol
t-Nonyl-	–
n-Decyl-	1-decanethi, decylmercaptan, 1-mercaptodecane
Undecyl-	1-undecanethi
n-Dodecyl-	m-dodecyl mercaptan, 1-dodecyl mercaptan, lauryl mercaptan,
t-Dodecyl-	t-dodecanethi, tert-dodecylthiol,
n-Hexadecyl-	1-hexadecanethi, cetyl mercaptan, hexadecanethiol-1, n-hexadecanethiol
n-Octadecyl-	1-octadecanethi, 1-mercaptooctadecane, octadecyl mercaptan, stearyl mercaptan

Mercaptan Compound	Common Synonyms
Phenyl-	thiophenol, benzenethiol, mercapto-benzenethiol, mercaptobenzene phenol, thio-phenylthiol
Benzyl-	alpha-toluolthiol, alpha-tolyl mercaptan, benzylhydrosulfide, benzylthiol, alpha-mercaptotoluene, phenylmethanethiol, phenylmethyl mercaptan, thiobenzyl alcohol
1,2-Ethylenedithiol-	1,2-dimercaptoethane, dithioethyleneglycol, ethylene dimercaptan, alpha-ethylene dimercaptan, ethylene dithioglycol, ethyl hydropersulfide
Perchloromethyl-	PCM, methanesulfenic acid, perchloromethanethiol, perchloro-methyl-mercaptan, thiocarbonyl tetrachloride, trichlormethyl sulfur chloride, Clairsit (war gas)

- Perchloromethyl mercaptan is an oily, yellow liquid and is neither flammable nor a serious fire hazard

- Mercaptans are generally slightly soluble in water; soluble in alcohol, ether and petroleum naphtha

- Mercaptans are generally extremely flammable liquids; when they are heated or decomposed mercaptans release oxides of sulphur; they react with water, steam or acids to produce toxic and flammable vapours; they can react violently with powerful oxidisers[2]

Summary of Human Toxicology

- Exposure is generally through inhalation; skin and eye absorption is minimal

- The gas is rapidly absorbed via the lungs

- Contact with the liquid or vapour may be irritating to the skin and eyes, mucous membranes, respiratory tract

- Severe toxicity is likely to occur with exposure to high concentrations

- Odour fatigue can occur quite rapidly with most mercaptans

- Methyl mercaptan inhibits the cell tissue mitochondrial respiration by interfering with cytochrome c oxidase[3,4]

- Perchloromethyl mercaptan is a severe pulmonary irritant and lacrimating agent, it has tear gas properties that has been of military interest; at high concentrations it has caused pulmonary oedema, liver and renal toxicity and death following both skin contact and direct inhalation[4]

Occupational Exposure Standards

Chemical	Long-term exposure limit	Short-term exposure limit
Methyl–	0.5 ppm (1 mg.m^{-3})	no data available
Ethyl–	0.5 ppm (1.3 mg.m^{-3})	2 ppm (5.2 mg.m^{-3})
Benzyl–	0.5 ppm (2.3 mg.m^{-3})	no data available

Toxicology of Mercaptans[1, 3]

Mercaptan compound	Concentration	Odour threshold
Methyl–	2 ppb	Odour threshold
	400 ppm	Immediately dangerous to life or health
	>10,000 ppm	Fatal after a few min
Ethyl–	0.1 ppb	Odour threshold
	6.6×10^{-10} to 4.6×10^{-4} mg.m^{-3}	Odour detection in air
	1.9×10^{-4} mg.l^{-1} to 43.5 ppm	Odour detection in water
	0.26–0.97 ppb	Odour threshold
	2.1 ppm	Noticeable in water
	50–112 ppm for 20 min	Decreased respiration, odour fatigue in min
	4 ppm for 1 h	Headache, abdominal pain, vomiting and diarrhoea
	4 ppm for 3 h daily for 5–10 d	Fatigue
Propyl–	0.0016 ppm	Odour threshold
	0.36 ppm	Easily noticeable
n–Butyl–	0.0001–0.001 ppm	Odour threshold
	0.1–1 ppm	Readily noticeable
	50–500 ppm for 1 h	Muscle weakness, sweating, nausea, vomiting and headache, confusion and coma for 20 min
Isobutyl–	0.84 ppb	Odour threshold
tert–Butyl–	0.8 ppb	Odour threshold
n–Pentyl–	0.8 ppb or 0.0005–0.0004 mg.m^{-3}	Odour threshold
Isoamyl–	0.0083 ppm	Odour threshold
Phenyl–	0.00025 ppm	Odour threshold
Benzyl–	0.0026 ppm	Odour threshold
	4.5 ppm	Nasal irritation
	7.5 ppm	Eye irritation
1,2-Ethylenedithiol–	0.031 ppm	Faint odour
	5.6 ppm	Easily noticeable
Perchloromethyl–	0.01 ppm	Odour threshold
	0.22 ppm	Irritation threshold
	1.3 ppm	Eye irritation
	8.8 ppm	Nausea and irritation to eyes, throat and respiratory tract

Acute Clinical Effects

Inhalation Effects

- *General symptoms:* Irritation to the nose, throat and lungs, coughing, dyspnoea and pulmonary oedema

- Nausea, vomiting, sweating, dizziness, ataxia, muscle weakness

- Headache, lethargy, weakness and drowsiness may be expected

- *Methyl mercaptan:* Hypertension, tachycardia, haematuria, proteinuria, haemolytic anaemia, methaemoglobinaemia, pulmonary embolism (may be due to glucose-6-phosphate dehydrogenase deficiency), bronchopneumonia and convulsions have been reported[4,5]

- *Butyl mercaptan:* Confusion, flushed face and mydriasis may be expected[4]

- *Perchloromethyl mercaptan:* severe pulmonary irritant and lacrimating agent; may cause pulmonary oedema, renal and liver damage; death has been reported

Dermal Effects

- Irritation on contact with the liquid will occur[6]

- Dermatitis and burns may occur with prolonged or large exposures

- Frostbite has resulted from methyl mercaptan exposure[3]

Eye Effects

- Irritation of the conjunctiva and lacrimation may be expected with direct contact or from the vapour

Oral Effects

- Gastric irritation, nausea, vomiting, abdominal pain and diarrhoea may occur

- Mucosal irritation and ulceration

Chronic Clinical Effects

Dermatitis has developed in workers exposed to chronic low levels of methyl mercaptan.

Management

- Casualties complaining of noxious smell with no other effects require reassurance only

- Those with signs and symptoms require general supportive measures only, although those who are particularly concerned about the smell may well benefit from removal of contaminated clothing, see dermal management below

Inhalation Management

- Maintain a clear airway, give humidified oxygen and ventilate if necessary

- If respiratory irritation occurs assess respiratory function and if necessary perform chest X-rays to check for chemical pneumonitis

- Consider the use of steroids to reduce the inflammatory response

- Treat pulmonary oedema with PEEP or CPAP ventilation

- Symptomatic and supportive care

- In severe exposures monitor blood pressure, hepatic and renal function and methaemoglobin level

- Some patients are likely to be more susceptible to the odour, such as asthmatics, and may present with respiratory distress even if exposure is very mild

Dermal Management

- *If frostbite has occurred:* remove clothing carefully, these may need to be soaked off with tepid water; irrigate the area

- Surgical referral may be necessary

- *If frostbite has not occurred:* remove any remaining contaminated clothing

- Irrigate with copious amounts of water

- Treat symptomatically

- Place any contaminated clothes in double, sealed, clear bags and label; store in a secure area away from patients and staff

Eye Management

- Irrigate thoroughly with running water or saline for 15 minutes

- Stain with fluorescein and refer to an ophthalmologist if there is any uptake of the stain

Oral Management

- Encourage oral fluids

- Symptomatic and supportive care

Summary of Environment Hazards

- In the atmosphere mercaptans generally react with photochemically produced hydroxyl radicals

- *Methyl mercaptan:* In the atmosphere it reacts with ozone to produce sulphur dioxide and methane sulphonic acid; is highly toxic to fish with a minimum lethal concentration of 0.9–1 ppm[3]

- *Butyl mercaptan:* In water it is mainly lost through evaporation with a half-life of 5 hours in a river; in the atmosphere it has a half-life of 1.6 days; it is harmful to aquatic life at very low concentrations and is toxic to fish[3]

- *sec-butyl mercaptan:* This is highly mobile in soil; in water it volatilises rapidly with a half-life of 2.9 hours and 3.8 days in rivers and lakes respectively; in the atmosphere it has a half-life of 9–10.8 hours

- *t-butyl mercaptan:* Reacts readily in the atmosphere with hydroxyl radicals with a half-life of 11.0–13.3 hours; it is moderately mobile in soil; in water it has a half-life of 2.9 hours and 3.8 days in rivers and lakes respectively

- *Phenyl mercaptan:* When released onto the soil it is moderately mobile; it has a half-life of 5–8 days in rivers; in the atmosphere its half-life is 8.8 hours; it is not expected to bioconcentrate in the aquatic system

- *Perchloromethyl mercaptan:* In the atmosphere the half-life is 8 days; lethal to fish at 4.5–5.0 mg.l^{-1}

- Drinking Water Standards: no data available

- Soil Guidelines: no data available

- Air Quality Standards: no data available

REFERENCES

1. Clayton GD & Clayton FE (eds.) *Patty's Industrial Hygiene and Toxicology*, 4th edn. John Wiley & Sons, Inc., New York, 1994

2. Sax NI. *Dangerous Properties of Industrial Materials*, 6th edn. Van Nostrand Reinhold, New York, 1984

3. Hall AH & Rumack BH (eds.) *TOMES System®* Micromedex, Englewood, Colorado. CD ROM. vol. 41 (exp. 31 July 1999)

4. Hathaway GJ, Proctor NH & Hughes JP. *Proctor and Hughes' Chemical Hazards of the Workplace*, 4th edn. Van Nostrand Reinhold, New York, 1996

5. Shults WT, Fountain EN & Lynch EC, 1970. Methanethiol poisoning: irreversible coma and hemolytic anemia following inhalation. *J Am Med Assoc*; 211(13): 2153–4

6. Grant MW & Schuman JS. *Toxicology of the Eye*, 4th edn. Charles C Thomas, Springfield, 1993

Henrietta Wheeler

MERCURY (ELEMENTAL)

Key Points

- There are many forms of mercury, which can be classified as elemental, inorganic (e.g., inorganic mercury salts), and organic; the following concerns only elemental mercury
- Elemental mercury is a silvery, heavy liquid which is slightly volatile at room temperature
- It is toxic both acutely and chronically by inhalation
- The vapour of elemental mercury may result in CNS and renal damage
- Due to the volatile nature of mercury, spills should be cleared up as promptly and thoroughly as possible

First Aid

- Terminate exposure and support vital functions
- The casualty should be moved to an uncontaminated area
- If the casualty is unconscious a clear airway should be established and maintained; give 100% oxygen if available
- **Dermal Exposure:** Remove contaminated clothing and place in double, sealed, clear bags and label; store the bags in a secure area away from patients and staff. Treatment is not required as long as the skin is intact
- **Eye Exposure:** Irrigate thoroughly with water or saline for 15 minutes
- **Oral Exposure:** Treatment is generally not required, unless large exposure and aspiration is suspected

Detailed Information

- Metallic mercury is an odourless, shiny silvery liquid; it is very mobile, easily divisible into globules, and readily volatilises on heating
- *Common synonyms* colloidal mercury, liquid silver, elemental mercury, quicksilver
- CAS 7439-97-6
- UN 2809
- NIOSH/RTECS OV 4550000
- Atomic symbol Hg
- Atomic weight 200.59
- Elemental mercury is no longer used as a therapeutic agent in the UK; it is used as a component of some dental amalgams and medical instruments
- Elemental mercury is insoluble in water, alcohol and ether; it is soluble in lipids, sulphuric acid (on boiling) and nitric acid
- It reacts with ammonia, amines, acetylene and oxalic acids to form compounds that are sensitive to mechanical shock; also reacts with most metals (except iron) to form amalgams
- Elemental mercury is non-combustible

Summary of Human Toxicity

- As elemental mercury is volatile it is absorbed by inhalation; acute toxicity is unlikely to occur from ingestion or dermal exposure
- Conditions which produce mercury containing vapour, dusts or particulate may also result in significant exposure: vacuuming of carpet contaminated with mercury, breaking of mercury containing fluorescent light bulbs, heating of mercury-gold amalgams in order to extract gold
- After inhalation, it is readily absorbed through the alveolar membrane and transported by blood to the brain and other tissues of the nervous system
- Elemental mercury is converted rapidly to mercuric ions (Hg^{++}), which are then excreted in the urine and faeces[1]
- Elimination of elemental mercury occurs primarily in the urine with a half-life of 30 days[2] but some is found in the faeces and saliva
- Fatalities due to inhalation of elemental mercury vapour have been reported, although specific airborne concentrations resulting in death have not been adequately documented; four adults died following exposure to mercury vapours while smelting silver from dental amalgam in their home, the concentration of mercury vapours was not known[1]
- Insertion or removal of dental amalgam can generate mercury vapour which may be inhaled, however the exposure is low and poses no health risk to the patient
- Ingestion or handling of liquid mercury following breakage of thermometers or other mercury-containing devices are also sources of exposure, but also pose no health risk if disposed of correctly
- Children less than 30 months of age may be at increased risk of adverse pulmonary effects following vapour exposure
- If the airborne concentration is less than $50\,\mu g\,.\,m^{-3}$ then elemental mercury is usually undetectable in the blood; if the airborne concentration is more than $50\,\mu g\,.\,m^{-3}$ elemental blood mercury is usually in the range of 0.4 to $6.1\mu g\,.\,l^{-1}$ [2]
- Occupational Exposure Standards :
  Long-term exposure limit: $0.025\,mg\,.\,m^{-3}$
  Short-term exposure limit: no data available
- Mercury thermometers and barometers are often the most likely sources of mercury in the home
- When mercury from broken thermometers or barometers is left in the home it will volatilise and could cause toxicity
- A spillage of as little as 5 ml of elemental mercury left on a rug has led to severe toxicity in a child[3]
- Attempts to remove the mercury often fail and may lead to dispersion of the mercury and further increase exposure; see mercury spills section, page 158

Acute Clinical Effects

Inhalation Effects

- Elemental mercury is readily and rapidly absorbed by inhalation

- Respiratory tract irritation may occur with coughing, shortness of breath, tight chest, cough, cyanosis, pneumonitis, necrotising bronchiolitis, pulmonary oedema, ARDS, and death

- Pyrexia and 'flu like symptoms (metal fume fever), tachycardia may occur[1,4]

- Rarely, acute renal failure, hepatocellular dysfunction and convulsions may occur[1]

Dermal Effects

- Elemental mercury is not significantly absorbed dermally and therefore is not a health risk

- Allergic contact dermatitis with papular erythema may result from skin exposure

Eye Effects

- Exposure to both the metal (e.g. splash) and vapour may cause ocular disturbances

- Vapours may cause conjunctivitis and eyelid tremor may occur

- External eye contact with liquid mercury has resulted in temporary mercury deposition beneath the corneal epithelium, without apparent adverse effects[5]

Oral Effects

- Ingestion of small quantities (e.g. from a mercury containing thermometer) does not cause acute toxicity

- Ingestion of larger quantities may lead to aspiration and thus absorption via the respiratory tract[6]

Chronic Clinical Effects

- Chronic exposure may lead to CNS disturbances such as personality changes, hallucinations, delirium, insomnia, decreased appetite, irritability, headache and memory loss

- Altered sense of taste and smell, and digestive disturbances may occur

- Bilateral fine tremors, eyelid tremor, ataxia, exaggerated reflexes, paraesthesias, excessive perspiration, and blushing[1,4]

- Nausea, vomiting, conjunctivitis and pyrexia

- Gingivitis and stomatitis are commonly described and dark spots or dark bluish lines along the gums may also occur[1]

- Renal dysfunction and acute renal failure; respiratory infiltrates and failure have been reported

- Thrombocytopenia is a rare occurrence[7]

- Fine glistening particulate opacities in the axial portions of the corneal stroma, in addition to mercurialentis,[6] usually indicates chronic exposure to elemental mercury rather than toxicity; band keratopathy, corneal opacity and impaired vision may occur[5]

- Children and some adults develop acrodynia, which is associated with severe leg cramps, irritability, maculopapular rash, and peeling erythematous skin on the fingers, hands, and feet; this is also called 'pink disease'

Management

Remove patients from any continuing mercury exposure until clean-up is complete.

Blood and urine analysis are important indicators of exposure and severity

Normal blood concentration:
$$< 4\,\mu g . l^{-1} \ (< 20\,nmol . l^{-1})$$

Normal urine concentration:
$$< 10\,\mu g/24\,hours \ (< 50\,nmol/24\,hours)$$

Acute Management

Inhalation Management

- Give supplement oxygen if required and observe closely for several hours for the development of acute pneumonitis

- Chest X-ray and arterial blood gases may be required

- Check renal function and mercury concentrations (these can be taken immediately due to the rapid absorption)

- Treat with chelation therapy according to results (see below)

Dermal Management

- Thoroughly wash the skin, no further treatment is generally required

Eye Management

- Irrigate thoroughly with running water or saline for 15 minutes

- Refer to an ophthalmologist if required

Oral Management

- Treatment is not required for ingestion of metallic mercury

- Abdominal X-ray may be of use to confirm even minimal exposure

- Patients with inflammation of the diverticula or other gastrointestinal fistulas are at risk of systemic toxicity

- Chest X-ray if aspiration is suspected and treat as for inhalation exposure

Chronic Management

- Chronic respiratory disease may occur as a result of mercury aspiration, radiographs may show mercury droplets at the lung bases

- Severe, acute exposures may result in residual restrictive pulmonary disease and fibrosis[2]

- Monitor renal function, respiratory function and WBC

- Mercury concentrations must be checked and chelation therapy commenced if indicated (see below)

Chelation Therapy

Contact a poisons information service for further guidance and paediatric doses.

DMPS (dimercaptopropanesulphonic acid) is the treatment of choice. It should be considered in symptomatic patients when blood concentrations are above $100 \, \mu g \cdot l^{-1}$ and is vital when blood concentrations are $200 \, \mu g \cdot l^{-1}$ or above. It should be given IV to seriously ill patients, or orally to those with less severe effects or chronic exposures.

DMPS

Parenteral: Slow IV injection over 3–5 minutes.

Day 1: $5 \, mg \cdot kg^{-1}$, 6 times daily

Day 2: $5 \, mg \cdot kg^{-1}$, 6 times daily

Day 3 and 4: $5 \, mg \cdot kg^{-1}$, 3 times daily

Day 5: oral if possible; if not then $5 \, mg \cdot kg^{-1}$ once daily

Oral: chronic or acute poisoning: $5 \, mg \cdot kg^{-1}$ daily in divided doses generally for 10 days

Other antidotes that are available are DMSA (succimer, 2-3 dimercaptosuccimic acid), dimercaprol and penicillamine – all of which have been used with varying degrees of success.

Mercury spills

Clinical thermometer

On carpets or porous surfaces

- The spill should **never** be vacuumed with an ordinary domestic cleaner, this may disperse liquid mercury and thus increase airborne concentrations

- Vacuuming should only ever be carried out with a specialised industrial vacuum cleaner

- Contaminated carpets may need to be cut out and disposed of

- Contact the local environmental health department for further assessment and advice

Non-absorbent, finished surfaces

- All jewellery should be removed prior to cleaning spill

- Heat sources in area of spill should be turned off

- Brooms must not be used to sweep up spill

- Slightly stiff paper should be used to gather the droplets into a jar or sealable plastic bag

- Disposal of the mercury at home should be discussed with the local environmental health department

- Most hospitals should have their own internal policies for clean-up (using specialised kits) and disposal of mercury

Larger devices/unknown amount

On carpets, porous surfaces

- Require immediate and specialised equipment, the local environmental health department must be contacted

Non-absorbent, finished surfaces

- See clinical thermometer non–absorbent, finished surfaces

- The local environmental health department must be contacted

Summary of Environmental Hazards

- The atmospheric residence time of mercury vapour is up to 3 years

- Mercury vapour is converted to soluble forms and deposited by rain into soil and water

- Elemental mercury can readily oxidise to inorganic mercury (II) under normal conditions, this may then combine with an organic fraction to form methyl mercury; both methyl mercury and inorganic mercury (II) can cause severe environmental contamination, they can bioaccumulate and concentrate in the food chain

- Eating fish contaminated with mercury has caused secondary poisoning in humans, e.g. Minamata Bay, Japan in the 1950s and Iraq during the 1970s

- Drinking Water Standards:
  mercury total: $1 \, \mu g \cdot l^{-1}$ (UK max)
  $1 \, \mu g \cdot l^{-1}$ (WHO guideline)

- Soil Guidelines:
  Dutch Criteria: $0.3 \, mg \cdot kg^{-1}$ (target)
  $10 \, mg \cdot kg^{-1}$ (intervention)

- Air Quality Standards:
  $1 \, \mu g \cdot m^{-3}$ averaging time 1 year, indoor air (WHO guideline)

REFERENCES

1. Ellenhorn MJ, Schonwalds S, Ordog G & Wasserberger J. *Ellenhorn's Medical Toxicology – Diagnosis and Treatment of Human Poisoning*, 2nd edn. Williams & Wilkins, London, 1997

2. Hall AH & Rumack BH (eds.) *TOMES System®* Micromedex, Englewood, Colorado. CD ROM. vol. 41 (exp. 31 July 1999)

3. Von Muhlendahl KE, 1990. Intoxication form mercury spilled on carpets. *Lancet*; 336 (8730): 1578

4. Hathaway GJ, Proctor NH & Hughes JP. *Proctor and Hughes' Chemical Hazards of the Workplace*, 4th edn. Van Nostrand Reinhold, New York, 1996

5. Grant MW & Schuman JS. *Toxicology of the Eye*, 4th edn. Charles C Thomas, Springfield, 1993

6. Wright N, Yeomen WB & Carter GF, 1980. Massive oral ingestion of mercury without poisoning. *Lancet*; 1(8161): 206

7. Fuortes LJ, Weisman DN, Graeff ML, Bale JF, Tannour R & Peters C, 1995. Immune thrombocytopenia and elemental mercury poisoning. *J Toxicol-Clin Toxicol*: 33(5): 449–55

Henrietta Wheeler

METHANE

Key Points

- Methane is a colourless, odourless gas
- Inhalation is the main route of exposure
- It is a simple asphyxiant
- Methane is extremely flammable and a major explosion hazard
- On the skin and in the eyes, liquefied methane can cause frostbite
- *In the event of a large spill, stay upwind and out of low areas. Ventilate closed spaces. Protective clothing, eye protection and breathing apparatus should be worn*

FIRST AID

- Terminate exposure and support vital functions
- The casualty should be moved to an uncontaminated area
- Rescuers should, ideally, be trained personnel and must be careful *not to put themselves at risk* and *so wear appropriate protective clothing and, if available, breathing apparatus*
- If the casualty is unconscious a clear airway should be established and maintained; give 100% oxygen if available
- **Inhalation Exposure:** If the patient stops breathing, expired air resuscitation should be started immediately using a pocket mask with a one way valve, if available. It is important where the face is contaminated that expired air resuscitation is NOT attempted unless an airway with rescuer protection is used
- **Dermal Exposure:** If frostbite occurs *do not* remove clothing, flush skin with water
- If frostbite is not present remove contaminated clothing, if possible under a shower and place in double, sealed, clear bags and label; store the bags in a secure area away from patients and staff
- Wash the skin thoroughly with copious amounts of water
- **Eye Exposure:** If the eye tissue is frozen seek medical advice as soon as possible
- If eye tissue is not frozen, irrigate thoroughly with water or saline for 15 minutes

Detailed Information

- Methane is a colourless, odourless gas
- *Synonyms* marsh gas, methyl hydride
- CAS 74-82-8
- UN 1971 (compressed)
- UN 1972 (liquid)
- NIOSH/RTECS PA 1490000
- Molecular formula CH_4
- Molecular weight 16.042
- Methane is prepared commercially from natural gas or by the fermentation of cellulose and sewage sludge; it is a constituent in cooking and illuminating gas; in the production of

ammonia, methanol and chlorohydrocarbons and in the synthesis of acetylene

- Methane is lighter than air, it accumulates at the top of an enclosed space
- Methane is relatively soluble in water; soluble in alcohol, ether and organic solvents[1]
- Methane forms explosive mixtures with air; air containing less than 5.53% methane does not explode[2]
- It is extremely flammable, and is easily ignited by heat, sparks, or flame
- Combustion produces carbon dioxide and water, or carbon monoxide and water when oxygen is limited

Summary of Human Toxicity

- When inhaled, the majority of the absorbed dose is exhaled unchanged; a small amount is converted to methanol and ultimately carbon dioxide[3]
- The liquefied gas may cause frostbite on skin contact
- Methane is practically inert and has no demonstrated physiological or toxicological effects[4]
- It acts as a simple asphyxiant at very high concentrations where the major factor in exposure is the lack of available oxygen
- Methane exposure can occur in coal miners when methane is trapped within coal seams; collapse or loss of consciousness may prove lifesaving as methane is lighter than air[4]
- Occupational Exposure Standards: simple asphyxiant (i.e. when present in high concentrations in air, reducing oxygen content by dilution to such an extent that life cannot be supported. Oxygen content of air should be monitored and not allowed to fall below 18% to ensure safety)

Acute Clinical Effects

Inhalation Effects

- *Initial effects:* tachycardia, tachypnoea, and decreased alertness
- *As dose increases:* disinhibition, euphoria, confusion, headache, dizziness, paraesthesiae, hyperventilation, drowsiness, ataxia, and decreased visual acuity
- *Large doses:* CNS depression leading to coma, convulsions, cyanosis, and acidosis
- Severe hypoxia may cause apnoea, cardiac or respiratory arrest

Dermal Effects

- Dermal exposure to liquid methane can cause burns or frostbite

Eye Effects

- Methane is not irritating to the eyes
- Compressed or liquefied gas may cause burns and frostbite

Oral Effects

• Unlikely route of exposure

Chronic Clinical Effects

No data available, unlikely to result in adverse health effects

Management

Measurement of blood concentration is of no clinical relevance other than to confirm exposure when this is in doubt.

Inhalation Management

• Recovery normally occurs quickly once exposure has ceased, but support the respiratory and cardiovascular systems if necessary

• Maintain a clear airway, give humidified oxygen and ventilate if necessary

• Diazepam may be used for convulsions

• Symptomatic and supportive care

Dermal Management

• *If frostbite has occurred:* remove clothing carefully, these may need to be soaked off with tepid water; irrigate the area

• Surgical referral may be necessary

• *If frostbite has not occurred:* remove any remaining contaminated clothing

• Irrigate with copious amounts of water

• Treat burns symptomatically

• Place any contaminated clothes in double, sealed, clear bags and label; store in a secure area away from patients and staff

Eye Management

• Irrigate thoroughly with running water or saline for 15 minutes

• Stain with fluorescein and refer to an ophthalmologist if there is any uptake of the stain

Oral Management

• Not applicable

Summary of Environmental Hazards

• Methane is expected to exist almost entirely in the vapour phase in ambient air; it is persistent in atmosphere due to very slow reaction with photochemically produced hydroxyl radicals, with a half-life of 1,908 days[5]

• In model river and model environmental pond the volatilisation half-lives for methane have been estimated to be 1.17 and 13.89 h, respectively[5]

• Drinking Water Standards:
hydrocarbon total $10 \, \mu g.l^{-1}$ (UK max)

• Soil Guidelines: no data available

• Air Quality Standards: no data available

REFERENCES

1. Snyder R (ed.) *Ethel Browning's Toxicity and Metabolism of Industrial Solvents*, 2nd Edn. Elsevier, Amsterdam, 1987

2. Budavari S, O'Neil MJ, Smith A, Heckelman PE & Kinneary JF (eds.) *The Merck Index*, 12th edn. Merck & Co., Inc., Whitehouse Station, 1996

3. Clayton GD & Clayton FE (eds.) *Patty's Industrial Hygiene and Toxicology*, 4th edn. John Wiley & Sons, Inc., New York, 1994

4. Hathaway GJ, Proctor NH & Hughes JP. *Proctor and Hughes' Chemical Hazards of the Workplace*, 4th edn. Van Nostrand Reinhold, New York, 1996

5. Hall AH & Rumack BH (eds.) *TOMES System®* Micromedex, Englewood, Colorado. CD ROM. vol. 41 (exp. 31 July 1999)

Robie Kamanyire

METHYL BROMIDE

Key Points

- Methyl bromide is a highly toxic, colourless gas or a colourless, transparent volatile liquid (under pressure)
- The effects of poisoning may be delayed and insidious in onset
- Inhalation is the major route of absorption, dermal absorption may occur but systemic effects are unlikely unless inhalation is significant
- The main effects of methyl bromide are on the central nervous system; initial effects are generally non-specific with nausea, vomiting and malaise
- Later, a variety of CNS effects occur, including ataxia, twitching, tremor, confusion, dizziness and weakness; convulsions are common and are often refractory to treatment
- Death may be due to circulatory failure, convulsions or pulmonary oedema
- *In the event of a large spill, stay upwind and out of low areas. Ventilate closed spaces. Protective clothing, eye protection and breathing apparatus should be worn*

FIRST AID

- Terminate exposure and support vital functions
- The casualty should be moved to an uncontaminated area
- Rescuers should, ideally, be trained personnel and must be careful *not to put themselves at risk,* and *so wear appropriate protective clothing and, if available, breathing apparatus*
- If the casualty is unconscious a clear airway should be established and maintained; give 100% oxygen if available
- **Inhalation Exposure:** If the patient stops breathing, expired air resuscitation should be started immediately using a pocket mask with a one way valve, if available. It is important where the face is contaminated that expired air resuscitation is NOT attempted unless an airway with rescuer protection is used
- **Dermal Exposure:** Remove contaminated clothing, if possible under a shower and place in double, sealed, clear bags and label; store the bags in a secure area away from patients and staff
- Wash the skin thoroughly with copious amounts of water
- **Eye Exposure:** Irrigate thoroughly with water or saline for 15 minutes
- **Oral Exposure:** Encourage small quantities of oral fluids (no more than 50–100 ml in total)

Detailed Information

- Methyl bromide is a highly toxic, colourless gas or a colourless, transparent volatile liquid (under pressure) with a burning taste and chloroform-like odour
- *Common synonyms* bromomethane, isobrome, monobromomethane, MBX
- CAS 74-83-9
- UN 1062

- NIOSH/RTECS PA 4900000
- Molecular formula CH_3Br
- Molecular weight 94.95
- Methyl bromide is used as a fumigant to sterilise soil and kill pests on fruit and vegetables, for degreasing wool, for extracting oils from seeds, nuts and flowers and as a methylating agent in the chemical industry
- Chloropicrin is often added to products containing methyl bromide as a warning agent
- It is three times heavier than air; is non-flammable in air, but burns in oxygen
- Methyl bromide penetrates leather, rubber and cloth
- Only qualified and competent personnel should use methyl bromide for fumigation; appropriate protective clothing and breathing apparatus should be worn
- Miscible with most organic solvents, it forms a voluminous crystalline hydrate with cold water
- It decomposes in flame or on hot surfaces giving off toxic and corrosive vapours including hydrogen bromide, bromine and carbon oxybromide
- Reacts violently with aluminium, with risk of fire and explosion; attacks zinc, magnesium and alkali metals
- Recovery may take weeks or months and patients may be left with permanent neurological or psychiatric impairment

Note: methyl bromide has poor olfactory warning properties; warning agents, such as chloropicrin (2%) or amyl acetate, may be added to overcome this problem

Summary of Human Toxicity

- Methyl bromide is readily absorbed by the lungs; significant dermal absorption is unlikely to occur
- Poisoning may occur from acute or chronic exposure
- Onset of effects may occasionally be delayed for up to 48 hours, although in most severe, but non-fatal cases, the effects occur within 14 hours; death generally occurs within 10–30 hours in fatal cases
- It should be noted that the presence or absence of chloropicrin (added as a warning agent) does not indicate the presence or absence of methyl bromide; chloropicrin may evaporate before the methyl bromide does
- Charcoal used in respirator protective devices has limited capacity to remove methyl bromide and may in fact preferentially remove chloropicrin
- Estimates of concentrations of methyl bromide that have caused fatalities range from 8,000 ppm for a few hours to 60,000 ppm for a brief exposure[1]
- Occupational Exposure Standards :
  Long term exposure limit 5 ppm (20 mg.m^{-3})
  Short term exposure limit 15 ppm (59 mg.m^{-3})

Acute Clinical Effects

Inhalation Effects

- Exposure leads to nausea, vomiting and headache

- Convulsions are common and are described as Jacksonian in character; status epilepticus may occur in severe cases

- Coma may be prolonged

- Behavioural and emotional disturbances may occur

- Also ataxia, diplopia, twitching, tremor, confusion, euphoria, nystagmus, slurred speech, hallucinations (visual and auditory), headache, malaise, dizziness, myoclonus, blurred vision, paraesthesia, weakness, drowsiness, confusion and extrapyramidal signs have been reported

- Bronchitis, pneumonitis and pulmonary oedema may occur in severe cases

Dermal Effects

- Erythema, oedema, itching and pain, numbness or a freezing sensation may occur

- Prolonged or repeated exposure may cause deep burns with delayed vesiculation (3–12 hours later)

- The most affected areas of the skin are those which are moist or subject to mechanical pressure[2,3]

Eye Effects

- Mild transient irritation and conjunctivitis may occur with delayed onset

Oral Effects

- No cases of ingestion have been reported, but nausea, vomiting, abdominal cramps, diarrhoea and ulceration of the gastrointestinal tract may be expected

Chronic Clinical Effects

Low level chronic exposure may result in neurological, behavioural and psychiatric abnormalities including confusion, anorexia, polyneuropathy, tremor, sensory abnormalities, slurred speech, visual disturbances and cognitive deficits. There may also be personality changes or psychosis.[4,5] Dizziness, numbness, paraesthesia, weakness of extremities, nightmares, and fatigue have been reported in chronically exposed individuals.

Dry, scaly skin with itching, bullae and red swollen hands may occur with chronic exposure.

Management

Plasma levels are not useful for methyl bromide. Measurement of blood bromide concentration correlates poorly with symptom severity and has no clinical relevance other than to confirm exposure where this is in doubt.

Inhalation Management

- Maintain a clear airway, give humidified oxygen and ventilate if necessary

- If respiratory irritation occurs assess respiratory function and if necessary perform baseline and serial chest X-rays to check for chemical pneumonitis

- Patients should be kept at rest

- Treat pulmonary oedema with PEEP or CPAP ventilation

- Symptomatic and supportive care

- See systemic management below

Dermal Management

- Remove any remaining contaminated clothing, place in double, sealed, clear bags and label; store in a secure area away from patients and staff

- Irrigate with copious amounts of water

- Treat burns or erythema symptomatically

Eye Management

- Irrigate thoroughly with running water or saline for 15 minutes

- Stain with fluorescein and refer to an ophthalmologist if there is any uptake of the stain

Oral Management

- NO GASTRIC LAVAGE OR EMETIC

- Encourage oral fluids

- Symptomatic and supportive care

- See systemic management below

Systemic Management

- Treatment is primarily supportive with attention to the CNS and respiratory effects

- All exposed patients will require observation for a minimum 48 hours post exposure

- The most important aspect of treatment is the control of fits which may not respond to anti-convulsant therapy

- It is usually necessary to paralyse and ventilate a severely poisoned patient

- Monitor EEG, blood gases and acid-base balance

- Rehabilitation may be required in patients suffering neurological sequelae

- Long-term neurological assessment and monitoring will be necessary to monitor the recovery of poisoned patients

Summary of Environmental Hazards

- In the atmosphere methyl bromide has a residence time of about 289 days; primary mechanism for removal is believed to be reaction with hydroxyl radicals

- Decomposition of methyl bromide in soil results in bromide ion production; the rate of production is influenced by the soil type; bromide in soil can be absorbed and concentrated by plants and may ultimately be ingested by humans and animals[6]

- Methyl bromide is moderately toxic to fish and is not expected to bioconcentrate in the food chain

- Drinking Water Standards:
  hydrocarbon total: $10\,\mu g.l^{-1}$ (UK max)
  pesticide: $0.1\,\mu g.l^{-1}$ (UK max)

- Soil Guidelines: no data available

- Air Quality Standards: no data available

REFERENCES

1. Hathaway GJ, Proctor NH & Hughes JP. *Proctor and Hughes' Chemical Hazards of the Workplace*, 4th edn. Van Nostrand Reinhold, New York, 1996

2. Hezemans-Boer M, Toonstra J, Meulenbelt J, Zwaveling J-H, Sangster B & van Vloten WA, 1988. Skin lesions due to exposure to methyl bromide. *Arch Dermatol*; 124: 917–21

3. Zwaveling JH, de Kort WLAM, Meulenbelt J, Hezemans-Boer M, van Vloten WA & Sangster B, 1987. Exposure of the skin to methyl bromide: a study of six cases occupationally exposed to high concentrations during fumigation. *Hum Toxicol*; 6: 491–5

4. Reidy TJ, Bolter JF & Cones JE, 1994. Neuropsychological sequelae of methyl bromide: a case study. *Brain Injury*; 8(1): 83–93

5. Kishi R, Itoh I, Ishizu S, Harabuchi I & Miyake H, 1991. Symptoms among workers with long-term exposure to methyl bromide. An epidemiological study. *Jpn J Ind Health*; 33: 241–50

6. Hall AH & Rumack BH (eds.) *TOMES System®* Micromedex, Englewood, Colorado. CD ROM. vol. 41 (exp. 31 July 1999)

Nicola Bates

METHYL ETHYL KETONE

Key Points

- Methyl ethyl ketone (MEK) is a clear colourless, volatile, highly flammable liquid with an acetone-like odour
- MEK is rapidly absorbed by inhalation; it is an irritant to eyes, skin and mucous membranes
- Clinical effects include headache, vomiting, dizziness, tiredness, confusion and numbness of extremities
- At high concentrations it may cause CNS depression, however because of its odour it can be safely detected well below recommended exposure limits
- MEK can potentiate the toxicity of other solvents
- *In the event of a large spill, stay upwind and out of low areas. Ventilate closed spaces. Protective clothing, eye protection and breathing apparatus should be worn*

First Aid

- Terminate exposure and support vital functions
- The casualty should be moved to an uncontaminated area
- Rescuers should, ideally, be trained personnel and must be careful *not to put themselves at risk* and *so wear appropriate protective clothing and, if available, breathing apparatus*
- If the casualty is unconscious a clear airway should be established and maintained; give 100% oxygen if available
- **Inhalation Exposure:** If the patient stops breathing, expired air resuscitation should be started immediately using a pocket mask with a one way valve, if available. It is important where the face is contaminated that expired air resuscitation is NOT attempted unless an airway with rescuer protection is used
- **Dermal Exposure:** Remove contaminated clothing, if possible under a shower and place in double, sealed, clear bags and label; store the bags in a secure area away from patients and staff
- Wash the skin thoroughly with copious amounts of water
- **Eye Exposure:** Irrigate thoroughly with water or saline for 15 minutes
- **Oral Exposure:** Encourage small quantities of oral fluids (no more than 50–100 ml in total)

Detailed Information

- MEK is a clear colourless, volatile, highly flammable liquid with an acetone-like odour

- *Common synonyms* butanone, 2-butanone, ethyl methyl ketone, MEK, methyl acetone

- CAS 78-93-3

- UN 1193

- NIOSH/RTECS EL 6475000

- Molecular formula C_4H_8O

- Molecular weight 72.10

- The main use for MEK is as a solvent for vinyl plastic used in coatings and moulding articles; other uses include varnishes, for degreasing metals, in the manufacture of magnetic tapes, inks and smokeless powder; it is also used in spray paints, paint removers, sealers and glues, and as an extraction solvent in the processing of foodstuffs and food ingredients

- MEK can form explosive mixtures with air

- It is very soluble in water and miscible with many organic solvents

- MEK is stable under ordinary conditions but can form peroxides following prolonged storage and these may be explosive[1]

- Reacts with oxidising agents such as peroxides, chlorinated solvents and strong alkalis

Summary of Human Toxicity

- MEK is rapidly absorbed by inhalation and may be absorbed dermally with prolonged contact[1]

- It is metabolised by the liver, mostly to carbon dioxide and water; excretion of MEK and its metabolites is mainly through the lungs

- MEK increases cytochrome P_{450} enzyme activity and this may be the mechanism by which MEK potentiates the toxic effects of other solvents

- It potentiates the neurotoxic effects of aliphatic hexacarbon compounds (n-hexane, methyl-n-butylketone and 2,5-hexanedione) and the liver and kidney toxicity of haloalkane solvents (carbon tetrachloride and trichloroethane)

- Following accidental ingestion of a glue containing MEK the plasma concentration was 13.2 mmol . l^{-1} [2]

Toxicity of MEK[3,4]

Concentration	Clinical effects
2–10 ppm	Odour threshold
100 ppm	Nose and throat irritation
200 ppm	Mild eye irritation
300 ppm	'Objectionable' with headache, throat irritation
300–600 ppm	Numbness of fingers, arms and legs

- Occupational Exposure Standards:
    Long-term exposure limit: 200 ppm (600 mg . m^{-3})
    Short-term exposure limit: 300 ppm (899 mg . m^{-3})

Acute Clinical Effects

Inhalation Effects

- Exposure may lead to irritation of nose and throat

- Headache, vomiting, dizziness, tiredness, confusion and numbness of extremities can occur

- At high concentrations MEK may cause CNS depression

- Myoclonic jerks, tremor and ataxia have been reported[5]

Dermal Effects

- MEK is an irritant to the skin
- It may defat the skin and cause dermatitis

Eye Effects

- Irritant with pain, lacrimation and blurred vision may occur
- No long-term effects are expected

Oral Effects

- Ingestion may lead to vomiting, drowsiness, coma, hypotension, hyperventilation and metabolic acidosis[2]

Chronic Clinical Effects

Peripheral neuropathy with numbness, weakness and tingling of extremities, and CNS depression with dizziness and loss of vision have all been reported from occupational exposures. Prolonged dermal exposure to MEK may cause contact dermatitis.[4]

Management

MEK can be detected in urine, blood and expired air.

Inhalation Management

- Maintain a clear airway, give humidified oxygen and ventilate if necessary
- If respiratory irritation occurs assess respiratory function and if necessary perform chest X-rays to check for chemical pneumonitis
- Treat pulmonary oedema with PEEP or CPAP ventilation
- Symptomatic and supportive care

Dermal Management

- Remove any remaining contaminated clothing, place in double, sealed, clear bags; store in a secure area away from patients and staff
- Irrigate with copious amounts of water
- Treat irritation symptomatically

Eye Management

- Irrigate thoroughly with running water or saline for 15 minutes
- Stain with fluorescein and refer to an ophthalmologist if there is any uptake of the stain

Oral Management

- Encourage oral fluids
- Consider gastric lavage within 1 hour of a substantial ingestion, ensuring airway is protected
- Contact a poisons information service for further guidance on gut decontamination
- Monitor blood gases following significant ingestion
- Symptomatic and supportive care

Summary of Environmental Hazards

- In the atmosphere MEK degrades by reacting with photochemically produced hydroxyl radicals with a half-life of 2.3 days; under photochemical smog, degradation may be slightly faster[6]
- It occurs naturally from volcanoes, forest fires and biological degradation of products
- High atmospheric MEK levels are associated with photochemical smog episodes although generally it is absent in ambient air
- In water MEK will evaporate with half-lives of 3 and 12 days in rivers and lakes, respectively[6]
- Spills on land will partially evaporate into the atmosphere and partially leach into ground water
- It has low aquatic toxicity and is not expected to bioaccumulate in the food chain
- Drinking Water Standards:
    hydrocarbons total: $10\,\mu g.l^{-1}$ (UK max)
- Soil Guidelines: no data available
- Air Quality Standards: no data available

REFERENCES

1. International Programme on Chemical Safety. *Environmental Health Criteria 143: Methyl ethyl ketone*. WHO, Geneva 1993

2. Kopelman PG & Kalfayan PY, 1983. Severe metabolic acidosis after ingestion of butanone. *Br Med J*; 286: 21–2

3. Clayton GD & Clayton FE (eds.) *Patty's Industrial Hygiene and Toxicology*, 4th edn. John Wiley & Sons, Inc., New York, 1994

4. Hathaway GJ, Proctor NH & Hughes JP. *Proctor and Hughes' Chemical Hazards of the Workplace*, 4th edn. Van Nostrand Reinhold, New York, 1996

5. Orit-Pareja M, Jiménez-Jiménez FJ, Miquel J, Montero E, Cabrera-Valdivia F, Benito A & García-Albea E, 1996. Reversible myoclonus, tremor, and ataxia in a patient exposed to methyl ethyl ketone. *Neurology*; 46: 272

6. Hall AH & Rumack BH (eds.) *TOMES System®* Micromedex, Englewood, Colorado. CD ROM. vol.41 (exp. 31 July 1999)

Nicola Bates

METHYL ETHYL KETONE PEROXIDE

Key Points

- Methyl ethyl ketone peroxide (MEKP) is a colourless liquid with an acetone-like odour
- MEKP is usually available as a 60% solution in dimethyl phthalate since it is explosive when pure
- MEKP is often stored in the refrigerator since a low temperature also reduces the risk of explosion
- MEKP is corrosive and causes liver and kidney damage
- *In the event of a large spill, stay upwind and out of low areas. Ventilate closed spaces. Protective clothing, eye protection and breathing apparatus should be worn*

FIRST AID

- Terminate exposure and support vital functions
- The casualty should be moved to an uncontaminated area
- Rescuers should, ideally, be trained personnel and must be careful *not to put themselves at risk* and *so wear appropriate protective clothing and, if available, breathing apparatus*
- If the casualty is unconscious a clear airway should be established and maintained; give 100% oxygen if available
- **Inhalation Exposure:** If the patient stops breathing, expired air resuscitation should be started immediately using a pocket mask with a one way valve, if available. It is important where the face is contaminated that expired air resuscitation is NOT attempted unless an airway with rescuer protection is used
- **Dermal Exposure:** Remove contaminated clothing, if possible under a shower and place in double, sealed, clear bags and label; store the bags in a secure area away from patients and staff
- Wash the skin thoroughly with copious amounts of water
- **Eye Exposure:** Irrigate thoroughly with water or saline for 15 minutes
- **Oral Exposure:** Encourage small quantities of oral fluids (no more than 50–100 ml in total)

Detailed Information

- MEKP, an organic peroxide, is a colourless liquid with an acetone-like odour
- *Common synonyms* 2-butanone peroxide, butanox, MEK peroxide, MEKP
- CAS 1338-23-4 (with approx. 10% active oxygen content)
- UN 3101 ($\leq$52%)
- UN 3105 ($\leq$45%)
- UN 3107 ($\leq$40%)
- NIOSH/RTECS EL 9450000 (with approx. 10% active oxygen content)
- Molecular formula $C_8H_{16}O_4$
- Molecular weight 176.24
- Ketone peroxides are used as catalysts for hardening fibreglass reinforced plastics

- MEKP is a strong oxidising agent and reacts violently with salts of heavy metals, tertiary amines, strong acids and alkalis; it decomposes above 80°C

Summary of Human Toxicity

- MEKP is corrosive and causes liver and kidney damage
- Few cases are reported in the literature but where they are, mortality and morbidity is high
- The mechanism of toxicity is unclear but may involve damage due to oxygen free radicals
- The dimethyl phthalate (see key points) is generally thought to be of relatively low toxicity; it is not corrosive
- Ingestion of 50–100 ml of 35–40% MEKP resulted in hepatic necrosis, respiratory failure and death after 4 days[1]
- Occupational Exposure Standards:
    Long-term exposure limit: no data available
    Short-term exposure limit: 0.2 ppm (1.5 mg.m^{-3})

Acute Clinical Effects

Inhalation Effects

- Limited information is available about inhalation
- The majority of respiratory effects occur as complications of ingestion, such as severe irritation to the respiratory tract

Dermal Effects

- There is limited information; exposure may cause irritation and burns

Eye Effects[2]

- *Mild* (e.g. finger to eye transfer): pain, lacrimation, photophobia, corneal epithelial damage; recovery 3–7 days
- *Moderate:* severe photophobia, perilimbal (junction of sclera and cornea) hyperaemia, epithelial and stromal oedema, recurrent erosions; recovery 6–7 months
- *Severe:* loss of corneal epithelium, full thickness corneal oedema and vascularisation, loss of vision or eye; deterioration over 1–3 years

Oral Effects

- Vomiting, haematemesis, oral and oesophageal burns, gastritis (may be haemorrhagic), drowsiness and coma may occur
- Exposure may lead to severe metabolic acidosis, respiratory distress, ARDS, aspiration pneumonitis, hypotension and haematuria
- Acute renal failure (secondary to rhabdomyolysis), liver damage with coagulopathy (raised INR, decreased haematocrit), myocarditis (gallop rhythm, tachycardia, inverted T-waves, congestive heart failure) and cardiac arrest have been reported
- Late sequelae include gastric perforation and oesophageal stricture

Chronic Clinical Effects

No data available.

Management

Inhalation Management

- Maintain a clear airway, give humidified oxygen and ventilate if necessary

- If respiratory irritation occurs assess respiratory function and if necessary perform chest X-rays to check for chemical pneumonitis

- Treat pulmonary oedema with PEEP or CPAP ventilation

- Symptomatic and supportive care

Dermal Management

- Remove any remaining contaminated clothing, place in double, sealed, clear bags and label; store in a secure area away from patients and staff

- Irrigate with copious amounts of water

- Treat irritation symptomatically

Eye Management

- Irrigate thoroughly with running water or saline for 15 minutes

- Stain with fluorescein and refer to an ophthalmologist if there is any uptake of the stain

Oral Management

- NO GASTRIC LAVAGE OR EMETIC

- Neutralising chemicals should never be given because heat is produced during neutralisation and this could exacerbate any injury

- Patients must be admitted for observation until the extent of the injury (if any) can be determined

- Monitor ECG, pH, respiratory, renal and liver function

- Encourage oral fluids unless there is evidence of severe injury

- Oral feeding should be maintained if the patient is able to tolerate it, otherwise tube feeding or parenteral nutrition should be provided

- The use of steroids in corrosive injury due to organic peroxides has not been studied, they may be given but are of no proven benefit

- Aggressive intervention is essential for severely affected patients. Urgent assessment of the airway is required

- If facilities are available, early gastro-oesophagoscopy should be undertaken within 12 hours of the event to assess the extent and severity of the injury

- Give plasma expanders/intravenous fluids for shock and check and correct the acid/base balance. Analgesia will almost certainly be needed

- A supraglottic-epiglottic burn with erythema and oedema is usually a sign that further oedema will occur which will lead to airway obstruction; it is an indication for early intubation or tracheostomy[3]

- Intubation and ventilation may be necessary for patients with respiratory distress

- On discharge all patients must be advised of the possibility of late onset sequelae and advised to return if necessary

- Strictures that prevent adequate nutritional intake and do not respond to dilatation require oesophagectomy and colonic interposition

- Oesophageal strictures which result in a lumen >10 mm do not impede normal life and should not require intervention.[4] Surgical intervention may also be required for gastrointestinal perforation or haemorrhage

Summary of Environmental Hazards

- If released to the atmosphere, MEKP degrades by reacting with photochemically produced hydroxyl radicals; the estimated half-life for this reaction is 1.7 days[5]

- If released to soil or water, MEKP may react with organic materials, possibly violently; leaching in soil may be possible in the absence of oxidation reactions or other degradation processes

- Aquatic volatilisation, adsorption to sediment, and concentration in the food chain are not expected[5]

- Drinking Water Standards:
    hydrocarbon total: $10 \, \mu g \cdot l^{-1}$ (UK max)

- Soil Guidelines: no data available

- Air Quality Standards: no data available

REFERENCES

1. Karhunen PJ, Ojanpera I, Lalu K & Vuori E. 1990. Peripheral zonal hepatic necrosis caused by accidental ingestion of methyl ethyl ketone peroxide. *Hum Exper Toxicol*; 9: 197–200

2. Grant MW & Schuman JS. *Toxicology of the Eye*, 4th edn. Charles C Thomas, Springfield, 1993

3. Meredith JW, Kon ND & Thompson JN, 1988. Management of injuries from liquid lye ingestion. *J Trauma*; 28(8): 1173–80

4. Sarfati E, Assens P & Celerier M, 1987. Management of caustic ingestion in adults. *Br J Surg*; 74: 146–8

5. Hall AH & Rumack BH (eds.) *TOMES System®* Micromedex, Englewood, Colorado. CD ROM. vol.41 (exp. 31 July 1999)

Nicola Bates

METHYL ISOBUTYL KETONE

Key Points

- Methyl isobutyl ketone (MIBK) is a clear liquid with a fruity odour
- Exposure to MIBK is generally via the dermal and respiratory systems
- MIBK causes respiratory, dermal and eye irritation
- *In the event of a large spill, stay upwind and out of low areas. Ventilate closed spaces. Protective clothing, eye protection and breathing apparatus should be worn*

FIRST AID

- Terminate exposure and support vital functions
- The casualty should be moved to an uncontaminated area
- Rescuers should, ideally, be trained personnel and must be careful *not to put themselves at risk* and *so wear appropriate protective clothing and, if available, breathing apparatus*
- If the casualty is unconscious a clear airway should be established and maintained; give 100% oxygen if available
- **Inhalation Exposure:** If the patient stops breathing, expired air resuscitation should be started immediately using a pocket mask with a one way valve, if available. It is important where the face is contaminated that expired air resuscitation is NOT attempted unless an airway with rescuer protection is used
- **Dermal Exposure:** Remove contaminated clothing, if possible under a shower and place in double, sealed, clear bags and label; store the bags in a secure area away from patients and staff
- Wash the skin thoroughly with copious amounts of water
- **Eye Exposure:** Irrigate thoroughly with water or saline for 15 minutes
- Oral Exposure: Encourage small quantities of oral fluids (no more than 50–100 ml in total)

Detailed Information

- MIBK is a clear liquid with a sweet, sharp, fruity odour
- *Common synonyms* hexanone, isopropyl–acetone, isobutylmethyl ketone, MIBK
- CAS 108-10-1
- UN 1245
- NIOSH/RTECS SA 9275000
- Molecular formula $C_6H_{12}O$
- Molecular weight 100.16
- MIBK is used in lacquers, such as cellulose and polyurethane lacquers, and as a minor component of paint solvents; it is also used as a solvent in some CS gas and other self defence spays
- MIBK occurs naturally in a wide range of foods
- It is soluble in water and miscible with alcohol, benzene and ether, and with most organic solvents

- MIBK reacts violently with oxidising agents such as peroxides, nitrates and perchlorates, reducing agents, or with potassium *tert*-butoxide; attacks many plastics
- It is easily ignited by heat, sparks or flames; when heated MIBK may form peroxides by auto-oxidation and under certain conditions these may explode spontaneously

Summary of Human Toxicity

- MIBK is absorbed dermally but is of low toxicity orally
- The low odour threshold of MIBK and its irritant effect should prevent over-exposure[1]
- Minimum human lethal exposure to MIBK has not been estimated
- Human response to ethanol may be affected by MIBK; simultaneous exposure to alcohol and MIBK should be avoided[2]

Toxicity of MIBK[3]

Concentrations	Clinical effects
0.10 ppm	Odour threshold concentration
1.3 ppm	Water odour threshold
100 ppm	Headache, nausea, vertigo, respiratory and eye irritation
200 ppm	Odour objectionable, discomfort and eye irritation
200–400 ppm	Eye and nasal irritation

- Occupational Exposure Standards:
  Long-term exposure limit: 50 ppm (208 mg.m^{-3})
  Short-term exposure limit: 100 ppm (416 mg.m^{-3})

Acute Clinical Effects

Inhalation Effects

- Inhalation may lead to sore throat, respiratory and eye irritation
- Larger exposures of MIBK may lead to systemic effects; see below

Dermal Effects

- Dermal exposure may produce minimal irritation unless sprayed in close proximity of skin[1]
- Prolonged dermal contact may cause irritation, defatting of the skin as well as flaking and peeling[2]
- Prolonged or excessive exposure can potentially lead to systemic effects; see below

Eye Effects

- Exposure to vapours may lead to eye irritation
- Direct splashes of MIBK in the eyes may cause pain and irritation

Oral Effects

- Ingestion may cause irritation to mucous membranes

- Coughing, wheezing and aspiration leading to chemical pneumonitis

Systemic Effects

- Dizziness, headache, drowsiness and ataxia have been reported

- Nausea, vomiting, weakness and loss of appetite may occur

- MIBK in high concentrations causes narcosis in animals, and it is expected that severe exposure to humans will produce the same effect[2]

Chronic Clinical Effects

Tolerance to MIBK can develop with chronic exposure, but it is quickly lost.[3] Excess long-term exposure may lead to weakness, dizziness, loss of appetite, headaches, eye irritation, abdominal pain, nausea, vomiting and sore throat. A few cases of peripheral neuropathy have been reported after exposure to spray paint or lacquer thinner that contained MIBK and other hydrocarbon solvents.[1] Prolonged or repeated skin contact may produce defatting dermatitis with dryness and cracking.

MIBK is an experimental teratogen in animals.[4]

Management

Inhalation Management

- Maintain a clear airway, give humidified oxygen and ventilate if necessary

- If respiratory irritation occurs assess respiratory function and if necessary perform chest X-rays to check for chemical pneumonitis

- Consider the use of IV steroids to reduce the inflammatory response

- Treat pulmonary oedema with PEEP or CPAP ventilation

- Symptomatic and supportive care

Dermal Management

- Remove any remaining contaminated clothing, place in double, sealed clear, bags and label; store in a secure area away from patients and staff

- Irrigate with copious amounts of water

- Treat skin irritation symptomatically

Eye Management

- Irrigate thoroughly with running water or saline for 15 minutes

- Stain with fluorescein and refer to an ophthalmologist if there is any uptake of the stain

Oral Management

- NO GASTRIC LAVAGE OR EMETIC

- Encourage oral fluids

- Maintain a clear airway, give humidified oxygen and ventilate if necessary

- If respiratory irritation occurs assess respiratory function and if necessary perform chest X-rays to check for chemical pneumonitis

- Symptomatic and supportive care

Summary of Environmental Hazards

- MIBK is not likely to persist in the environment

- It is moderately soluble in water and volatilises slowly from soil and surface waters with a half-life of 15 to 33 hours[3,4]

- MIBK is readily biodegraded in fresh and salt water

- MIBK has low toxicity to aquatic organisms; there is no evidence that MIBK accumulates in the food chain

- Drinking Water Standards: no data available

- Soil Guidelines: no data available

- Air Quality Standards: no data available

REFERENCES

1. Clayton GD & Clayton FE (eds.) *Patty's Industrial Hygiene and Toxicology*, 4th edn. John Wiley & Sons, Inc., New York, 1994

2. Hathaway GJ, Proctor NH & Hughes JP. Proctor and Hughes' *Chemical Hazards of the Workplace*, 4th edn. Van Nostrand Reinhold, New York, 1996

3. International Programme on Chemical Safety. *Environmental Health Criteria 117: Methyl Isobutyl Ketone*. WHO, Geneva 1995

4. Hall AH & Rumack BH (eds.) *TOMES System®* Micromedex, Englewood, Colorado. CD ROM. vol.41 (exp. 31 July 1999)

Henrietta Wheeler

METHYL METHACRYLATE

Key Points

- Methyl methacrylate is a colourless, volatile liquid with an acrid, fruity odour
- It is irritating to the eyes, skin, respiratory tract and mucous membranes
- Methyl methacrylate is moderately toxic by inhalation and mildly toxic by ingestion
- Effects following inhalation include narcosis, excitation, anorexia and hypotension
- Dermal contact may cause contact dermatitis and sensitisation
- The vapours (gas) are heavier than air and spread at ground level, with a risk of ignition at a distance
- *In the event of a large spill, stay upwind and out of low areas. Ventilate closed spaces. Protective clothing, eye protection and breathing apparatus should be worn*

FIRST AID

- Terminate exposure and support vital functions
- The casualty should be moved to an uncontaminated area
- Rescuers should, ideally, be trained personnel and must be careful *not to put themselves at risk* and *so wear appropriate protective clothing and, if available, breathing apparatus*
- If the casualty is unconscious a clear airway should be established and maintained; give 100% oxygen if available
- **Dermal Exposure:** Remove contaminated clothing, if possible under a shower and place in double, sealed, clear bags and label; store the bags in a secure area away from patients and staff
- Wash the skin thoroughly with copious amounts of water
- **Eye Exposure:** Irrigate thoroughly with water or saline for 15 minutes
- **Oral Exposure:** Encourage small quantities of oral fluids (no more than 50–100 ml in total)

Detailed Information

- Methyl methacrylate is a colourless, volatile liquid with an acrid, fruity odour
- *Common synonyms* 2-methylpropenoic acid methyl ester, methacrylic acid methyl ester, methyl-2-methyl-2-propenoate, MMA
- CAS 80-62-6
- UN 1247
- NIOSH/RTECS OZ 5075000
- Molecular formula $C_5H_8O_2$
- Molecular weight 100.12
- Methyl methacrylate polymers or copolymers are widely used in the manufacture of plastics, coatings, dental restorations and surgical implants; methyl methacrylate monomers readily polymerise to form polymethyl methacrylate (hard plastic) which are used as sheets,

moulding, extrusion powders, surface coating resins, emulsions polymers, fibres, inks and films

- Commercial forms contain a small amount of hydroquinone or hydroquinone monomethyl ether to inhibit spontaneous polymerisation[1]
- Methyl methacrylate vapour (gas) is heavier than air and may spread along the ground
- Methyl methacrylate can be copolymerised with other methacrylate esters and many other monomers
- It is slightly soluble in water, soluble in most organic solvents
- Methyl methacrylate is able to polymerise when moderately heated or exposed to light or due to contaminants
- It reacts with many substances, violently with strong oxidants, with a risk of fire and explosion

Summary of Human Toxicity

- Methyl methacrylate is absorbed by inhalation, orally and dermally
- Toxic effects are due to the monomer; the polymer is inert and non-toxic
- The severity of the effects seems to be inversely proportional to the degree of polymerisation[1]
- Toxic effects on the nervous tissues may be due to diffusion into nerve cells causing lysis of the membrane lipids and destruction of the myelin sheath[1]
- Methyl methacrylate is metabolised to methacrylic acid which is a normal intermediate in the Krebs' cycle
- The lethal oral dose in laboratory animals 6–9 g.kg^{-1} [2]

Toxicity of Methyl Methacrylate [1,2,3]

Concentration	Clinical effects
170–250 ppm	Maximum tolerated dose; irritation observed
2,300 ppm	Intolerable
0.5–50 ppm	Irritability, fatigue, headache and mucous membrane irritation
150 mg.m^{-3}	CNS effects

- Occupational Exposure Standards:
  Long-term exposure limit: 50 ppm (208 mg.m^{-3})
  Short-term exposure limit: 100 ppm (416 mg.m^{-3})

Acute Clinical Effects

Inhalation Effects

- Mucous membrane irritation, nausea and anorexia may occur
- Headache, fatigue, dizziness, irritability or excitation have been reported

- Respiratory depression, pulmonary oedema, emphysema and atelectasis have occurred in animals following large doses[3]

- Systemic effects may occur; see below

Dermal Effects

- Contact dermatitis and eczematous reactions may occur

- Methyl methacrylate is a dermal sensitiser and may cause systemic effects; see below

Eye Effects

- Methyl methacrylate is moderately irritating to the eyes

- Lacrimation and corneal irritation may occur

Oral Effects

- Limited information available; irritation to the mucous membranes may occur

- Kidney lesions have been reported in humans and animals following ingestion, although there are no details available[3]

Systemic Effects

- Systemic vasodilation and transient hypotension have been reported[3]

- Acute hypertension, erythroderma and dyspnoea may occur[4]

- AV block has been reported following insertion of methyl methacrylate cement into the femoral cavity; cardiac arrest, fat and bone marrow embolism have been reported due to heat and pressure in the femoral canal while the polymer is curing[3]

- Transient, but dose related, elevation of serum gamma glutamyl transferase has been reported following a hip replacement using polymethyl methacrylate

Chronic Clinical Effects

Brief but repeated high concentrations of methyl methacrylate has been associated with the development of asthma.[5]

Sensitisation and allergic dermatitis has been reported; contact dermatitis is characterised by paraesthesia and tenderness that outlasts the eruptions. Prolonged or repeated contact may lead to numbness, coldness and pain; sensorimotor neuropathy has also been reported including loss of manual dexterity, impaired peripheral sensation, muscle wasting, hand and limb weakness and depression or absence of reflexes.[6]

Management

Methyl methacrylate concentrations are not of use; methanol concentrations in blood and urine may indicate exposure.[1]

Inhalation Management

- Maintain a clear airway, give humidified oxygen and ventilate if necessary

- If respiratory irritation occurs assess respiratory function and if necessary perform chest X-rays to check for chemical pneumonitis

- See systemic management below

Dermal Management

- Remove any remaining contaminated clothing, place in double, sealed, clear bags and label; store in a secure area away from patients and staff

- Irrigate with copious amounts of water

- An emollient may be required

- For prolonged or large exposure see systemic management below

Eye Management

- Irrigate thoroughly with running water or saline for 15 minutes

- Stain with fluorescein and refer to an ophthalmologist if there is any uptake of the stain

Oral Management

- NO GASTRIC LAVAGE OR EMETIC

- Encourage oral fluids

- See systemic management below

Systemic Management

- For symptomatic patients monitor cardiac, renal and liver functions

- Symptomatic and supportive care

Summary of Environmental Hazards

- In the atmosphere methyl methacrylate will photodegrade with a half-life of 2.7 hours in urban areas and >3 hours in rural areas[3]

- In water methyl methacrylate will principally be lost by volatilisation with a half-life of 6.3 hours in a typical river[3]

- In soil it will volatilise and leach into the groundwater where its fate is unknown

- It is not expected to bioconcentrate in fish

- Drinking Water Standards:
  hydrocarbon total: $10 \, \mu g \cdot l^{-1}$ (UK max)

- Soil Guidelines: no data available

- Air Quality Standards: no data available

REFERENCES

1. Hathaway GJ, Proctor NH & Hughes JP. *Proctor and Hughes' Chemical Hazards of the Workplace*, 4th edn. Van Nostrand Reinhold, New York, 1996

2. Gosselin RE, Smith RP & Hodge HC. *Clinical Toxicology of Commercial Products*, 5th edn. Williams & Wilkins, Baltimore, 1984

3. Hall AH & Rumack BH (eds.) *TOMES System®* Micromedex, Englewood, Colorado. CD ROM. vol.41 (exp. 31 July 1999)

4. Scolnick B & Collins J, 1986. Systemic reaction to methylmethacrylate in an operating room nurse. *J Occup Med*; 28:196–8

5. Pickering CAC, Bainbridge D, Birtwistle IH & Griffiths DL, 1986. Occupational asthma due to methyl methacrylate in an orthopaedic theatre sister. *Brit Med J*; 292: 1362–3

6. Ellenhorn MJ, Schonwalds S, Ordog G & Wasserberger J. *Ellenhorn's Medical Toxicology - Diagnosis and Treatment of Human Poisoning*, 2nd edn. Williams & Wilkins, London, 1997

Henrietta Wheeler

MUSTARD GAS

Key Points

- Mustard gas is a colourless, odourless, oily liquid
- Mustard gas is a vesicant chemical warfare agent; it is a powerful irritant and symptoms may be delayed for several hours
- Mustard gas (either vapour or liquid) causes damage to skin, eyes and respiratory system
- Following inhalation pulmonary damage may occur
- Keratitis can be delayed for years following ocular exposure
- Mustard gas is a human carcinogen
- Its use as a chemical warfare agent was prohibited by the Geneva Protocol in 1925
- *Mustard gas vapour is 5.5 times heavier than air and may accumulate at ground level[1]*
- *In the event of a large spill, stay upwind and out of low areas. Ventilate closed spaces. Protective clothing, eye protection and breathing apparatus should be worn*
- *Mustard gas vapours may evaporate from heavily contaminated casualties and so there is a risk that emergency personnel may become contaminated*
- *Casualties should be transported in such a way that there is no risk to the drivers of emergency vehicles becoming contaminated by the fumes*

FIRST AID

- *Rescuers must not enter a contaminated area without full personal protective equipment and self-contained breathing apparatus*
- Terminate exposure and support vital functions
- The casualty should be moved to an uncontaminated area
- If the casualty is unconscious a clear airway should be established and maintained; give 100% oxygen if available
- **Inhalation Exposure:** If the patient stops breathing, expired air resuscitation should be started immediately using a pocket mask with a one way valve, if available. It is important where the face is contaminated that expired air resuscitation is NOT attempted unless an airway with rescuer protection is used
- **Dermal Exposure:** Rapid decontamination is vital; remove contaminated clothing, if possible under a shower and place in double, sealed, clear bags and label; store the bags in a secure area away from patients and staff
- Skin exposed to liquid should initially be decontaminated with large amounts of Fuller's earth (or activated charcoal if Fuller's earth is not available)
- Wash the skin thoroughly with organic solvents such as paraffin for 30 minutes followed by copious amounts of soap and water
- **Eye Exposure:** Irrigate thoroughly with water or saline for 15 minutes
- **Oral Exposure:** Encourage small quantities of oral fluids (no more than 50–100 ml in total)

Detailed Information

- Mustard gas is a colourless, odourless, oily liquid; it is an organic sulphide which acts as a strong vesicant[2]
- *Common synonyms* bis(2-chloroethyl)sulphide, distilled mustard, H, HD, sulphur mustard gas, yperite
- CAS 505-60-2
- NIOSH/RTECS WQ 0900000
- Molecular formula $C_4H_8Cl_2S$
- Molecular weight 159.08
- Mustard gas was used as a vesicant in chemical warfare, initially as an offensive by the German Army in Ypres, Belgium in 1917; from 1917 to the end of World War I British casualties from mustard gas were at least 125,000 with about 1,859 deaths[3]
- 1945–1948 mustard gas was dumped in the Baltic Sea, fishermen in this area have been increasingly exposed; it was used by Italian forces against Ethiopian forces in 1936, by Japanese forces against Chinese troops during World War II and in the Iran/Iraq conflict in the 1980s[1]
- Mustard gas should only be handled under controlled conditions; it may penetrate clothing including leather[4]
- Rubber gloves and overalls should be worn and provide dermal protection for several hours
- Mustard gas is only slightly soluble in water but is soluble in fat solvents and other common organic solvents[1,5]
- Oxidising agents react with mustard gas to produce the corresponding sulphone and sulphoxide; the sulphone is produced by stronger oxidising agents, e.g. hypochlorite, which is itself a vesicant and may produce lacrimation and sneezing; the sulphoxide is not a vesicant[1]
- Increased temperature and pH increase the rate of hydrolysis in water
- Mustard gas is combustible when exposed to heat or flame and can be ignited by a large explosive charge[2]

Summary of Human Toxicity

- Toxic effect of mustard gas is primarily due to its alkylating ability
- Vesicant chemical warfare agents incapacitate far more people than they kill[3]
- 80% of liquid mustard gas placed on the skin evaporates; of the remainder, 10% is fixed to the skin and the remainder is absorbed systemically[1]
- Lowest published lethal concentration for humans by inhalation is 23 ppm for 10 min[4]
- Lowest published lethal dose for humans by the dermal route is 60 to 64 $mg.kg^{-1}$ for 1 hour contact time[4]

Estimated Concentrations and Clinical Effects of Mustard Gas[1]

Type of exposure	Concentration	Clinical effects
Vapour in eyes	50 mg.min.m^{-3}	Maximum safe exposure
	70 mg.min.m^{-3}	Mild reddening of the eyes
	100 mg.min.m^{-3}	Partial incapacitation due to eye effects
	200 mg.min.m^{-3}	Total incapacitation due to eye effects
Liquid on skin	50 μg.cm^{-2} for 5 min	Slight erythema
	250–500 μg.cm^{-2} for 5 min	Blistering
Vapour on skin	100–400 mg.min.m^{-3}	Erythema of skin
	200–1000 mg.min.m^{-3}	Blistering
	750–1000 mg.min.m^{-3}	Severe, incapacitating skin burns

- Occupational Exposure Standards: no data available

Acute Clinical Effects

Mustard gas does not usually cause pain at the time of exposure; symptoms may be delayed for 4 to 6 hours.[6] Keratitis can be delayed for years following ocular exposure. The higher the concentration the shorter the time for symptoms to develop.

Development of Clinical Effects for Mustard Gas[1,3]

Time after vapour exposure	Clinical effects
20–60 min	Nausea, retching, vomiting and eye smarting occasionally reported, sometimes no initial symptoms
1 h	First appearance of erythema
2–6 h	Nausea, fatigue, headache, inflammation of eyes with intense pain, lacrimation, blepharospasm, photophobia and rhinorrhoea; erythema of face and neck; sore throat, hoarse voice or total loss; tachycardia and increased respiration; definite erythema
8–12 h	Raised erythema (oedema)
13–22 h	Inflammation in areas where tight clothing was worn and inner thighs, genitalia, perineum, buttocks and axillae followed by blister formation which may be pendulous and filled with clear, yellow fluid; death within 24 hours is rare and extremely unlikely under civilian conditions
42–72 h	Maximum blisters or necrosis; coughing appears: muco pus and necrotic slough may be expectorated; intense itching of skin; increase in skin pigmentation
6–9 d	Possible complete skin surface denudation
20–28 d	Removal of scab
22–29 d	Usually complete skin healing

Inhalation Effects

- Coughing (which may be worse at night and become productive), wheezing, dyspnoea, paroxysmal coughing, and pulmonary oedema may be delayed for 1 to 12 hours

- Fever, headache, hoarseness or loss of voice may be delayed for 24 hours[3]

- Symptoms may persist for 1 or more years[3]

- Long-term effects include asthma, chronic bronchitis, emphysema and chronic laryngitis

Dermal Effects

- Mustard gas is primarily a vesicant with blisters being formed following either liquid or vapour contact

- Symptoms may be delayed for 4 to 6 hours

- Contaminated skin may become inflamed and blistered with severe pain (Note: the blister fluid is not dangerous to health carers)

- Oedema, hyperaemia and irritation

Eye Effects

- The eyes are the organs most sensitive to mustard gas although no clinical indication of injury becomes evident until a latent period of several hours

- The corneal epithelium may become oedematous; lids and conjunctiva become red and swollen

- Burning, discomfort, photophobia, lacrimation, blepharospasm

- Exposure to vapour induces extreme discomfort and temporary disablement but in most cases recovery is complete

- In more severe cases, injuries have involved not only the epithelium but also deeper layers; corneas may become cloudy and infiltrated, and in extreme cases eyes may become totally opaque

- Mustard gas may cause late recurrences of keratitis; patients with burns to the eyes which recovered may develop repeated corneal ulceration sometimes up to 40 years later[7]

- Long term effects include corneal opacities, chronic conjunctivitis and ocular keratitis

Oral Effects

- Ingestion of food or water contaminated with mustard gas may cause nausea and vomiting, pain, bloody diarrhoea and, in severe cases, dehydration

Systemic Effects

- Dizziness, generalised malaise, anorexia and lethargy can occur after acute exposure

- CNS excitation with convulsions may occur, followed by CNS depression; AV-block and cardiac arrythmias

- Bone marrow depression may occur

Chronic Clinical Effects

Chronic exposure has been associated with an increased risk of respiratory tract cancer (nasopharyngeal, laryngeal and lung), and skin cancer, especially in ammunition factory workers; also chronic bronchitis, pigmentation abnormalities, chronic skin ulceration and scar formation; bone marrow depression and sexual dysfunction due to scarring of the scrotum and penis.[3]

IARC has determined that mustard gas is a Class I human carcinogen and experimental teratogen.[4,8]

Management

NB: Medical staff must protect themselves. Full chemical warfare suits and respirators should be worn if highly contaminated patients require treatment. Dermal decontamination should be conducted outside, not in a casualty department.

Mustard gas or its metabolite, thiodiglycol, can be detected in urine up to a week after acute exposure.

Inhalation Management

* Maintain a clear airway, give humidified oxygen and ventilate if necessary

* TIME IS OF THE ESSENCE: a nebulisation mist of 2.5 % solution of sodium thiosulphate may have some value in neutralisation if exposure has occurred in the past 15 minutes[4]

* If respiratory tract irritation occurs assess respiratory function and if necessary perform chest X-rays to check for chemical pneumonitis; monitor arterial blood gases

* Treat pulmonary oedema with PEEP or CPAP ventilation

* Consider the use of steroids to reduce the inflammatory response

* Monitor WBC for severe exposures

Dermal Management

* Initially cover the area with Fuller's earth or activated charcoal; wash with paraffin for 30 minutes then copious amounts of soap and water

* For erythema and blisters treat with emollients

* Silver sulphadiazine 1% cream was used for casualties from the Iran/Iraq conflict and they benefited by reduced infection[1]

* Patients may develop a dermal hypersensitivity reaction which may require treatment with systemic or topical corticosteroids or antihistamines

* Pain will require analgesia

* Topical antiseptic solutions, and a regimen of oral vitamin E may be beneficial[4]

* Observation is advised for a minimum of 12 hours

* Treat blisters as burns which may require long healing periods

* Large full-thickness burns will not heal satisfactorily without grafting[1]

* Monitor WBC for severe exposures

Eye Management

* Irrigate thoroughly with running water or saline

* Immediate referral to ophthalmologist

* **For liquid contamination:** Attempts to irrigate eyes 5 minutes after liquid contamination is not likely to be of value and may increase the severity of the injury[1]

Oral Management

* NO GASTRIC LAVAGE OR EMETIC, activated charcoal may be of use

* Encourage oral fluids

* Give IV fluids if dehydrated; analgesics for pain

* Symptomatic and supportive care

Summary of Environmental Hazards

* WHO reports a persistence of 12 to 48 hours at 10 °C with rain and a moderate wind, 2 to 7 days at 15 °C with sun and a light breeze, and 2 to 8 weeks at -10 °C with sun, no wind, and a snow cover[4]

* Mustard gas may be persistent when placed on the ground in large quantities, if protected from wind and rain

* It may be very persistent in soil, especially when bulk quantities are involved. Soldiers have been burned while digging in areas where mustard gas has not been used for 10 years and it has persisted for several decades in land dumps

* At 0 °C, 25 °C, and 40 °C the estimated half-lives of mustard gas when dissolved in large amounts of water are 1.75 h, 4 min, and 43 s, respectively[4]

* Mustard gas in water, at very low concentrations so that it can dissolve, will rapidly hydrolyse to form mustard chlorohydrin (CH) and thiodiglycol (TDG)

* Hydrolysis in seawater will be a factor of 2.5 slower because of the common-ion effect exerted by the chloride ion[4]

* Drinking Water Standards: no data available

* Soil Guidelines: no data available

* Air Quality Standards: no data available

REFERENCES

1. Marrs TC, Maynard RL & Sidell FR. *Chemical Warfare Agents*. John Wiley & Sons, Chichester, 1996

2. Sax NI. *Dangerous Properties of Industrial Materials*. Van Nostrand Rheinhold Company, New York 1984

3. Ellenhorn MJ, Schonwalds S, Ordog G & Wasserberger J. *Ellenhorn's Medical Toxicology – Diagnosis and Treatment of Human Poisoning*, 2nd edn. Williams & Wilkins, London, 1997

4. Hall AH & Rumack BH (eds.) *TOMES System®* Micromedex, Englewood, Colorado. CD ROM. vol.41 (exp. 31 July 1999)

5. Budavari S, O'Neil MJ, Smith A, Heckelman PE & Kinneary JF (eds.) *The Merck Index*, 12th edn. Merck & Co., Inc., Whitehouse Station, 1996

6. Hathaway GJ, Proctor NH & Hughes JP. *Proctor and Hughes' Chemical Hazards of the Workplace*, 4th edn. Van Nostrand Reinhold, New York, 1996

7. Grant MW & Schuman JS. *Toxicology of the Eye*, 4th edn. Charles C Thomas, Springfield, 1993

8. Goldfrank LR, Flomenbaum NE, Lewin NA, Weisman RS, Howland MA & Hoffman RS. *Goldfrank's Toxicologic Emergencies*, 5th edn. Appleton & Lange, Norwalk, 1994

Henrietta Wheeler

NICKEL AND NICKEL COMPOUNDS

Key Points

- Elemental nickel is a silver-white metal, nickel salts are generally crystalline; nickel carbonyl is a colourless, volatile liquid
- Nickel and its inorganic compounds can be absorbed via the gastrointestinal and respiratory tracts; under certain circumstances nickel is also absorbed through the skin
- The most acutely toxic nickel compound is nickel carbonyl, with the acute toxic effects occurring in two phases:
  - **Phase 1:** headache, nausea, vomiting, vertigo and irritability
  - **Phase 2:** delayed pulmonary oedema, resembling viral pneumonia
  - There may be an asymptomatic interval between the phases
- Contact with nickel and nickel compounds may cause dermatitis and sensitisation
- Nickel carbonyl vapours are heavier than air and may spread along the ground or confined spaces with the risk of ignition
- *In the event of a large spill of nickel carbonyl, stay upwind and out of low areas. Ventilate closed spaces. Protective clothing, eye protection and breathing apparatus should be worn*

FIRST AID

- Exposure to most nickel compounds is a minor risk in terms of acute effects

Nickel carbonyl

- Terminate exposure and support vital functions
- The casualty should be moved to an uncontaminated area
- Rescuers should, ideally, be trained personnel and must be careful *not to put themselves at risk* and *so wear appropriate protective clothing and, if available, breathing apparatus*
- If the casualty is unconscious a clear airway should be established and maintained; give 100% oxygen if available
- **Inhalation Exposure:** If the patient stops breathing, expired air resuscitation should be started immediately using a pocket mask with a one way valve, if available. It is important where the face is contaminated that expired air resuscitation is NOT attempted unless an airway with rescuer protection is used
- **Dermal Exposure:** Remove contaminated clothing, if possible under a shower and place in double, sealed, clear bags and label; store the bags in a secure area away from patients and staff
- Wash the skin thoroughly with copious amounts of water
- **Eye Exposure:** Irrigate thoroughly with water or saline for 15 minutes
- **Oral Exposure:** Encourage small quantities of oral fluids (no more than 50–100 ml in total)

Detailed Information

- Elemental nickel is a silver-white metal, nickel salts are generally crystalline; nickel carbonyl is a colourless, volatile liquid
- CAS 7440-02-0
- UN specific number not available
- NIOSH/RTECS QR 5950000
- Atomic symbol Ni
- Atomic weight 58.69
- Nickel is insoluble in water

Identification Numbers of Nickel Compounds

Compound	CAS	UN	NIOSH/RTECS
Nickel acetate	373–02–4	n.a.	QR 6125000
Nickel bromide	13462–88–9	n.a.	n.a.
Nickel carbonate	3333–67–3	n.a.	QR 6200000
Nickel carbonyl	13463–39–3	1259	QR 6300000
Nickel chloride	7718–54–9	9139	QR 6475000
Nickel chloride hexahydrate	7791–20–0	n.a.	QR 6480000
Nickel fluoride	10028–18–9	n.a.	QR 6825000
Nickel hydroxide	12054–48–7	n.a.	QR 7040000
Nickel monoxide	1313–99–1	n.a.	n.a.
Nickel nitrate	13138–45–9	2725	QR 7200000
Nickel phosphate	10381–36–9	n.a.	n.a.
Nickel sulphate	7786–81–4	n.a.	QR 9400000
Nickel subsulphide	12035–72–2	n.a.	QR 9800000

n.a. = number not available

UN 1325 Nickel catalyst, spent (flammable solid, not otherwise specified)

UN 2881 Nickel catalyst, dry

Molecular Formula and Weight of Nickel Compounds

Compound	Molecular formula	Molecular weight
Nickel acetate	$C_4H_6O_4.Ni$	176.81
Nickel bromide	$NiBr_2$	218.50
Nickel carbonate	$CH_2O_3.Ni$	118.70
Nickel carbonyl	$Ni(CO)_4$	170.73
Nickel chloride	$NiCl_2$	129.60
Nickel chloride hexahydrate	$NiCl_2.6H_2O$	237.70
Nickel fluoride	NiF_2	96.69
Nickel hydroxide	H_2NiO_2	92.72
Nickel monoxide	NiO	74.69
Nickel nitrate	$N_2O_6.Ni$	182.73
Nickel phosphate	$Ni_3O_8P_2$	366.02
Nickel sulphate	$NiSO_4$	154.77
Nickel subsulphide	Ni_3S_2	240.19

Appearance and Solubility of Nickel Compounds[1]

Compound	Appearance	Solubility
Nickel acetate	Green crystalline mass or powder	Soluble in water and alcohol
Nickel bromide	Yellow-green deliquescent crystals	Soluble in water and alcohol
Nickel carbonate	Light green crystals	Insoluble in water; soluble in acids
Nickel carbonyl	Colourless, volatile liquid	Insoluble in water; soluble in organic solvents
Nickel chloride	Green, deliquescent crystals	Soluble in water
Nickel chloride hexahydrate	Green, monoclinic, deliquescent crystals	Soluble in water and alcohol
Nickel fluoride	Yellow-green, tetragonal crystals	Slightly soluble in water
Nickel hydroxide	Green powder	Insoluble in water; soluble in acids and ammonia
Nickel monoxide	Green or black powder	Insoluble in water; soluble in acid
Nickel nitrate	Green, deliquescent crystals	Soluble in water and alcohol
Nickel phosphate	Light green powder	Insoluble in water; soluble in acid
Nickel sulphate	α: blue-green tetragonal crystals β: green, monoclinic, transparent crystals	Soluble in water
Nickel subsulphide	Pale, yellowish-bronze metallic lustre	Insoluble in water; soluble in nitric acid

- Nickel compounds are used in ink production, ceramics, jewellery, electroplating, storage batteries; manufacturing coins, stainless steel, cooking utensils and electrical parts; corrosion-resistant alloys

- Nickel carbonyl is used as a catalyst in the petroleum, plastic and rubber industries

- Nickel carbonyl is flammable and explosive at 60°C; it reacts violently with oxidants with risk of fire or explosion

- Nickel carbonyl decomposes in contact with acids giving off carbon monoxide

- Nickel sulphate decomposes when heated giving of corrosive sulphur oxides

Summary of Human Toxicity

- Nickel and its inorganic compounds can be absorbed via the gastrointestinal and respiratory tracts; under certain circumstances nickel is also absorbed through the skin[1]

- The amount absorbed is determined by the physical and chemical characteristics of the nickel compound; solubility is an important factor in all routes of exposure

- Soluble salts of nickel dissociate readily in the aqueous environment of biological membranes, thus allowing their transport as metal ions[1]

- Nickel is distributed to the lungs, kidneys and erratically into the blood; it is also found in the brain, stomach and intestinal tissue

- Absorbed nickel is excreted mainly in urine; the half-life of nickel in serum is 11 hours; in the body the half-life is several days with little tissue accumulation

- 32 workers ingested water contaminated with nickel sulphate and nickel chloride ($1.63\,g.l^{-1}$ of nickel); 20 workers rapidly developed nausea, vomiting, abdominal discomfort, diarrhoea, headache, dizziness, lassitude, cough and shortness of breath which lasted a few hours but persisted 1–2 days in 7 cases. Nickel doses ingested in symptomatic individuals were estimated to range from 0.5–2.5 g with blood concentrations ranging from $13–1,340\,\mu g.l^{-1}$ and urine concentrations ranging from $0.15–12\,mg.g^{-1}$ creatine on day 1 [2]

- Ingestion of 15g of nickel sulphate was fatal in a child ($220\,mg.kg^{-1}$) [3]

- Nickel compounds may cause contact dermatitis and sensitisation, presenting with 'nickel itch'

- Nickel contact dermatitis is the most common reaction to nickel; approximately 5% of all eczemas are nickel reactions[4]

- Ingestion of nickel preparations or swallowing nickel coins has caused or exacerbated dermal reactions

- Once acquired, nickel sensitivity usually persists and may aggravate atopic dermatitis[5]

Nickel carbonyl

- Inhalation exposure to nickel carbonyl is acutely more dangerous than exposure to nickel metals or nickel salts

- Nickel carbonyl can pass across the alveolar membrane in either direction without alteration[4]

- Once absorbed, nickel carbonyl is slowly broken down to carbon monoxide and nickel

- Inhalation of nickel carbonyl for several hours resulted in headache, shortness of breath, chest pain and weakness at 24 hours; tachypnoea, tight chest and paraesthesia subsequently developed. Chest X-ray showed ARDS. Following treatment with dilsulfiram and dithiocarb, respiratory impairment resolved over 3 months[4]

- Treatment should continue until urine concentration is below $100\,\mu g.l^{-1}$ [4]

- Cigarette tobacco may contain approximately $1.3–4.0\,mg.kg^{-1}$ of nickel; about $0.04–0.58\,\mu g$ is released with the smoke of one cigarette.[1] There is no agreement on the chemical nature of nickel in tobacco smoke and its health significance[4]

- Occupational Exposure Standards (organic compounds):
  Long-term exposure limit: $1\,mg.m^{-3}$ (as Ni)
  Short-term exposure limit: $3\,mg.m^{-3}$

Acute Clinical Effects

Inhalation Effects

- Acute toxicity is a minor risk from nickel or its compounds; sore throat and hoarseness may be observed, with the exception of nickel carbonyl where systemic effects are likely; see below

Dermal Effects

- Nickel contact dermatitis is characterised by a burning sensation and pruritis followed by erythema and nodular eruptions

- Nodules may ulcerate which may discharge and become crusted or eczematous

- Eruptions can spread to areas related to primary activity site, such as elbow flexure, eyelids or face[6]

- Pigmentation and depigmentation may occur

- Asthma has been reported following dermal exposure to nickel sulphide

Eye Effects

- Nickel metal foreign bodies in the eye may cause mechanical injury

Oral Effects

- Large exposure may lead to nausea, vomiting, abdominal pain and diarrhoea

- Ingestion of nickel carbonyl may result in cough, shortness of breath and dizziness

- Renal changes including vacuolisation of the proximal convoluted tubules has been reported[4]

- Systemic effects may occur; see below

Systemic Effects

Nickel carbonyl

- **Phase 1:** nausea, vomiting, non-productive cough, headache, vertigo, weakness, dizziness, chills and sweating

- Anxiety, restlessness, insomnia, numbness have been reported

- **Phase 2:** delayed effects of tightness of chest, cough, tachypnoea, dyspnoea, abdominal pain and blurred vision may occur

- Tachycardia or bradycardia, sinus irregularities with or without heart block, premature ventricular complexes and toxic myocarditis with changes in S–T and T-waves, and QT prolongation waves have been reported[4]

- Convulsions, ARDS, cerebral oedema, pulmonary oedema and interstitial fibrosis may be delayed[7]

- Weakness and somnolence can persist for 3–6 months post exposure

- Between phases there may be an asymptomatic interval

Chronic Clinical Effects

Chronic exposure to nickel dust may cause eczematous dermatitis, asthma and Loefflar's syndrome (pulmonary eosinophilia).[4] Nasal irritation, damage of nasal mucosa, perforation of the nasal septum and loss of smell have occasionally been reported following chronic exposure to nickel aerosols and other contaminants.[6]

Pneumoconiosis has been reported following occupational exposure to nickel dust, although exposure to other known fibrogenic substances could not be excluded.[6] Chronic exposure may lead to nasal carcinomas, nasal epithelial dysplasia, lung cancer and potentially cancer of the larynx. IARC classifies nickel compounds as human carcinogens and metallic nickel as a possible carcinogen.

Chronic exposure to nickel carbonyl may lead to decreased serum monoamineoxidase and EEG abnormalities.

Management

Blood and urine concentrations may be determined (normal range for blood is 2.6–4.6 $\mu g\,l^{-1}$; urine 0.5–6.0 μg / 24 hours).

Inhalation Management

- Maintain a clear airway, give humidified oxygen and ventilate if necessary

- If respiratory irritation occurs assess respiratory function and if necessary perform chest X-rays to check for chemical pneumonitis

- Consider the use of steroids to reduce the inflammatory response

- Treat pulmonary oedema with PEEP or CPAP ventilation

- **Note:** Phase 1 of nickel carbonyl exposure may lack severe symptoms, all exposures should be monitored closely

- Monitor ECG and renal function

- Symptomatic and supportive care; for chelation therapy see below

Dermal Management

- Remove any remaining contaminated clothing, place in double, sealed, clear bags and label; store in a secure area away from patients and staff

- Irrigate with copious amounts of water

- Dermatitis should be treated symptomatically, referral to a dermatologist may be necessary

- Various creams have been used for the treatment of nickel-sensitised cases with varying effects; disulfiram has been shown to be effective in clearing nickel dermatitis; 50–100 mg.d^{-1} may be of use[4]

- Recovery usually occurs after a week but may be delayed for several weeks

Eye Management

- Irrigate thoroughly with running water or saline for 15 minutes

- Stain with fluorescein and refer to an ophthalmologist if there is any uptake of the stain

- Surgical removal of a foreign body if appropriate

Oral Management

- Encourage oral fluids

- Consider gastric lavage followed by 50 g activated charcoal within 1 hour of a substantial ingestion

- Contact a poisons information service for further guidance on gut decontamination

- Monitor ECG and renal function

- Symptomatic and supportive care; see chelation therapy below

Chelation Therapy

Contact a poisons information service for further guidance and paediatric doses

- **DITHIOCARB**

  - **Mild exposure** (urine nickel concentration $<10\,\mu g.dl^{-1}$): 2 g orally in divided doses

  - **Moderate exposure** (urine nickel concentration $>10\,\mu g.dl^{-1}$):

  - **Day 1:** 2.0 g at 0 hours; 1.0 g at 4 hours; 0.6 g at 8 hours; 0.4 g at 16 hours

  - **Subsequent days:** 0.4 g every 8 hours

- Therapy should be continued until the urine nickel concentration is $< 10\,\mu g.dl^{-1}$

- Dithiocarb therapy may cause a metallic taste and abdominal discomfort

- **DISULFIRAM**

- *Only where dithiocarb is not available:* disulfiram, which is metabolised to two dithiocarb molecules, is hypothetically of value; a suitable dose, however, has not been established

- Alcohol should be avoided because of the disulfiram effect

Summary of Environmental Hazards

- Transport and distribution of nickel particulates between different environmental compartments is strongly influenced by particle size; fine particulate matter has a longer residence time in the atmosphere and is carried a long distance from source; larger particles are deposited near emission sources[1]

- Atmospheric residence time for nickel particulates is estimated to be 5.4 to 7.9 days

- Water solubility and bioavailability is affected by soil pH; decreases in pH generally mobilises nickel, thus acid rain mobilises nickel from soil and increases nickel concentrations in groundwater

- Nickel bioaccumulates in the food chain but is not bioconcentrated

- Drinking Water Standards:
  Nickel: $50\,\mu g.l^{-1}$ (UK max)
  $20\,\mu g.l^{-1}$ (WHO guideline)

  Chloride: $400\,mg.l^{-1}$ (UK max)
  $250\,mg.l^{-1}$ (WHO guideline)

  Sulphate: $250\,mg.l^{-1}$ (UK max)

- Soil Guidelines:
  Dutch Criteria: $35\,mg.kg^{-1}$ (target)
  $210\,mg.kg^{-1}$ (intervention)

- Air Quality Standards: no safe level recommended because of carcinogenic properties (WHO guideline)

REFERENCES

1. International Programme on Chemical Safety. *Environmental Health Criteria 108: Nickel.* WHO, Geneva, 1991

2. Sunderman FW, Dingle B, Hopfer SM & Swift T, 1988. Acute nickel toxicity in electroplating workers who accidentally ingested a solution of nickel sulfate and nickel chloride. *Am J Ind Med*; 14(3):257–66

3. Ellenhorn MJ, Schonwalds S, Ordog G & Wasserberger J. *Ellenhorn's Medical Toxicology – Diagnosis and Treatment of Human Poisoning*, 2nd edn. Williams & Wilkins, London, 1997

4. Hall AH & Rumack BH (eds.) *TOMES System®* Micromedex, Englewood, Colorado. CD ROM. vol.41 (exp. 31 July 1999)

5. Veien NK, 1994. Nickel sensitivity and occupational skin disease. *Occ Med*; 9(1):81–95

6. Hathaway GJ, Proctor NH & Hughes JP. *Proctor and Hughes' Chemical Hazards of the Workplace*, 4th edn. Van Nostrand Reinhold, New York, 1996

7. Jones CC, 1973. Nickel carbonyl poisoning. Report of a fatal case. *Arch Environ Health*; 26(5):245–8

Henrietta Wheeler

NITRATES

Key Points

- Nitrate and nitrite are naturally occurring ions that are part of the nitrogen cycle
- Nitrates are absorbed by ingestion and inhalation
- Nitrate is mostly excreted unchanged in the urine; some is converted *in vivo* to nitrites
- Nitrites induce methaemoglobinaemia which may result in cyanosis which is not responsive to oxygen therapy
- *In the event of a large spill, stay upwind and out of low areas. Ventilate closed spaces. Protective clothing, eye protection and breathing apparatus should be worn*

FIRST AID

- Terminate exposure and support vital functions
- The casualty should be moved to an uncontaminated area
- Rescuers should, ideally, be trained personnel and must be careful *not to put themselves at risk* and *so wear appropriate protective clothing and, if available, breathing apparatus*
- If the casualty is unconscious a clear airway should be established and maintained; give 100% oxygen if available
- **Inhalation exposure:** If the patient stops breathing, expired air resuscitation should be started immediately using a pocket mask with a one way valve, if available. It is important where the face is contaminated that expired air resuscitation is NOT attempted unless an airway with rescuer protection is used
- **Dermal exposure:** Remove contaminated clothing, if possible under a shower, and place in double, sealed, clear bags and label; store the bags in a secure area away from patients and staff
- Wash the skin thoroughly with copious amounts of water
- **Eye exposure:** Irrigate thoroughly with water or saline for 15 minutes
- **Oral exposure:** Encourage small quantities of oral fluids (no more than 50–100 ml in total)

Detailed Information

Nitrates and Common Synonyms

Compound	Synonyms
Ammonium nitrate	Nitrate of ammonia, ammonium saltpeter, German saltpeter
Potassium nitrate	Saltpeter, niter
Sodium nitrate	Chile saltpeter, cubic niter, soda niter

Identification Numbers of Common Nitrates

Compound	CAS	UN	NIOSH/RTECS
Ammonium nitrate	6484-52-2	1942	BR 9050000
Potassium nitrate	7757-79-1	1486	TT 3700000
Sodium nitrate	7631-99-4	1498	WC 5600000

Molecular Formula and Weight of Common Nitrates

Compound	Molecular formula	Molecular weight
Ammonium nitrate	NH_4NO_3	80.04
Potassium nitrate	KNO_3	101.10
Sodium nitrate	$NaNO_3$	84.99

- Nitrates are used in fertilisers, pharmaceuticals and in the manufacture of explosives
- Nitrates are used in the building, printing and chemical industries

Summary of Human Toxicity

- Nitrates are well absorbed from the gastro-intestinal tract; they are generally not absorbed in toxic amounts across intact skin; absorption occurs across skin burns[1]
- Absorbed nitrate is excreted mostly unchanged in the urine; only some is reduced to nitrite[2]
- The extent of conversion of nitrate to nitrite under various conditions is not known
- Nitrates are converted to nitrites by bacteria in the gastro-intestinal tract; toxicity is due to the nitrites produced[1,3]
- Nitrites induce methaemoglobinaemia; cyanosis may result which is not responsive to oxygen therapy[4,5,6]
- Nitrate-induced methaemoglobinaemia is delayed in onset and prolonged due to ongoing conversion to nitrites
- The minimal toxic dose is extremely variable; assessment of severity of toxicity should be based on clinical condition of patient
- Infants are more susceptible to nitrate-induced methaemoglobinaemia[6]
- Chronic daily ingestion of more than 5 mg . kg^{-1} of nitrate is considered unacceptable[1]
- Occupational Exposure Standards: no data available

Acute Clinical Effects

Inhalation Effects

- Irritation of the mucous membranes of the nose, headache and nausea
- Systemic effects may develop; see below

Dermal Effects

- No data available

Eye Effects

- Irritation ranging from mild to severe depending on the nitrate salt involved and the physical form (solution, crystals etc.)

Oral Effects

- Irritation of the mucous membranes causing nausea, vomiting, diarrhoea, gastritis and abdominal pain

- Gastrointestinal haemorrhage has been reported

- Systemic effects may develop; see below

Systemic Effects

- Once the nitrate has been converted *in vivo* to nitrite, dizziness, fatigue, shortness of breath, syncope, tachycardia or bradycardia, and throbbing headache may develop[6]

- Warm flushed skin which later becomes cold and cyanotic[6]

- ECG changes, including atrial fibrillation, cardiac ischaemia, frequent or occasional ventricular premature beats and bigeminy have been reported

- Vasodilatation may cause hypotension, decreased peripheral resistance, cardiovascular collapse, convulsions, and coma in severe cases[1]

- Methaemoglobinaemia may develop, which may cause cyanosis and dyspnoea[6]

Chronic Clinical Effects

Some tolerance to headaches develops following chronic exposure to nitrates; this tolerance is usually lost after a few days without exposure. An increase in the incidence of stomach cancer has been reported following chronic occupational exposure to nitrate fertilisers.[1]

Management

Plasma nitrate concentrations are not clinically useful. Urine nitrate concentration depends on the conversion of nitrate to nitrite by bacterial action; there should be a minimum of four hours between ingestion and testing. Methaemoglobin concentrations are the best indicator of toxicity.

Inhalation Management

- Maintain a clear airway and give 100% oxygen

- Determine the blood methaemoglobin concentration

- For further treatment, see systemic management below

Dermal Management

- Remove any remaining contaminated clothing, place in double, sealed, clear bags and label; store in a secure area away from patients and staff

- Irrigate with copious amounts of water

Eye Management

- Irrigate thoroughly with running water or saline for 15 minutes

- Stain with fluorescein and refer to an ophthalmologist if there is any uptake of the stain

Oral Management

- Consider gastric lavage followed by 50 g activated charcoal within 1 hour of ingestion

- Contact a poisons information service for further guidance on gut decontamination

- Encourage oral fluids

- Give 100% oxygen

- Determine the blood methaemoglobin concentration

- For further treatment, see systemic management below

Systemic Management

- Give 100% oxygen

- Determine the blood methaemoglobin concentration

- If the patient is very drowsy, unconscious or the methaemoglobin concentration exceeds 30%, methylene blue should be given at the dosage below

- If methaemoglobinaemia is suspected but levels can not be measured, methylene blue (1% solution, i.e. 10mg.ml^{-1}) 1–2 mg.kg^{-1} should be given IV over 5 minutes and repeated in 1 hour if there is no response

- Suspect methaemoglobinaemia if the arterial blood is chocolate brown in appearance and remains dark on aeration, if cyanosis is unresponsive to oxygen therapy, or pO$_2$ is normal in the presence of a decreased oxygen saturation

- For methaemoglobinaemia unresponsive to methylene blue, exchange transfusion and/or haemodialysis could be considered[4,6]

- Give plasma expanders/blood or IV fluids for shock

- Symptomatic and supportive care

Summary of Environmental Hazards

- Water soluble nitrates are persistent in water

- Liquid nitrates can infiltrate the soil and migrate downward towards the groundwater system; the rate depends on the soil type and water content

- Nitrate degradation is fastest in anaerobic conditions

- Drinking water standards:
  Nitrate (as NO$_3^-$): 50 mg.l^{-1} (UK max)
  50 mg.l^{-1} (WHO guideline)

  Nitrite (as NO$_2^-$): 0.1 mg.l^{-1} (UK max)
  3 mg.l^{-1} (WHO provisional guideline)

- Soil Guidelines: no data available

- Air Quality Standards: no data available

REFERENCES

1. Hall AH & Rumack BH (eds.) *TOMES System®* Micromedex, Englewood, Colorado. CD ROM. vol.41 (exp. 31 July 1999)

2. Stringer DA. *Technical Report No. 27: Nitrate and Drinking Water*. European Chemical Industry Ecology and Toxicology Centre, Belgium, 1988

3. Baselt RC & Cravey RH. *Disposition of Toxic Drugs and Chemicals in Man*, 4th edn. Chemical Toxicology Institute, 1995

4. Walley T, Flanagan M, 1987. Nitrite-induced methaemoglobinaemia. *Clin Toxicol*; 63:643–4

5. Kempster PL, 1981. Nitrite, Iron Deficiency Anaemia and Methaemoglobinemia. *SA Water*; 7(1):61

6. Kaplan A, Smith C, Promnitz DA, Joffe BI, Seftel HC, 1990. Methaemoglobinaemia due to accidental sodium nitrite poisoning. Report of 10 cases. *SAMJ*; 77:300–1

Catherine Farrow

NITRIC ACID

Key Points

- Nitric acid is a colourless to yellow, corrosive liquid
- Exposure may be fatal following inhalation, ingestion and dermal exposure
- Nitric acid causes corrosion or irritation to the lungs, eyes, skin, mucous membranes, oesophagus or any tissue in direct contact
- It can cause severe burns and ulceration or necrosis, and immediate corneal opacification
- Pulmonary oedema may result from acute inhalation
- *In the event of a large spill, stay upwind and out of low areas. Ventilate closed spaces. Protective clothing and breathing apparatus should be worn*

First Aid

- Terminate exposure and support vital functions
- The casualty should be moved to an uncontaminated area
- Rescuers should, ideally, be trained personnel and must be careful *not to put themselves at risk* and *so wear appropriate protective clothing and, if available, breathing apparatus*
- If the casualty is unconscious a clear airway should be established and maintained; give 100% oxygen if available
- **Inhalation Exposure:** If the patient stops breathing, expired air resuscitation should be started immediately using a pocket mask with a one way valve, if available. It is important where the face is contaminated that expired air resuscitation is NOT attempted unless an airway with rescuer protection is used
- **Dermal Exposure:** Remove contaminated clothing, if possible under a shower and place in double, sealed, clear bags and label; store the bags in a secure area away from patients and staff
- Wash the skin thoroughly with copious amounts of water
- **Eye Exposure:** Irrigate thoroughly with water or saline for 15 minutes
- **Oral Exposure:** Encourage small quantities of oral fluids (no more than 50–100 ml in total) unless perforation is suspected

Detailed Information

- Nitric acid is a colourless to yellow or browny-red, extremely corrosive, non-flammable, fuming liquid with a pungent odour
- *Common synonyms* aqua fortis, engraver's acid, hydrogen nitrate, nitric acid (red fuming), nitric acid (white fuming), nitrous fumes
- CAS 7697-37-2
- UN 2031 (other than fuming, with more than 40% acid)
- UN 2032 (fuming, red fuming)
- NIOSH/RTECS QU 5775000
- Molecular formula HNO_3
- Molecular weight 63.01

- Nitric acid is used in the manufacture of ammonium nitrate for fertilisers and explosives, organic synthesis (dyes, drugs, explosives, cellulose nitrate, nitrate salts), metallurgy, photoengraving, etching steel, ore flotation, rubber chemicals and reprocessing spent nuclear fuel
- Nitric acid is a solution of nitrogen dioxide in water and is commercially available in several forms
- Nitric acid is heavier than air with a characteristic choking odour and is extremely corrosive
- Nitric acid is highly soluble in water and may form nitrate at neutral pH; decomposes in alcohol
- Strong oxidant which reacts violently with combustible substances and reducing agents; reacts violently with bases, dispersed metals and many organic compounds to produce nitrous vapours
- Decomposes when heated or exposed to light, giving off nitrous vapours

Summary of Human Toxicity

- Nitric acid may be fatal following inhalation, ingestion and dermal exposure
- It exerts its effects by virtue of its strong acidity
- Nitric acid can denature proteins and alter the acid–base balance in localised regions
- If nitric acid has been in contact with organic materials it is likely to release nitric oxide; methaemoglobin may be formed as a result of exposure to nitric oxide
- Extensive injury to the jejunum occurred after ingestion of 28–56 ml of nitric acid
- A patient ingested 'half a cupful' of unknown concentration and developed burns to the lips and mouth, perforated stomach and oesophageo-tracheal fistula and respiratory distress; death occurred after 19 days[1]
- Occupational Exposure Standards:
    Long-term exposure limit: 2 ppm (5.2 mg . m^{-3})
    Short-term exposure limit: 4 ppm (10 mg . m^{-3})

Acute Clinical Effects

Inhalation Effects

- Nitric acid is a severe respiratory tract irritant
- Initially there may be a dry mouth with sore throat and eyes, a cough, tight chest, headache, ataxia, and confusion
- Dyspnoea may develop between 3 to 30 hours post exposure
- Hypoxia, cyanosis and methaemoglobinaemia may potentially occur
- Increased concentrations may cause laryngeal spasm, pneumonitis and pulmonary oedema which may be fatal

Dermal Effects

- Mild exposures to a dilute form may cause irritation and erythema to the skin

- The liquid or concentrated vapours may cause immediate, severe and penetrating burn

- Concentrated solutions on the skin may lead to deep ulcers and stain the skin bright yellow or yellowy-brown[2]

Eye Effects

- Nitric acid vapours and fumes may cause irritation and conjunctivitis and even necrosis of the conjunctiva at low concentrations

- Liquid nitric acid causes severe pain, corneal ulcers, corneal clouding, or severe burns of the corneal epithelium if splashed in the eye

- Reduced visual fields or blindness may occur from direct contact with the eye

- Perforation of the globe and loss of ocular contents may occur

- Permanent damage and visual impairment can occur from nitric acid splashes in the eye[2]

Oral Effects

- Small exposures lead to local mucosal irritation, epigastric pain, nausea and vomiting, which may be mucoid and coffee ground in nature[3]

- Larger exposures may lead to oesophageal corrosion, stricture, necrosis and perforation of the stomach, especially the pylorus and occasionally the small bowel

- Severe metabolic acidosis and shock may occur

- Extensive injury to the jejunum occurred after ingestion of 28–56 ml of nitric acid[3]

- Pyloric stenosis often occurs after several weeks and cancer may develop years later in patients who survive severe acute oral exposures[4]

Chronic Clinical Effects

Prolonged low concentrations of nitric acid may lead to chronic bronchitis, pulmonary fibrosis and chemical pneumonitis. Conjunctivitis and overt symptoms may resemble acute viral respiratory tract infections. Discoloration and erosion of dental enamel may occur.

Management

Inhalation Management

- Maintain a clear airway, give humidified oxygen and ventilate if necessary

- If respiratory irritation occurs assess respiratory function and if necessary perform chest X-rays to check for chemical pneumonitis

- Consider the use of steroids to reduce the inflammatory response

- Treat pulmonary oedema with PEEP or CPAP ventilation

- Symptomatic and supportive care

- If cyanosis develops or methaemoglobinaemia is suspected, determine methaemoglobin concentration

- See Nitrates (page 181) for management of methaemoglobinaemia

Dermal Management

- Remove any remaining contaminated clothing, place in double, sealed, clear bags and label; store in a secure area away from patients and staff

- Irrigate with copious amounts of water

- Treat burns symptomatically

Eye Management

- Irrigate thoroughly with running water or saline for 15 minutes

- Stain with fluorescein and refer to an ophthalmologist if there is any uptake of the stain

Oral Management

- NO GASTRIC LAVAGE OR EMETIC

- Encourage oral fluids, unless perforation is suspected

- Give plasma expanders/blood or IV fluids for shock and analgesics for pain

- Consider the use of steroids to reduce the inflammatory response

- Take abdominal X-ray to check for perforation

- If facilities are available, early gastro-oesophagoscopy should be undertaken within 12 hours of the event to assess the extent and severity of the injury

- Symptomatic and supportive care

- Long term follow up is necessary following severe oral exposure due to sequelae

Summary of Environmental Hazards

- Nitric acid is formed in photochemical smog from the reaction between nitric oxide and hydrocarbons; individuals living in heavily polluted areas may receive chronic inhalation exposures to nitric acid

- Nitric acid will be gradually neutralised by hardness minerals (calcium and magnesium) in water; the nitrate ion may persist longer but will ultimately be consumed as a plant nutrient[4]

- Elevated nitrate levels will stimulate plankton and aquatic weed growth

- During transport through the soil, nitric acid will dissolve some of the soil material, in particular the carbonate based materials; the acid will be neutralised to some degree with adsorption of the proton also occurring on clay materials although significant amounts of acid are expected to remain for transport down toward the ground water table

- Drinking Water Standards: no data available

- Soil Guidelines: no data available

- Air Quality Standards: no data available

REFERENCES

1. *National Poisons Information Service (London).* Internal Case; 77/17494

2. Hathaway GJ, Proctor NH & Hughes JP. *Proctor and Hughes' Chemical Hazards of the Workplace,* 4th edn. Van Nostrand Reinhold, New York, 1996

3. Adams JT & Skucas J, 1980. Corrosive jejunitis due to ingestion of nitric acid. *Am J Surgery;* 139(2): 282–5

4. Hall AH & Rumack BH (eds.) *TOMES System®* Micromedex, Englewood, Colorado. CD ROM. vol.41 (exp. 31 July 1999)

Henrietta Wheeler

NITROGEN

Key Points

- Nitrogen is a colourless gas; liquid nitrogen is a colourless to faint yellow liquefied gas with no odour or taste
- Liquid nitrogen causes cryogenic effects and may be fatal following inhalation, ingestion and dermal exposure
- Contact with liquid may cause frostbite and damage to the lungs, eyes, skin, mucous membranes, oesophagus or any tissue with which it is directly in contact
- Nitrogen may cause dizziness or suffocation
- Fire may produce irritating or poisonous gases
- *In the event of a large spill, stay upwind and out of low areas. Ventilate closed spaces. Protective clothing, eye protection and breathing apparatus should be worn*

FIRST AID

- Terminate exposure and support vital functions
- The casualty should be moved to an uncontaminated area
- Rescuers should, ideally, be trained personnel and must be careful *not to put themselves at risk* and *so wear appropriate protective clothing and, if available, breathing apparatus*
- If the casualty is unconscious a clear airway should be established and maintained; give 100% oxygen if available
- **Inhalation Exposure:** If the patient stops breathing, expired air resuscitation should be started immediately using a pocket mask with a one way valve, if available. It is important where the face is contaminated that expired air resuscitation is NOT attempted unless an airway with rescuer protection is used
- **Dermal Exposure:** If frostbite occurs *do not* remove clothing, flush skin with water
- If frostbite is not present remove contaminated clothing, if possible under a shower and place in double, sealed, clear bags and label; store the bags in a secure area away from patients and staff
- Wash the skin thoroughly with copious amounts of water
- **Eye Exposure:** If the eye tissue is frozen seek medical advice as soon as possible
- If eye tissue is not frozen, irrigate thoroughly with water or saline for 15 minutes
- **Oral Exposure:** Encourage small quantities of oral fluids (no more than 50–100 ml in total)

Detailed Information

- Nitrogen is a colourless gas; liquid nitrogen is a colourless to faint yellow liquefied gas with no odour or taste, stored under compression in metal containers
- *Common synonyms* liquid nitrogen: nitrogen compress, nitrogen refrigerated liquid
- CAS 7727-37-9
- UN 1066 compressed nitrogen
- UN 1977 refrigerated liquid nitrogen
- UN 1981 rare gases and nitrogen mixtures

- NIOSH/RTECS QW 9700000
- Molecular formula N_2
- Molecular weight 28.02
- Liquid nitrogen is used in medicine and biology for quick freezing of tissues and micro-organisms, in cryosurgery, to euthanase animals and chilling metals to alter physical state
- Nitrogen exists in a liquid phase between -209.86°C (melting point) and -195°C (boiling point)
- Vapours from liquid nitrogen are heavier than air and may spread along the ground
- Nitrogen is slightly soluble in water; soluble in alcohol and liquid ammonia
- Can react violently with lithium, neodynium and titanium
- Liquid nitrogen does not react with common materials, but the low temperature may cause rubber and plastics to become brittle
- Nitrogen gas is non-flammable and does not support combustion

Summary of Human Toxicity

- Inhalation of liquid nitrogen may cause injury to the pharynx; dermal exposure may lead to frostbite
- Within a confined space or following a very high dose, nitrogen is a simple asphyxiant
- Signs of asphyxia are noted if evolved nitrogen gas displaces atmospheric oxygen such that the oxygen concentration is <15%; unconsciousness occurs when oxygen is <6–8% [1]
- The release of nitrogen from solution in the blood may cause formation of small bubbles potentially leading to gas embolism formation [2]
- Cardiac arrest, neuropathies, syncope, and a variety of dermal injuries may occur following topical application
- Occupation Exposure Standards: simple asphyxiant (i.e. when in high concentrations in air, reducing oxygen content by dilution to such an extent that life can not be supported. Oxygen content of air should be monitored and not allowed to fall below 18% to ensure safety)

Acute Clinical Effects

Inhalation Effects

- Inhalation of liquid nitrogen may lead to acute burns to lips and oropharynx with signs and symptoms of acute upper airway distress
- Large mucosal ulcers may develop in the posterior hypopharynx and hard palate [3,4]
- Nitrogen gas is a simple asphyxiant; high concentrations may lead to dizziness, respiratory distress, coma and death
- Anoxic symptoms may also include decreased visual acuity, night vision, or visual fields (tunnel vision)

- Increased respiratory rate, air hunger and tachycardia may occur

- Decrease in performance and alertness, fatigue, dizziness, headache, belligerence, euphoria, numbness and tingling in the extremities, sleepiness, mental confusion, poor judgement, loss of memory, cyanosis, unconsciousness, and death may occur

Dermal Effects

- Liquid nitrogen may cause immediate, severe and penetrating burns

- Mild exposures may cause irritation, dermatitis and erythema

- Dermal injury may include bullae, oedema and necrosis

- Cardiac arrest has occurred following topical application for cryotherapy[4]

- Escaping compressed gas may cause frostbite

Eye Effects

- Exposure may lead to frostbite of the eye

- Redness, pain and blisters may occur

Oral Effects

- Not applicable

Management

Inhalation Management

- Maintain a clear airway, give humidified oxygen and ventilate if necessary

- If respiratory irritation occurs assess respiratory function and if necessary perform chest X-rays to check for chemical pneumonitis

- Consider the use of steroids to reduce the inflammatory response

- Treat pulmonary oedema with PEEP or CPAP ventilation

- Symptomatic and supportive care

Dermal Management

- *If frostbite has occurred*: remove clothing carefully, these may need to be soaked off with tepid water; irrigate the area

- Surgical referral may be necessary

- *If frostbite has not occurred*: remove any remaining contaminated clothing

- Irrigate with copious amounts of water

- Treat burns symptomatically

- Place any contaminated clothes in double, sealed, clear bags and label; store in a secure area away from patients and staff

Eye Management

- Irrigate thoroughly with running water or saline for 15 minutes

- Stain with fluorescein and refer to an ophthalmologist if there is any uptake of the stain

Oral Management

- not applicable

Summary of Environmental Hazards

- See nitrogen oxide page 189

- Drinking Water Standards: no data available

- Soil Guidelines: no data available

- Air Quality Standards: no data available

REFERENCES

1. Kizer KW, 1984. Toxic inhalation. *Emerg Med Clin North Am*; 2: 649–66

2. Sax NI. *Dangerous Properties of Industrial Materials*, 6th edn. Van Nostrand Reinhold, New York, 1984

3. Rockswold G & Buran BJ, 1982. Inhalation of liquid nitrogen vapour. *Ann Emerg Med*; 11(10): 553–5

4. Hall AH & Rumack BH (eds.) *TOMES System®* Micromedex, Englewood, Colorado. CD ROM. vol.41 (exp. 31 July 1999)

Henrietta Wheeler

NITROGEN OXIDES

Key Points

- Oxides of nitrogen are toxic by inhalation; the majority are very irritating to the respiratory system, and nitrous oxide is narcotic in high concentrations
- Two phases of toxicity – acute respiratory injury followed by possible delayed bronchiolitis
- Skin contact with these gases in compressed form can cause frostbite
- They can cause methaemoglobinaemia in high concentrations
- *In the event of a large spill, stay upwind and out of low areas. Ventilate closed spaces. Protective clothing, eye protection and breathing apparatus should be worn*

FIRST AID

- Terminate exposure and support vital functions
- The casualty should be moved to an uncontaminated area
- Rescuers should, ideally, be trained personnel and must be careful *not to put themselves at risk* and *so wear appropriate protective clothing and, if available, breathing apparatus*
- If the casualty is unconscious a clear airway should be established and maintained; give 100% oxygen if available
- **Inhalation Exposure:** If the patient stops breathing, expired air resuscitation should be started immediately using a pocket mask with a one way valve, if available. It is important where the face is contaminated that expired air resuscitation is NOT attempted unless an airway with rescuer protection is used
- **Dermal Exposure:** If frostbite occurs *do not* remove clothing, flush skin with water
- If frostbite is not present remove contaminated clothing, if possible under a shower and place in double, sealed, clear bags and label; store the bags in a secure area away from patients and staff
- Wash the skin thoroughly with copious amounts of water
- **Eye Exposure:** If the eye tissue is frozen seek medical advice as soon as possible
- If eye tissue is not frozen, irrigate thoroughly with water or saline for 15 minutes
- **Oral Exposure:** *Do not* give oral fluids

Detailed information

- There are seven oxides of nitrogen that may be found in ambient air: NO; NO_2; N_2O; NO_3; N_2O_3; N_2O_4 and N_2O_5
- The main source of oxides of nitrogen in both indoor and outdoor air is from combustion; 90–95% in the form of NO, 5–10% is released as NO_2 from this source

Description of Individual Oxides of Nitrogen

Compound	Description
Nitrogen dioxide	Colourless solid, a yellow liquid, or a brown gas
Nitric oxide	At room temperature, colourless gas with a sweet to acrid odour; deep blue when liquid, and a bluish–white 'snow' texture when solid
Nitrous oxide	Colourless gas with a slightly sweetish odour and taste; may exist in the form of a liquid or cubic crystals
Nitrogen trioxide	Blue liquid that is partially dissociated upon vaporisation into nitric oxide and nitrogen dioxide

Common Synonyms of Oxides of Nitrogen

Compound	Synonyms
Nitric oxide	nitrogen monoxide
Nitrogen dioxide	nitrogen peroxide
Nitrous oxide	nitrogen oxide, laughing gas
Nitrogen trioxide	dinitrogen trioxide, nitrogen sesquioxide, nitrous anhydride
Nitrogen tetroxide	dinitrogen tetroxide, nitrogen tetroxide (liquid), nitrogen peroxide, nitrogen di-dioxide
Dinitrogen pentoxide	nitric anhydride, nitrogen pentoxide

Identification Numbers of Oxides of Nitrogen

Compound	CAS	UN	NIOSH/RTECS
Nitric oxide	10102-43-9 90452-29-2 90880-94-7 10104-43-9	1660	QX 0525000
Nitrogen dioxide	10102-44-0	1067	QW 9800000
Nitrous oxide	10024-97-2	1070 (compressed) 1070 (refrigerated liquid) 2201	QX 1350000
Nitrogen trioxide	10544-73-7 (16529-92-3: Deleted Registry number)	2421	QX 1960000
Nitrogen tetroxide	10544-72-6	1067	QX 1575000
Dinitrogen pentoxide	10102-03-1		

Molecular Formula and Weight of Oxides of Nitrogen

Compound	Molecular formula	Molecular weight
Nitric oxide	NO	30.01
Nitrogen dioxide	NO_2	46.01
Nitrous oxide	N_2O	44.02
– (unstable)	NO_3	
Nitrogen trioxide	N_2O_3	76.02
Nitrogen tetroxide	N_2O_4	92.02
Dinitrogen pentoxide	N_2O_5	108.01

- NO and NO_2 are the most prevalent forms and are present in the greatest concentrations in urban and industrial air

- When the other oxides come into contact with atmospheric oxygen or are heated they at least partly dissociate into NO and NO_2

- NO and NO_2 form nitric and nitrous acid in contact with water

- NO_3 is highly reactive and unstable and only exists for short periods of time

- NO and NO_2 are not flammable but will enhance the flammability of other substances during a fire (as they are oxidising agents)

Uses for Oxides of Nitrogen[1,2,3]

Oxide	Use
Nitric oxide	Manufacture of nitric acid, bleaching of rayon, stabiliser for propylene and methyl ether
Nitrogen dioxide	Intermediate in nitric and sulphuric acid production; nitration of organic compounds and explosives; manufacture of oxidised cellulose compounds; bleaching flour; oxidising agent in rocket propulsion
Nitrous oxide	Oxidising organic compounds at temperatures above 300°C; manufacture of nitrites from alkali metals at their boiling points; with carbon disulphide in rocket fuel formulations; in the preparation of whipped cream; inhalation anaesthetic/analgesic
Nitrogen trioxide	Oxidant in special fuel systems; for the identification of terpenes; in the preparation of pure alkali nitrites
Nitrogen tetroxide	Titan II weapons systems are charged with hypergolic liquid rocket propellants (hydrazine fuel and nitrogen tetroxide oxidiser); same propellants are used in support of other systems, including the MX missile and the space shuttle
Dinitrogen pentoxide	Nitrating agent in chloroform solution

Solubility and Interactions of Oxides of Nitrogen

Oxide	Solubility	Interactions
Nitric oxide	Slightly soluble in water; solubility decreases with increasing temperature. Soluble in sulphuric acid, carbon disulphide and iron sulphate	Combines with O_2 to form NO_2 and with chorine and bromine to form nitrosyl halides
Nitrogen dioxide	Soluble in concentrated sulphuric and nitric acids; reacts with water forming nitrous and nitric acids (HNO_2 and HNO_3)	Reacts violently with cyclohexane, fluorine, formaldehyde and alcohol, nitrobenzene, petroleum, toluene. Reacts with combustible material, water (to form nitric acid), chlorinated hydrocarbons, carbon disulfide and ammonia
Nitrous oxide	Slightly soluble in water; freely soluble in sulphuric acid; soluble in alcohol, ether, oils	No data available
Nitrogen trioxide	Reacts with water forming nitrous acid (HNO_2)	Strong oxidising agent, reacts explosively with many materials including fuels
Nitrogen tetroxide	Reacts with water forming nitrous acid (HNO_2) and nitric acid (HNO_3)	Forms mixtures of nitric and nitrous acids on contact with water producing nitrous and nitric acids; reacts violently with most chlorinated hydrocarbons; liquid ammonia reacts explosively with the solid N_2O_4 at –80°C, while aqueous ammonia reacts vigorously with the gas at ambient temperature; reduced iron, potassium and pyrophoric manganese ignite in N_2O_4 at ambient temperature; slightly warm sodium ignites when in contact with N_2O_4 gas; reacts explosively with calcium
Dinitrogen pentoxide	Freely soluble in chloroform, less soluble in carbon tetrachloride	Reacts with water, forming nitrous acid (HNO_2)

Combustion and Flammability of Oxides of Nitrogen

Oxide	Combustion/ pyrolysis	Flammability
Nitric oxide	When heated to decomposition, it emits highly toxic fumes of NO_x	Burns only when heated with hydrogen
Nitrogen dioxide		Does not burn but supports combustion of carbon, phosphorus and sulphur
Nitrous oxide	Strong oxidising agent when heated to above 300 °C	Not flammable but supports combustion when in proper conc. with a flammable anaesthetic; can form explosive mixture with air; decomposes explosively at high temperatures; mixtures of nitrous oxide and phosphine can be exploded by a spark
Nitrogen trioxide	On vaporising partially dissociates into NO and NO_2	Does not burn but will support combustion
Nitrogen tetroxide	On vaporising partially dissociates into NO and NO_2	Does not burn but will support combustion

Summary of Human Toxicity

- NO_2 acts as a strong oxidising agent which can act on cell membrane proteins and lipids, resulting in the loss of control of cell permeability

- It increases airway responsiveness to chemicals (and other provoking agents e.g. cold air) causing bronchoconstriction, this effect is more pronounced in asthmatics and those with pre-existing pulmonary disease than in healthy subjects

- It can cause bronchial inflammation with increased numbers of mast cells and lymphocytes; alveolar macrophages recovered after NO_2 exposure have been shown to have impaired activity (as measured by ability to phagocytose viruses and yeasts)

- There have been a few reports that high concentrations of NO_2 (above 4 ppm) can decrease arterial oxygen partial pressure and systemic blood pressure

- Both NO and NO_2 react with water to form nitric and nitrous acids, which mediates some of their toxicity

- The mechanism of action of the other oxides of nitrogen is not well delineated, but most of them will be at least partly dissociated into NO and NO_2 under conditions in which they might be encountered (e.g. spills and fires/explosions)

- Nitrous oxide produces a euphoric and anxiolytic effect and has been demonstrated to be a partial agonist at μ, κ, and σ receptors of the endogenous opioid system; this may explain the emetic and addictive properties of nitrous oxide

- Naloxone appears to partially reverse nitrous oxide-induced anaesthesia[3]

Toxicity of Nitrogen Dioxide[6]

Concentration	Clinical effects
1 ppm for 60 min	Equivocal respiratory function effects and impaired dark adaptation of vision
5 ppm for 60 min	Acute reversible respiratory function effects
25 ppm for 60 min	Immediate respiratory irritation with chest pain
50 ppm for 60 min	Immediate respiratory and eye irritation with possible subacute and chronic pulmonary lesions
100 ppm for 60 min	Immediate respiratory and eye irritation with progressive respiratory injury and death
1000 ppm for 15 min	Immediate incapacitation with respiratory and eye injury followed by death

- Occupational Exposure Standards:

    NO: Long-term exposure limit: 25 ppm (31 mg.m^{-3})
    Short-term exposure limit: 35 ppm (44 mg.m^{-3})

    NO_2: Long-term exposure limit: 3 ppm (5.7 mg.m^{-3})
    Short-term exposure limit: 5 ppm (9.6 mg.m^{-3})

    N_2O: Long-term exposure limit: 100 ppm (183 mg.m^{-3})
    Short-term exposure limit: no data available

    Exposure limits for other nitrogen oxides: no data available

Acute Clinical Effects

Inhalation Effects

Oxides of nitrogen (other than N^2O) are toxic as follows:

- Severely irritating to the respiratory tract

- Initial symptoms can be delayed for between 1–5 hours

- Coughing, haemoptysis, chest pain, vomiting, shortness of breath, bronchospasm, laryngospasm and respiratory arrest may occur

- Acute pulmonary oedema may be delayed and can be fatal in severe exposure

- Methaemoglobinaemia with cyanosis from severe exposures; this will aggravate hypoxaemia

- Bronchiolitis obliterans may develop 2–3 weeks post exposure

Nitrous oxide

- Nausea and vomiting may occur

- Hypotension and cardiac arrhythmias have been reported

- Respiratory irritation may occur; interstitial emphysema and pneumomediastinum have been reported following deliberate inhalation

- At high concentration it is an asphyxiant and can cause narcosis; symptoms may include headache, dizziness and excitation leading to CNS depression

- Intracranial pressure may be elevated[3]

Dermal Effects

- Contact with pressurised liquid forms of nitrogen oxides can cause frostbite

Eye Effects

- Very irritating to the eyes

- Exposure to the vapours can cause conjunctival irritation, lacrimation, burns

- Direct contact with the liquid forms can cause severe burns to the cornea and eyelids

Oral Effects

- Unlikely to occur

- On ingestion the pressurised liquid forms of nitrogen oxides could cause frost injury to the oropharynx, leading to glottal oedema and suffocation

Chronic Clinical Effects

Prolonged inhalation of nitrogen oxides has been associated with decreased respiratory function in children, increased lower respiratory tract morbidity in children, higher incidence of acute respiratory disease in adolescents, increased airway resistance in asthmatics and people with chronic obstructive airways disease.[4]

Leucopenia, thrombocytopenia and severe megaloblastic anaemia have been reported following chronic intermittent inhalation of N_2O[3]

Management

Blood concentrations are unlikely to be of clinical use. Monitor full blood count for chronic exposures to N_2O.

Inhalation Management

- Remove from exposure, maintain a clear airway, give oxygen and ventilate if necessary

- If respiratory symptoms are present assess respiratory function and if necessary perform serial chest X-rays to check for chemical pneumonitis

- Consider the use of steroids to reduce the inflammatory response

- Treat pulmonary oedema with CPAP or PEEP ventilation

- If cyanosis is present or methaemoglobinaemia is suspected, determine methaemoglobin concentration

- See Nitrates (page 181) for management of methaemoglobinaemia

- If patients are unconscious they should be cardiac monitored, particularly if secondary to N_2O exposure

- Naloxone may partially reverse nitrous oxide-induced anaesthesia

- If early respiratory symptoms have been seen, patients should be reviewed up to three weeks post exposure in case they subsequently develop bronchiolitis obliterans

Dermal Management

- *If frostbite has occurred:* remove clothing carefully, these may need to be soaked off with tepid water; irrigate the area

- Surgical referral may be necessary

- *If frostbite has not occurred:* remove any remaining contaminated clothing

- Irrigate with copious amounts of water

- Treat burns symptomatically

- Place any contaminated clothes in double, sealed, clear bags and label; store in a secure area away from patients and staff

Eye Management

- Irrigate thoroughly with running water or saline for 15 minutes

- Stain with fluorescein and refer to an ophthalmologist if there is any uptake of the stain

Oral Management

- NO GASTRIC LAVAGE OR EMETIC

- Intubate to protect the airway, if burns or swelling are apparent

- Consider use of steroids to reduce inflammatory response

Summary of Environmental Hazards

- NO is the most prevalent oxide of nitrogen in the atmosphere, mainly derived from combustion, 90–95% of nitrogen oxides from this source are in the form of NO with 5–10% released as NO_2

- Oxides of nitrogen can enter the environment from other sources e.g. bacterial decomposition of nitrogenous compounds (main natural source of N_2O) and volcanic emission; they may also be released from industrial plants manufacturing nitric acid or explosives

- NO is converted to NO_2, rapidly by reacting with ozone, and more slowly by photochemical reactions (which require both the presence of sunlight and reactive organic compounds)

- NO and NO_2 are then progressively oxidised to other nitrates, particularly HNO_3, which is removed from the atmosphere mainly by wet or dry deposition

- N_2O is a so-called 'greenhouse gas'; emissions of NOx are thought to be leading to an increase in tropospheric concentrations of ozone[7]

- N_2O is a product of soil denitrification

- Drinking Water Standards: no data available

- Soil Guidelines: no data available

- Air Quality Standards: Nitrogen dioxide:
     150 ppb averaging time 1 hour (UK)
     150 $\mu g \cdot m^{-3}$ averaging time 24 hours (WHO guideline)

REFERENCES

1. Budavari S, O'Neil MJ, Smith A, Heckelman PE & Kinneary JF (eds). *The Merck Index*, 12th edn. Merck & Co., Inc., Whitehouse Station, 1996

2. Sax NI & Lewis RJ. *Hawley's Condensed Chemical Dictionary*, 11th edn. Van Nostrand Reinhold Company, New York, 1987

3. Hall AH & Rumack BH (eds.) *TOMES System* ® Micromedex, Englewood, Colorado. CD ROM. vol.41 (exp. 31 July 1999)

4. International Programme on Chemical Safety. *Environmental Health Criteria 188: Nitrogen oxides*. WHO, Geneva 1997

5. Hardman JG & Limbird LE (ed.) *Goodman and Gilman's The Pharmacological Basics of Therapeutics*, 9th Edn. McGraw-Hill, New York, 1996

6. Mayorga MA 1994. Overview of nitrogen dioxide effects on the lung with emphasis on military relevance. *Toxicology*; 89: 175–92

7. Harrison RM. *Understanding Our Environment: an Introduction to Environmental Chemistry and Pollution*, 2nd edn. Royal Society of Chemistry, Cambridge, 1992

Sarah McCrea

ORGANOPHOSPHATE INSECTICIDES

Key Points

- Organophosphate insecticides (OPs) are usually esters, amides or thiol derivatives of phosphoric, phosphonic, phosphorthioic or phosphonothioic acids
- OPs are generally highly lipid-soluble agents and are well absorbed from the skin, oral mucous membranes, conjunctiva, gastrointestinal and respiratory routes
- Many OP preparations contain organic solvents with the attendant risk of aspiration and increased irritation
- OPs act by inhibiting cholinesterase enzymes
- The onset, severity and duration of poisoning is determined by the degree and route of exposure, the lipid solubility and rate of metabolism of that particular organophosphate and whether transformation in the liver is required before the compound is active: the onset of clinical effects may be as early as 5 minutes post exposure but may be delayed for 12 hours or more and always occur within 24 hours
- Clinical effects may be divided into three types and the clinical picture may be a mixture of all three:

  Muscarinic effects: Bradycardia, bronchospasm, bronchorrhoea, sweating, salivation, lacrimation, vomiting, diarrhoea and miosis

  Nicotinic effects: Tachycardia, hypertension, muscle fasciculation and cramps, weakness and respiratory paralysis

  Central effects: CNS depression, agitation, confusion, psychosis, delirium, coma and convulsions; following severe exposures prolonged neurological manifestations may occur which may be irreversible or only slowly reversible

- *In the event of a large spill, stay upwind and out of low areas. Ventilate closed spaces. Protective clothing, eye protection and breathing apparatus should be worn*
- *Organophosphate vapours may evaporate from heavily contaminated casualties and so there is a risk that emergency personnel may become contaminated*
- *Casualties should be transported in such a way that there is no risk to the drivers of the emergency vehicles becoming contaminated by the fumes*

FIRST AID

- *Rescuers must not enter a contaminated area without full personal protective equipment and self-contained breathing apparatus*
- Terminate exposure and support vital functions
- The casualty should be moved to an uncontaminated area
- If the casualty is unconscious a clear airway should be established and maintained; give 100% oxygen if available
- **Inhalation Exposure:** If the patient stops breathing, expired air resuscitation should be started immediately using a pocket mask with a one way valve, if available; if the face is contaminated, it is important that expired air resuscitation is NOT attempted unless an airway with rescuer protection is used

---

- **Dermal Exposure:** Remove contaminated clothing, if possible under a shower, place in double, sealed, clear bags and label; store the bags in a secure area away from patients and staff
- Wash the skin thoroughly with copious amounts of water
- **Eye Exposure:** Irrigate thoroughly with water or saline for 15 minutes
- **Oral Exposure:** Encourage small quantities of oral fluids (no more than 50–100 ml in total)

Detailed Information

- OPs are used as agricultural pesticides, insecticides in domestic and public health applications and as chemical warfare agents

- Chemical warfare agents are not discussed in this entry (see Sarin on page 232)

- Most OPs are only slightly soluble in water and most organic solvents[1]

Common Synonyms for Organophosphates

See appendix i, page 197

Identification Numbers for Organophosphates

See appendix ii, page 201

Chemical Formulae and Molecular Weights for Organophosphates

See appendix iii, page 202

Summary of Human Toxicity

- Most OPs are highly lipid soluble agents which are well absorbed from the skin, oral mucous membranes, conjunctiva, gastrointestinal and respiratory routes

- Many OP preparations contain organic solvents with the attendant risk of aspiration and increased irritation

- The onset, severity and duration of poisoning is determined by the degree and route of exposure, the lipid solubility and rate of metabolism of that particular organophosphate and whether transformation in the liver is required before the compound is active: the onset of clinical effects may be as early as 5 minutes post exposure but may be delayed for 12 hours or more and always occurs within 24 hours

- The major toxicity of organophosphate insecticides is the covalent binding of phosphate radicals to the active site of cholinesterase enzymes, transforming them into enzymatically inert proteins; this inhibition of cholinesterase activity leads to the accumulation of acetylcholine at synapses, causing overstimulation and subsequent disruption of transmission in both the central and peripheral nervous systems. Exposure to OPs therefore interferes with synaptic transmission peripherally at muscarinic neuroeffector junctions and nicotinic receptors within sympathetic ganglia and at skeletal myoneural junctions and disruption of transmission will also occur at the acetylcholine receptor sites in the central nervous system[2]

- Antidotes: diazepam, atropine and pralidoxime are used for the treatment of OP poisoning

- *Diazepam* may be used to control twitching and convulsions

- *Atropine* is primarily effective for muscarinic effects as it blocks the action of acetylcholine at muscarinic receptors. *Hypoxia must be corrected before atropine is given*

- *Oximes* (pralidoxime) act by reactivating acetylcholinesterase; for greatest efficacy, pralidoxime should be given within 24 hours of exposure, although treatment commenced after this time may be effective

Toxicity of Organophosphates

See appendix iv, page 203

Occupational Exposure Standards

See appendix v, page 203

Acute Clinical Effects

Inhalation Effects

- OPs and their solvent carriers may cause irritation to the upper respiratory tract leading to coughing, wheezing and shortness of breath

- Systemic effects may occur; see below

Dermal Effects

- OPs and their solvent carriers may cause irritation and erythema

- Systemic poisoning may occur particularly after heavy dermal contamination, prolonged exposure or if patients have not been decontaminated effectively; see below

Eye Effects

- Contact with the eye may cause hyperaemia of the conjunctiva, constriction of the pupil and spasm of accommodation with aching pain in and about the eye and temporary blurring of vision[3]

- Systemic toxicity is unlikely

Oral Effects

- Ingestion may lead rapidly to severe poisoning

- See systemic effects below

Systemic effects

- Nausea, vomiting, abdominal pain, constricted pupils, sweating, salivation, muscle weakness, pyrexia, drowsiness, tachycardia, muscle fasciculation, profuse urinary and faecal incontinence may occur

- Tracheobronchial oversecretion with bronchoconstriction leading to pulmonary oedema occurs in severe cases

- In severe cases, coma, convulsions, cyanosis and hypoxia

- The main cause of death is respiratory depression (due to muscular paralysis)

- Hyperglycaemia and glycosuria indicating acute pancreatitis may also be present; neutrophil leucocytosis and renal damage have been reported

- Cardiac effects include bradycardia, cardiac arrhythmias including prolonged QT interval, in rare cases torsade de pointes and cardiac arrest

- Prolonged neurological manifestations including peripheral neuropathy, depression, poor memory and insomnia may develop following severe exposures

Intermediate Syndrome

- Some patients, after apparent recovery, develop acute respiratory failure 24 to 96 hours after the cholinergic phase of poisoning, with paralysis of the proximal limb muscles, motor cranial muscles and respiratory muscles

- The Intermediate Syndrome is refractory to atropine and pralidoxime treatment and ventilation is required; it is thought that the Intermediate Syndrome may not occur in patients who have received adequate pralidoxime therapy during the acute cholinergic phase[4]

Chronic Effects

Repeated exposure via inhalation or through the skin may eventually lead to clinical effects. Several chronic central nervous system disturbances due to acute or chronic OP exposure have been described. These are widely variable and include parkinsonian and pseudobulbar signs, alterations in affect, libido and memory, psychiatric and neuropsychological dysfunction and a cerebellar syndrome. It is not known whether chronic low dose exposure to OPs can result in neurological effects in the absence of acute poisoning and this area is controversial. Some studies have shown evidence of neurotoxicity in pesticide workers [5,6]

Management

All medical staff should wear full personal protective equipment when decontaminating patients; if they also become symptomatic they should be treated as below.

Monitor plasma and erythrocyte cholinesterase activity in every symptomatic case. Although the correlation between cholinesterase values and clinical effects is poor, depression in excess of 50% of red blood cell cholinesterase is associated with severe effects; for less severe exposures, correlation is very poor. Depression of cholinesterase activity is a good indicator of exposure.

Inhalation Management

- See systemic management below

Dermal Management

- Remove any remaining contaminated clothing, place in double, sealed, clear bags and label; store in a secure area away from patients and staff

- Wash thoroughly with soap and water

- See systemic management below

Eye Management

- Irrigate thoroughly with saline or water for at least 15 minutes

- Stain with fluorescein and refer to an ophthalmologist if there is any uptake of the stain

Oral Management

- DO NOT INDUCE EMESIS

- Consider gastric lavage (using a cuffed ET tube if an organic solvent is involved) followed by 50 g activated charcoal within 1 hour of ingestion

- Contact a poisons information service for further guidance on gut decontamination

- See systemic management below

Systemic Management

- Maintain respiration, give oxygen if required, ventilate if necessary

- Remove excess bronchial secretions by suction

- Observe for 24 hours with ECG monitoring

- Monitor red blood cell cholinesterase concentrations in every symptomatic case 4 to 6 hourly until recovery

- Administer antidotes in moderate or severe cases; see below

Antidotes

Contact a poisons information service for further guidance and paediatric doses

- **Diazepam** may have an overall benefit as well as controlling twitching and convulsions

- **Atropine** NOTE *hypoxia must be corrected before atropine is given*, Adult: 2 mg repeatedly SC or IV until atropinisation is achieved and maintained (atropinisation is characterised by decreased bronchial secretions, heart rate >100 bpm, dry mouth, dilated pupils)

- **Pralidoxime** NOTE *pralidoxime should be given as an adjunct to, not as a replacement for, atropine and should be given in every case where atropine therapy is deemed necessary*
Traditional dose: 1 g (or 2 g in very severe cases) by slow IV injection over 5–10 minutes. 1–2 g 4 hourly (maximum dose 12 g in 24 hours) until clinical and analytical recovery is achieved and maintained

Organophosphate spills

Wearing the appropriate personal protective equipment, absorb spilled liquid and cover contaminated areas with a 1:3 mixture of sodium carbonate crystals and damp sawdust, lime, sand or earth; sweep up and place in a closeable impervious container; ensure that the container is tightly closed and suitably labelled before transport to a safe place for disposal.

Summary of Environmental Hazards

- During application, pesticides are carried by drift which is enhanced by aerial spraying

- OPs also enter the atmosphere by volatilisation and are subsequently washed out by wet deposition

- OPs are degraded by hydrolysis, yielding water–soluble products that are believed to be in non–toxic concentrations[7]

- The toxic hazard is essentially short–lived in contrast to the persistent organochlorine pesticides; the half–life in neutral water varies from a few hours for dichlorvos to weeks for parathion

- In slightly acidic soils the half–lives may be extended to many hours

- Accumulation in aquatic organisms is low; however persistence in body tissues causes inhibition of acetylcholinesterase[8]

- Drinking Water Standards: Pesticide: 0.1 µg.l[-1] (UK max)

- Soil Guidelines: no data available

- Air quality Standards: no data available

REFERENCES

1. Al-Saleh IA, 1994. Pesticides: A review article. *J Environ Path Toxicol & Oncol*; 13(3): 151–61

2. Tafuri J, & Roberts J, 1987. Organophosphate poisoning; *Ann Emerg Med*; 16: 193–202

3. Grant MW & Schuman JS. *Toxicology of the Eye*, 4th edn. Charles C Thomas, Springfield, 1993

4. Senanayake N & Karalliedde L, 1987. Neurotoxic effects of organophosphorus insecticides. An intermediate syndrome. *N Eng J Med*; 316: 761–3

5. Steenland K, 1996. Chronic neurological effects of organophosphate pesticides. *Brit Med J*; 312: 1312–3

6. Stokes L, Stark A, Marshall E & Narang A, 1995. Neurotoxicity among pesticide applicators exposed to organophosphates. *Occup Environ Med*; 52: 648–53

7. International Programme on Chemical Safety. Environmental Health Criteria 63: *Organophosphorus Insecticides: A general Introduction*. WHO, Geneva, 1986

8. Ballentyne B & Marrs TC. *Clinical & Experimental Toxicology of Organophosphates and Carbamates*. Butterworth-Heinemann Ltd., Oxford, 1992

9. Tomlin CDS, 1997. *The Pesticide Manual. A World Compendium*, 11th edn. British Crop Protection Council

10. IPCS 1998. *The WHO recommended classification of pesticides by hazard and guidelines to classification 1998–99*

Grainne Cullen

Appendix i Common Synonyms and Trade Names of Organophosphates[9]

Organophosphate	Use	Synonyms and Trade Names
Acephate	Insecticide	O,S-dimethyl acetylphosphoramidothioate; N-[methoxy(methylthio)phosphinoyl]acetamide; Orthene (Tomen, Valent); Ortran (Tomen, Valent); Acevol (Voltas); Amithene (Amico); Asataf (Rallis); Cekucefate (Cequisa); Lancer (United Phosphorus); Racet (Rotam); Rythane (Ramcides); Saphate (Sanonda); Starthene (Shaw Wallace); Torpedo (Searle India); Vital (Productos OSA)
Azamethiphos	Insecticide	S-6-chloro-2,3-dihydro-2-oxo-1,3-oxazolo[4,5-b]pyridin-3-ylmethyl O,O-dimethylphosphorothioate; 6-chloro-3-dimethoxyphosphinoylthiomethyl-1,3-oxazolo[4,5-b]pyridin-2(3H)-one; S-[(6-chloro-2-oxooxazolo[4,5-b]pyridin-3(2H)-yl) methyl] O,O-dimethylphosphorothioate; Alfacron (Novartis)
Azinphos-ethyl	Insecticide and acaricide	Azinphosethyl; Triazotion; S-(3,4-dihydro-4-oxobenzo[d]-[1,2.3]-triazin-3-ylmethyl) O,O-diethylphosphorodithioate; O,O-diethyl S-[(4-oxo-1,2,3-benzotriazin-3(4H)-yl) methyl]phoshorodithioate; Azinugec E (Sipcam Phyteurop); Cotnion-Ethyl (Makhteshim-Agan)
Azinphos-methyl	Insecticide	Azinphosmethyl; Metiltriazotion; S-(3,4-dihydro-4-oxobenzo[d]-[1,2,3]-triazin-3-ylmethyl) O,O-dimethylphosphorodithioate; O,O-dimethyl S-[(4-oxo-1,2,3-benzotriazin-3(4H)-yl) methyl] phosphorodithioate; Gusathion M (Bayer); Acifon (General Quimica); Azinugec (Sipcam Phyteurop); Cotnion-Methyl (Makhteshim-Agan); Valefos (Productos OSA)
Cadusafos	Nematicide and insecticide	Ebufos; S,S-di-sec-butyl O-ethylphosphorodithioate; O-ethyl S.S-bis (1-methylpropyl)phosphorodithioate; Rugby (FMC); Apache (FMC)
Chlorfenvinphos	Insecticide and acaricide	2-chloro-1-(2,4-dichlorophenyl)vinyl diethyl phosphate; 2-chloro-1-(2,4-dichlorophenyl)ethenyl diethyl phosphate; Birlane (Cyanamid); Supona (Cyanamid); Apachlor (Rhone-Poulenc)
Chlormephos	Insecticide	S-chloromethyl O,O-diethyl phosphorodithioate; S-(chloromethyl) O,O-diethyl phosphorodithioate; Dotan (Rhone-Poulenc)
Chlorpyrifos	Insecticide	Chlorpyriphos; Chlorpyriphos-ethyl; O,O-diethyl O-3,5,6-trichloro-2-pyridyl phosphorothioate; O,O-diethyl O-(3,5,6-trichloro-2-pyridinyl) phosphorothioate; Dursban (DowElanco); Lorsban (DowElanco); Agromil (Westrade); Bullet (Mitsu); Chlorfos (Griffin); Destroyer (Agriphar); Dhanvan (Dhanuka); Dorsan (Luxembourg); Omexan (Aimco); Pyrifoz (Sanoda); Pyrinex (Makhteshim-Agan); Pyrivol (Voltas); Silrifos (Siapa); Spannit (Pan Brittanica); Strike (Wockhardt); Tafaban (Rallis); Talon (FCC); Terraguard (Gharda); Tricel (Excel)
Chlorpyrifos-methyl	Insecticide and acaricide	Chlorpyriphos-methyl; O,O-dimethyl O-3,5,6-trichloro-2-pyridyl phosphorothioate; O,O-dimethyl O-(3,5,6-trichloro-2-pyridinyl) phosphorothioate; Reldan (DowElanco); Pyriban-M (Amico)
Coumaphos	Insecticide	O-3-chloro-4-methyl-2-oxo-2H-chromen-7-yl O,O-diethyl phosphorothioate; 3-chloro-7-diethoxyphosphinothioyloxy-4-methylcoumarin; O-(3-chloro-4-methyl-2-oxo-2H-1-benzopyran-7-yl) O,O-diethyl phosphorothioate; Asuntol (Bayer); Perizin (Bayer)
Cyanophos	Insecticide	O-4-cyanophenyl O,O-dimethyl phosphorothioate; 4-(dimethoxyphosphinothioyloxy)benzonitrile; O-(4-cyanophenyl) O,O-dimethyl phosphorothioate; Cyanox (Sumitomo)
Demeton-S-methyl	Insecticide and acaricide	Methyl demeton; Methylmercaptofositol; S-2-ethylthioethyl O,O-dimethyl phosphorothioate; S-[2-(ethylthio)ethyl] O,O-dimethyl phosphorothioate; Metasystox (I) (Bayer); DSM (MTM); Mifatox (FCC); Metaphor (United Phosphorus Ltd.)
Diazinon	Insecticide and acaricide	Dimpylate; O,O-diethyl O-2-isopropyl-6-methylpyrimidin-4-yl phosphorothioate; O,O-diethyl O-[6-methyl-2-(1-methylethyl)-4-pyrimidinyl] phosphorothioate; Basudin (Novartis); Cekuzinon (Cequisa); Dianon (Nippon Kayaku); Dianozyl (Agriphar); Diazol (Makhteshim-Agan); Ectoban (Agropharm); Efdiazon (Efthymiadis); Knox-out (Elf Atochem)
Dichlorvos	Insecticide and acaricide	2,2-dichlorovinyl dimethyl phosphate; 2,2-dichloroethenyl dimethyl phosphate; Dedevap (Bayer); Nuvan (Novartis); Vapona (Cyanamid); Amidos (Aimco); Charge (Sanonda); Didivane (Diachem); Denkavepon (Denka); Divipan (Makhteshim-Agan); Phosvit (Nippon Soda); Doom (United Phosphorus); Rupini (Ramcides); Swing (Siapa); Uniphos (Florin); Vantaf (Rallis)
Dicrotophos	Insecticide and acaricide	(E)-2-dimethylcarbamoyl-1-methylvinyl dimethyl phosphate; 3-dimethoxyphosphinoyloxy-N,N-dimethylisocrotonamide; (E)-3-(dimethylamino)-1-methyl-3-oxo-1-propenyl dimethyl phosphate; Bidrin (Cyanamid); Dicron (Hui Kwang)

Organophosphate	Use	Synonyms and Trade Names
Dimethoate	Insecticide and acaricide	*O,O*-dimethyl *S*-methylcarbamoylmethyl phosphorodithioate; 2-dimethoxyphosphinothioylthio-*N*-methylacetamide; *O,O*-dimethyl *S*-[2-(methylamino)-2-oxoethyl] phosphorodithioate; Cygon (Wilbur-Ellis); Perfekthion (BASF); Rogor (Isagro, Rallis); Roxion (Wilbur-Ellis); Cekutoate (Cequisa); Champ (Searle India);Chimigor (Diachem); Danadim (Cheminova); Diadhan (Dhanuka); Dicentra (Sanonda); Dimezyl (Agriphar); Efdacon (Efthymiadis); Robgor (Ramcides); Romethoate (Rotam); Tara 909 (Shaw Wallace)
Dimethylvinphos	Insecticide	Chlorfenvinphos-methyl; (Z)-2-chloro-1-(2,4-dichlorophenyl)vinyl dimethyl phosphate; 2-chloro-1-(2,4-dichlorophenyl)ethenyl dimethyl phosphate; Rangado (Cyanamid)
Disulfoton	Insecticide and acaricide	Ethylthiodemeton; *O,O*-diethyl *S*-2-ethylthioethyl phosphorodithioate; *O,O*-diethyl *S*-[2-(ethylthio)ethyl] phosphorodithioate; Disyston (Bayer); Frumin AL (Novartis); Solvirex (Novartis)
EPN	Insecticide and acaricide	*O*-ethyl *O*-4-nitrophenyl phenylphosphonothioate; *O*-ethyl *O*-(4-nitrophenyl) phenylphosphonothioate; EPN (Nissan)
Ethion	Insecticide and acaricide	Diethion; *O,O,O',O'*-tetraethyl *S,S'*-methylene bis(phosphorodithioate); *S,S'*-methylene bis(*O,O*-diethyl) phosphorodithioate); Cekuetion (Cequisa); Cethion (Cheminova); Dhanumit (Dhanuka); Ethiol (Rhone Poulenc); MIT 505 (Shaw Wallace); Rayethion (Krishi Rasayan); Rhodocide (Rhone-Poulenc); Tafethion (Rallis)
Ethoprophos	Nematicide and insecticide	Ethoprop; *O*-ethyl *S,S*-dipropyl phosphorodithioate; Mocap (Rhone-Poulenc)
Famphur	Insecticide	Famophos; *O,4*-dimethylsulfamoylphenyl *O,O*-dimethyl phosphorothioate; 4-dimethoxyphosphinothioyloxy-*N,N*-dimethylbenzenesulfonamide; *O*-[4-[(dimethylamino)sulfonyl]phenyl] *O,O*-dimethyl phosphorothioate; Bo-Ana (Cyanamid); Warbexol (AgrEvo)
Fenamiphos	Nematicide	Methaphenamiphos; ethyl 4-methylthio-*m*-tolyl isopropylphosphoramidate; ethyl 3-methyl-4-(methylthio)phenyl (1-methylethyl) phosphoramidate; Nemacur (Bayer)
Fenitrothion	Insecticide	*O,O*-dimethyl *O*-4-nitro-*m*-tolyl phosphorothioate; *O,O*-dimethyl *O*-(3-methyl-4-nitrophenyl) phosphorothioate; Folithion (Bayer); Sumithion (Sumitomo); Dicofen (PBI); Cekutrothion (Cequisa); Farmathion (Sanonda); Fentron (Efthymiadis); Shaminliulin (Shenzhen Jiangshan)
Fenthion	Insecticide	*O,O*-dimethyl *O*-4-methylthio-*m*-tolyl phosphorothioate; *O,O*-dimethyl *O*-[3-methyl-4-(methylthio)phenyl] phosphorothioate; Lebaycid (Bayer); Beiliulin (Shenzhen Jiangshan); Faster (Sanonda); Pilartex (Pilarquim)
Fonofos	Insecticide	*O*-ethyl *S*-phenyl (*RS*)-ethylphosphonodithioate; (±)-*O*-ethyl *S*-phenyl ethylphosphonodithioate; Dyfonate (Zeneca); Capfos (Zeneca)
Formothion	Insecticide and acaricide	*S*-[formyl(methyl)carbamoylmethyl] *O,O*-dimethyl phosphorodithioate; 2-dimethoxyphosphinothioylthio-*N*-formyl-*N*-methylacetamide; *S*-[2-formylmethylamino)-2-oxoethyl] *O,O*-dimethyl phosphorodithioate; Anthio (Novartis)
Fosthiazate	Nematicide	(*RS*)-*S-sec*-butyl *O*-ethyl 2-oxo-1,3-thiazolidin-3-ylphosphonothioate; (*RS*)-3-[*sec*-butylthio(ethoxy)phosphinoyl]-1,3-thiazolidin-2-one; *O*-ethyl *S*-(1-methylpropyl) (2-oxo-3-thiazolidinyl) phosphonothioate; Nemathorin (Ishihara Sangyo)
Heptenophos	Insecticide	7-chlorobicyclo[3.2.0]hepta-2,6-dien-6-yl dimethyl phosphate; Hostaquick (AgrEvo); Ragadan (Hoechst)
Isazofos	Nematicide and insecticide	*O*-5-chloro-1-isopropyl-1*H*-1,2,4-triazol-3-yl *O,O*-diethyl phosphorothioate; *O*-[5-chloro-1-(1-methylethyl)-1*H*-1,2,4-triazol-3-yl] *O,O*-diethyl phosphorothioate; Miral (Novartis)
Isofenphos	Insecticide	Isophenphos; *O*-ethyl *O*-2-isopropoxycarbonylphenyl isopropylphosphoroamidothioate; isopropyl *O*-[ethoxy-*N*-isopropylamino(thiophosphoryl)]salicylate; 1-methylethyl 2-[[ethoxy[(1-methylethyl)amino]phosphinothioyl]oxy]benzoate; Oftanol (Bayer)
Isoxathion	Insecticide	*O,O*-diethyl *O*-5-phenylisoxazol-3-yl phosphorothioate; *O,O*-diethyl *O*-(5-phenyl-3-isoxazolyl) phosphorothioate; Karphos (Sankyo)
Malathion	Insecticide and acaricide	Maldison; Malathon; Mercaptothion; Carbofos; Mercaptotion; diethyl (dimethoxythiophosphorylthio)succinate; *S*-1,2-bis(ethoxycarbonyl)ethyl *O,O*-dimethyl phosphorodithioate; diethyl [(dimethoxyphosphinothioyl)thio]butanedioate; Cekumal (Cequisa); Celthion (Excel); Fyfanon (Cheminova); Malathane (Agriphar); Malatox (Pesticides India, Siapa); Malixol (Rhone-Poulenc); MLT (Sumitomo); Maltox (Aimco); White Star (Sanonda)
Mecarbam	Insecticide and acaricide	*S*-(*N*-ethoxycarbonyl-*N*-methylcarbamoylmethyl) *O,O*-diethyl phosphorodithioate; ethyl *N*-(diethoxythiophosphorylthio)acetyl-*N*-methylcarbamate; ethyl (diethoxyphosphinothioylthio)acetyl(methyl)carbamate; ethyl 6-ethoxy-2-methyl-3-oxo-7-oxa-5-thia-2-aza-6-phosphanonanoate 6-sulfide; Murfotox (Efthymiadis)

Organophosphate	Use	Synonyms and Trade Names
Methacrifos	Insecticide and acaricide	methyl (E)-3-(dimethoxyphosphinothioyloxy)-2-methylacrylate; (E)-O-2-methoxycarbonylprop-1-enyl O,O-dimethyl phosphorothioate; (E)-methyl 3-[(dimethoxyphosphinothioyl)oxy]-2-methyl-2-propenoate; Damfin (Novartis)
Methamidophos	Insecticide and acaricide	O,S-dimethyl phosphoramidothioate; Monitor (Bayer, Tomen, Valent); Tamaron (Bayer); Cekumidofos (Cequisa); Giant (Sanonda); Jiaanlin (Shenzhen Jiangshan); Methaphos (Efthymiadis); MTD-600 (Westrade); Patrole (Productos OSA); Pilaron (Pilarquim)
Methidathion	Insecticide and acaricide	S-2,3-dihydro-5-methoxy-2-oxo-1,3,4-thiadiazol-3-ylmethyl O,O-dimethyl phosphorodithioate; 3-dimethoxyphosphinothioylthiomethyl-5-methoxy-1,3,4-thiadiazol-2(3H)-one; S-[(5-methoxy-2-oxo-1,3,4-thiadiazol-3(2H)-yl)methyl] O,O-dimethyl phosphorodithioate; Supracide (Novartis); Suprathion (Makhteshim-Agan)
Mevinphos	Insecticide and acaricide	2-methoxycarbonyl-1-methylvinyl dimethyl phosphate; methyl 3-(dimethoxyphosphinoyloxy)but-2-enoate; methyl 3-[(dimethoxyphosphinyl)oxy]-2-butenoate; Phosdrin (Cyanamid, Amvac); Duraphos (Amvac); Mevindrin (Hui Kwang)
Monocrotophos	Insecticide and acaricide	dimethyl (E)-1-methyl-2-(methylcarbamoyl)vinyl phosphate; 3-dimethoxyphosphinoyloxy-N-methylisocrotonamide; (E)-dimethyl 1-methyl-3-(methylamino)-3-oxo-1-propenyl phosphate; Azodrin (Cyanamid); Nuvacron (Novartis); Apadrin (Rhone-Poulenc); Balwan (Rallis); Croton (Searle India); Crotos (Siapa); Efacron (Efthymiadis); Macabre (Sanonda); Monocron (Makhteshim-Agan); Monodhan (Dhanuka); Monodrin (Hui Kwang); Monostar (Shaw Wallace); Monovol (Voltas); Phoskill (United Phosphorus); Pilardrin (Pilarquim)
Naled	Insecticide and acaricide	Bromchlophos; Dibrom; 1,2-dibromo-2,2-dichloroethyl dimethyl phosphate; Dibrom (Valent); Bromex (Makhteshim-Agan)
Omethoate	Insecticide and acaricide	O,O-dimethyl S-methylcarbamoylmethyl phosphorothioate; O,O-dimethyl S-[2-(methylamino)-2-oxoethyl] phosphorothioate; Folimat (Bayer)
Oxydemeton-methyl	Insecticide	S-2-ethylsulfinylethyl O,O-dimethyl phosphorothioate; S-[2-(ethylsulfinyl)ethyl] O,O-dimethyl phosphorothioate; Metasystox R (Bayer, Gowan); Aimcosystox (Aimco); Dhanusystox (Dhanuka)
Parathion	Insecticide and acaricide	Parathion-ethyl; Thiophos; Ethyl parathion; O,O-diethyl O-4-nitrophenyl phosphorothioate; O,O-diethyl O-(4-nitrophenyl) phosphorothioate; E605 (Bayer); Fostox E (Siapa); Chimac Par H (Agriphar); Fighter (Sanonda)
Parathion-methyl	Insecticide	Metaphos; Methyl parathion; O,O-dimethyl O-4-nitrophenyl phosphorothioate; O,O-dimethyl O-(4-nitrophenyl) phosphorothioate; Folidol-M (Bayer); Metacide (Bayer); Cekumethion (Cequisa); Dhanuman (Dhanuka); Fostox metil (Siapa); Jiajiduiliulin (Shenzhen Jiangshan); Morfos Methyl (Efthymiadis); Parartaf (Rallis); Paraxox (Aimco); Pencap-M (Elf Atochem); Sweeper (Sanonda); Thionyl (Agriphar)
Phenthoate	Insecticide and acaricide	Dimephenthoate; S-α-ethoxycarbonylbenzyl O,O-dimethyl phosphorodithioate; ethyl dimethoxyphosphinothioylthio(phenyl)acetate; ethyl α-[(dimethoxyphosphinothioyl)thio]benzeneacetate; Elsan (Nissan); Cidial (Isagro); Aimsan (Aimco); Papthion (Sumitomo)
Phorate	Insecticide, acaricide and nematicide	O,O-diethyl S-ethylthiomethyl phosphorodithioate; O,O-diethyl S-[(ethylthio)methyl] phosphorodithioate; Thimet (Cyanamid); Cekuforatox (Cequisa); Dhan (Dhanuka); Kurunai (Ramcides); Umet (United Phosphorus); Volphor (Voltas); Warrant (Searle India)
Phosalone	Insecticide and acaricide	Benzphos; Benzofos; S-6-chloro-2,3-dihydro-2-oxobenzoxazol-3-ylmethyl O,O-diethylphosphorodithioate; S-[(6-chloro-2-oxo-3(2H)-benzoxazolyl)methyl] O,O-diethylphosphorodithioate; Zolone (Rhone-Poulenc)
Phosmet	Insecticide and acaricide	Phtalofos; O,O-dimethyl S-phthalimidomethyl phosphorodithioate; N-(dimethoxyphosphinothioylthiomethyl)phthalimide; S-[(1,3-dihydro-1,3-dioxo-2H-isoindol-2-yl)methyl] O,O-dimethyl phosphorodithioate; Cekumet (Cequisa); Fosdan (General Quimica); Imidan (Gowan); Inovat (Productos OSA); Inovitan (Efthymiadis); Prolate (Gowan)
Phosphamidon	Insecticide and acaricide	2-chloro-2-diethylcarbamoyl-1-methylvinyl dimethyl phosphate; 2-chloro-3-dimethoxyphosphinoyloxy-N,N-diethylbut-2-enamide; 2-chloro-3-(diethylamino)-1-methyl-3-oxo-1-propenyl dimethyl phosphate; Dimecron (Novartis); Aimphon (Aimco); Kinadon (United Phosphorus); Phosron (Hui Kwang); Rilan (Rallis); Rimdon (Ramcides)
Phoxim	Insecticide	Phoxime; O,O-diethyl α-cyanobenzylideneamino-oxyphosphonothioate; 2-(diethoxyphosphinothioyloxyimino)-2-phenylacetonitrile; α-[[(diethoxyphosphinothioyl)oxy]imino]benzene-acetonitrile; Baythion (Bayer); Volaton (Bayer)

Organophosphate	Use	Synonyms and Trade Names
Pirimiphos-ethyl	Insecticide	*O,O*-diethyl *O*-2-diethylamino-6-methylpyrimidin-4-yl phosphorothioate; *O*-[2-(diethylamino)-6-methyl-4-pyrimidinyl] *O,O*-diethyl phosphorothioate; Primicid (Zeneca)
Pirimiphos-methyl	Insecticide and acaricide	*O,O*-dimethyl *O*-2-diethylamino-6-methylpyrimidin-4-yl phosphorothioate; *O*-[2-(diethylamino)-6-methyl-4-pyrimidinyl] *O,O*-dimethyl phosphorothioate; Actellic (Zeneca); Actellifog (Hortichem)
Profenofos	Insecticide and acaricide	*O*-4-bromo-2-chlorophenyl *O*-ethyl *S*-propyl phosphorothioate; *O*-(4-bromo-2-chlorophenyl) *O*-ethyl *S*-propyl phosphorothioate; Curacron (Novartis); Sanofos (Sanonda)
Propaphos	Insecticide	Propafos; 4-(methylthio)phenyl dipropyl phosphate; Kayaphos (Nippon Kayaku)
Propetamphos	Insecticide and acaricide	Prometamfos; (*E*)-*O*-2-isopropoxycarbonyl-1-methylvinyl *O*-methylethylphosphoramidothioate; 1-methylethyl (*E*)-3-[[(ethylamino)methoxyphosphinothioyl]oxy]-2-butenoate; Safrotin (Novartis)
Prothiofos	Insecticide	*O*-2,4-dichlorophenyl *O*-ethyl *S*-propyl phosphorodithioate; *O*-(2,4-dichlorophenyl) *O*-ethyl *S*-propyl phosphorodithioate; Tokuthion (Bayer)
Pyraclofos	Insecticide	(*RS*)-[*O*-1-(4-chlorophenyl)pyrazol-4-yl *O*-ethyl *S*-propyl phosphorothioate] (±)-*O*-[1-(4-chlorophenyl)-1*H*-pyrazol-4-yl] *O*-ethyl *S*-propyl phosphorothioate; Boltage (Takeda); Voltage (Takeda)
Pyridaphenthion	Insecticide and acaricide	*O*-(1,6-dihydro-6-oxo-1-phenylpyridazin-3-yl) *O,O*-diethyl phosphorothioate; *O*-(1,6-dihydro-6-oxo-1-phenyl-3-pyridazinyl) *O,O*-diethyl phosphorothioate; Ofunack (Mitsui Toatsu); Oreste (Sipcam Phyteurop)
Quinalphos	Insecticide and acaricide	Chinalphos; *O,O*-diethyl *O*-quinoxalin-2-yl phosphorothioate; *O,O*-diethyl *O*-2-quinoxalinyl phosphorothioate; Ekalux (Novartis); Danulux (Dhanuka); Hubelux (Sanonda); Quinaal (Ramicides); Quinatox (Aimco); Smash (Searle India); Starlux (Shaw Wallace)
Sulfotep	Insecticide and acaricide	Dithio; Dithione; Thiotep; *O,O,O',O'*-tetraethyl dithiopyrophosphate; tetraethyl thiodiphosphate; Bladafum (Bayer)
Sulprofos	Insecticide	*O*-ethyl *O*-4-(methylthio)phenyl *S*-propyl phosphorodithioate; *O*-ethyl *O*-[4-(methylthio)phenyl] *S*-propyl phosphorodithioate; Bolstar (Bayer)
Temephos	Insecticide	*O,O,O',O'*-tetramethyl *O,O'*-thiodi-*p*-phenylene bis(phosphorothioate); *O,O,O',O'*-tetramethyl *O,O'*-thiodi-*p*-phenylene diphosphorothioate; *O,O'*-(thiodi-4,1-phenylene) bis(*O,O*-dimethyl phosphorothioate); Abate (Cyanamid); Temeguard (Gharda)
Terbufos	Insecticide and nematicide	*S-tert*-butylthiomethyl *O,O*-diethyl phosphorodithioate; *S*-[[(1,1dimethylethyl)thio]methyl] *O,O*-diethyl phosphorodithioate; Contraven (Cyanamid); Counter (Cyanamid, BASF, DowElanco); Cyanater (Siapa); Hunter (United Phosphorus); Pilarfox (Pilarquim); Terborox (Rotam); Tertin (Sanonda)
Tetrachlorvinphos	Insecticide and acaricide	Stirofos; (Z)-2-chloro-1-(2,4,5-trichlorophenyl)vinyl dimethyl phosphate; (Z)-2-chloro-1-(2,4,5-trichlorophenyl)ethenyl dimethyl phosphate; Rabon (Cyanamid, Du Pont); Debantic (SDS Biotech); Gardona (Cyanamid)
Thiometon	Insecticide and acaricide	Dithiometon; *S*-2-ethylthioethyl *O,O*-dimethyl phosphorodithioate; *S*-[2-(ethylthio)ethyl] *O,O*-dimethyl phosphorodithioate; Ekatin (Novartis); Mavrik (Novartis)
Triazophos	Insecticide, acaricide and nematicide	*O,O*-diethyl *O*-1-phenyl-1*H*-1,2,4-triazol-3-yl phosphorothioate; *O,O*-diethyl *O*-(1-phenyl-1*H*-1,2,4-triazol-3-yl) phosphorothioate; Hostathion (AgrEvo); Trelka (AgrEvo); Spark (AgrEvo); Try (Sanonda)
Trichlorfon	Insecticide	Trichlorphon; Chlorophos; Metriphonate; dimethyl 2,2,2,-trichloro-1-hydroxyethylphosphonate; dimethyl (2,2,2,-trichloro-1-hydroxyethyl)phosphonate; Dipterex (Bayer); Cekufon (Cequisa); Danex (Makhteshim-Agan); Denkaphon (Denka); Saprofon (Sanonda)
Vamidothion	Insecticide and acaricide	*O,O*-dimethyl *S*-2-(1-methylcarbamoylethylthio)ethyl phosphorothioate; 2-(2-dimethoxyphosphinoylthioethylthio)-*N*-methylpropionamide; *O,O*-dimethyl *S*-[2-[[1-methyl-2-(methylamino)-2-oxoethyl]thio]ethyl] phosphorothioate; Kilval (Rhone-Poulenc)

Appendix ii Identification Numbers for Organophosphates

Organophosphate	CAS	UN	NIOSH/RTECS
Acephate	30560-19-1	2783, 2784, 3017, 3018	TB 4760000
Azamethiphos	35575-96-3	not available	TE 8070000
Azinphos-ethyl	2642-71-9	2783, 2784, 3017, 3018	TD 8400000
Azinphos-methyl	86-50-0	2783	TE 1925000
Cadusafos	95465-99-9	not available	TE 3800000
Chlorfenvinphos	470-90-6 formerly 2701-86-2	2783, 2784, 3017, 3018	TB 8750000
Chlormephos	24934-91-6	3018	TD 5170000
Chlorpyrifos	2921-88-2	2783	TF 6300000
Chlorpyrifos-methyl	5598-13-0	2783	TG 0700000
Coumaphos	56-72-4	2783	GN 6300000
Cyanophos	2636-26-2	2783, 2784, 3017, 3018	TF 7600000
Demeton-S-methyl	919-86-8	3018	TG 1750000
Diazinon	333-41-5	2783	TF 3325000
Dichlorvos	62-73-7	2783, 2784, 3017, 3018	TC 0350000
Dicrotophos	141-66-2	2783, 2784, 3017, 3018	TC 3850000
Dimethoate	60-51-5	2783	TE 1750000
Dimethylvinphos	2274-67-1	not available	TB 8805000
Disulfoton	298-04-4	2783	TD 9275000
EPN	2104-64-5	2783, 2784, 3017, 3018	TB 1925000
Ethion	563-12-2	3018	TE 4550000
Ethoprophos	13194-48-4	2783, 2784, 3017, 3018	TE 4025000
Famphur	52-85-7	2783, 2784, 3017, 3018	TF 7650000
Fenamiphos	22224-92-6	2783, 2784, 3017, 3018	TB 3675000
Fenitrothion	122-14-5	3018	TG 0350000
Fenthion	55-38-9	2783, 2784, 3017, 3018	TF 9625000
Fonofos	944-22-9	2783, 2784, 3017, 3018	TA 5950000
Formothion	2540-82-1	2783, 2784, 3017, 3018	TE 1050000
Fosthiazate	98886-44-3	not available	TB 1707000
Heptenophos	23560-59-0	not available	TB 8545000
Isazofos	42509-80-8	not available	TE 7760000
Isofenphos	25311-71-1	not available	VO 4395500
Isoxathion	18854-01-8	not available	TF 5600000
Malathion	121-75-5	3279	WM 8400000
Mecarbam	2595-54-2	2783, 2784, 3017, 3018	FB 3850000
Methacrifos	62610-77-9	not available	UD 3335000
Methamidophos	10265-92-6	2783, 2784, 3017, 3018	TB 4970000
Methidathion	950-37-8	2783, 2784, 3017, 3018	TE 2100000
Mevinphos	26718-65-0 formerly 298-01-1	not available	GQ 5250100
Monocrotophos	6923-22-4 formerly 919-44-8	2783, 2784, 3017, 3018	TC 4375000
Naled	300-76-5	2783, 2784, 3017, 3018	TB 9450000
Omethoate	1113-02-6	2783, 2784, 3017, 3018	TF 8050000
Oxydemeton-methyl	301-12-2	not available	TG 1420000
Parathion	56-38-2	2783	TF 4550000
Parathion-methyl	298-00-0	2783, 3018	TG 0175000
Phenthoate	2597-03-7	2783	AI 7875000
Phorate	298-02-2	3018	TD 9450000
Phosalone	2310-17-0	2783, 2784, 3017, 3018	TD 5175000
Phosmet	732-11-6	2783	TE 2275000
Phosphamidon	13171-21-6	3018	TC 2800000
Phoxim	14816-18-3	not available	MD 4740000
Pirimiphos-ethyl	23505-41-1	2783, 2784, 3017, 3018	TF 1610000

Organophosphate	CAS	UN	NIOSH/RTECS
Pirimiphos–methyl	29232-93-7	not available	TF 1410000
Profenofos	41198-08-7	not available	TE 6975000
Propaphos	7292-16-2	not available	TO 6586000
Propetamphos	31218-83-4	not available	GQ 4750000
Prothiofos	34643-46-4	not available	TD 5680000
Pyraclofos	77458-01-6	not available	TE 8346000
Pyridaphenthion	119-12-0	not available	TF 2275000
Quinalphos	13593-03-8	not available	TF 6125000
Sulfotep	3689-24-5	1703, 1704	XN 4375000
Sulprofos	35400-43-2	not available	TE 4165000
Temephos	3383-96-8	2783, 2784, 3017, 3018	TF 6890000
Terbufos	13071-79-9	2783, 2784, 3017, 3018	TD 7200000
Tetrachlorvinphos	22248-79-9	2783	TB 9100000
Thiometon	640-15-3	not available	TE 4375000
Triazophos	24017-47-8	2783, 2784, 3017, 3018	TE 5635000
Trichlorfon	52-68-6	2783, 2784, 3017, 3018	TA 0700000
Vamidothion	2275-23-2	2783, 2784, 3017, 3018	TF 7900000

UN 1703 Tetraethyl dithiopyrophosphate and compressed gas mixture

UN 1704 Tetraethyl dithiopyrophosphate liquid or mixtures

UN 2783 OP solid, toxic, NOS

UN 2784 OP liquid, flammable, toxic, NOS, flashpoint <23°C

UN 3017 OP liquid, toxic, flammable

UN 3018 OP liquid, toxic, NOS

UN 3279 Organophosphorus compound, poisonous, flammable, NOS

Appendix iii Chemical Formulae and Molecular Weights for Organophosphates

Organophosphate	Formula	Molecular Weight
Acephate	$C_4H_{10}NO_3PS$	183.2
Azamethiphos	$C_9H_{10}ClN_2O_5PS$	324.7
Azinphos–ethyl	$C_{12}H_{16}N_3O_3PS_2$	345.4
Azinphos–methyl	$C_{10}H_{12}N_3O_3PS_2$	317.3
Cadusafos	$C_{10}H_{23}O_2PS_2$	270.4
Chlorfenvinphos	$C_{12}H_{14}Cl_3O_4P$	359.6
Chlormephos	$C_5H_{12}ClO_2PS_2$	234.7
Chlorpyrifos	$C_9H_{11}Cl_3NO_3PS$	350.6
Chlorpyrifos–methyl	$C_7H_7Cl_3NO_3PS$	322.5
Coumaphos	$C_{14}H_{16}ClO_5PS$	362.8
Cyanophos	$C_9H_{10}NO_3PS$	243.2
Demeton-S-methyl	$C_6H_{15}O_3PS_2$	230.3
Diazinon	$C_{12}H_{21}N_2O_3PS$	304.3
Dichlorvos	$C_4H_7Cl_2O_4P$	221.0
Dicrotophos	$C_8H_{16}NO_5P$	237.2
Dimethoate	$C_5H_{12}NO_3PS_2$	229.3
Dimethylvinphos	$C_{10}H_{10}Cl_3O_4P$	331.5
Disulfoton	$C_8H_{19}O_2PS_3$	274.4
EPN	$C_{14}H_{14}NO_4PS$	323.3
Ethion	$C_9H_{22}O_4P_2S_4$	384.5
Ethoprophos	$C_8H_{19}O_2PS_2$	242.3

Organophosphate	Formula	Molecular Weight
Famphur	$C_{10}H_{16}NO_5PS_2$	325.3
Fenamiphos	$C_{13}H_{22}NO_3PS$	303.4
Fenitrothion	$C_9H_{12}NO_5PS$	277.2
Fenthion	$C_{10}H_{15}O_3PS_2$	278.3
Fonofos	$C_{10}H_{15}OPS_2$	246.3
Formothion	$C_6H_{12}NO_4PS_2$	257.3
Fosthiazate	$C_9H_{18}NO_3PS_2$	283.3
Heptenophos	$C_9H_{12}ClO_4P$	250.6
Isazofos	$C_9H_{17}ClN_3O_3PS$	313.7
Isofenphos	$C_{15}H_{24}NO_4PS$	345.4
Isoxathion	$C_{13}H_{16}NO_4PS$	313.3
Malathion	$C_{10}H_{19}O_6PS_2$	330.3
Mecarbam	$C_{10}H_{20}NO_5PS_2$	329.4
Methacrifos	$C_7H_{13}O_5PS$	240.2
Methamidophos	$C_2H_8NO_2PS$	141.1
Methidathion	$C_6H_{11}N_2O_4PS_3$	302.3
Mevinphos	$C_7H_{13}O_6P$	224.1
Monocrotophos	$C_7H_{14}NO_5P$	223.2
Naled	$C_4H_7Br_2Cl_2O_4P$	380.8
Omethoate	$C_5H_{12}NO_4PS$	213.2
Oxydemeton–methyl	$C_6H_{15}O_4PS_2$	246.3
Parathion	$C_{10}H_{14}NO_5PS$	291.3
Parathion–methyl	$C_8H_{10}NO_5PS$	263.2
Phenthoate	$C_{12}H_{17}O_4PS_2$	320.4
Phorate	$C_7H_{17}O_2PS_3$	260.4
Phosalone	$C_{12}H_{15}ClNO_4PS_2$	367.8
Phosmet	$C_{11}H_{12}NO_4PS_2$	317.3
Phosphamidon	$C_{10}H_{19}ClNO_5P$	299.7
Phoxim	$C_{12}H_{15}N_2O_3PS$	298.3
Pirimiphos–ethyl	$C_{13}H_{24}N_3O_3PS$	333.4
Pirimiphos–methyl	$C_{11}H_{20}N_3O_3PS$	305.3
Profenofos	$C_{11}H_{15}BrClO_3PS$	373.6
Propaphos	$C_{13}H_{21}O_4PS$	304.3
Propetamphos	$C_{10}H_{20}NO_4PS$	281.3

Organophosphate	Formula	Molecular Weight
Prothiofos	$C_{11}H_{15}Cl_2O_2PS_2$	345.2
Pyraclofos	$C_{14}H_{18}ClN_2O_3PS$	360.8
Pyridaphenthion	$C_{14}H_{17}N_2O_4PS$	340.3
Quinalphos	$C_{12}H_{15}N_2O_3PS$	298.3
Sulfotep	$C_8H_{20}O_5P_2S_2$	322.3
Sulprofos	$C_{12}H_{19}O_2PS_3$	322.4
Temephos	$C_{16}H_{20}O_6P_2S_3$	466.5
Terbufos	$C_9H_{21}O_2PS_3$	288.4
Tetrachlorvinphos	$C_{10}H_9Cl_4O_4P$	366.0
Thiometon	$C_6H_{15}O_2PS_3$	246.3
Triazophos	$C_{12}H_{16}N_3O_3PS$	313.3
Trichlorfon	$C_4H_8Cl_3O_4P$	257.4
Vamidothion	$C_8H_{18}NO_4PS_2$	287.3

Appendix iv Toxicity of Organophosphates[10]

Organophosphate	Toxicity
Chlormephos	Extremely hazardous
Coumaphos	Extremely hazardous
Disulfoton	Extremely hazardous
EPN	Extremely hazardous
Ethoprofos	Extremely hazardous
Fenamiphos	Extremely hazardous
Fonophos	Extremely hazardous
Mevinphos	Extremely hazardous
Parathion	Extremely hazardous
Parathion–methyl	Extremely hazardous
Phorate	Extremely hazardous
Phosphamidon	Extremely hazardous
Sulfotep	Extremely hazardous
Terbufos	Extremely hazardous
Azinphos–ethyl	Highly hazardous
Azinphos–methyl	Highly hazardous
Cadusafos	Highly hazardous
Chlorfenvinphos	Highly hazardous
Demeton–S–methyl	Highly hazardous
Dichlorvos	Highly hazardous
Dicrotophos	Highly hazardous
Famphur	Highly hazardous
Heptenophos	Highly hazardous
Isazofos	Highly hazardous
Isoxathion	Highly hazardous
Mecarbam	Highly hazardous
Methamidophos	Highly hazardous
Methidathion	Highly hazardous
Monocrotophos	Highly hazardous
Omethoate	Highly hazardous
Oxydemeton–methyl	Highly hazardous
Pirimiphos–ethyl	Highly hazardous
Propaphos	Highly hazardous
Propetamphos	Highly hazardous
Thiometon	Highly hazardous
Triazophos	Highly hazardous
Vamidothion	Highly hazardous
Chlorpyrifos	Moderately hazardous

Organophosphate	Toxicity
Cyanophos	Moderately hazardous
Diazinon	Moderately hazardous
Dimethoate	Moderately hazardous
Ethion	Moderately hazardous
Fenitrothion	Moderately hazardous
Fenthion	Moderately hazardous
Formothion	Moderately hazardous
Methacrifos	Moderately hazardous
Naled	Moderately hazardous
Phenthoate	Moderately hazardous
Phosalone	Moderately hazardous
Phosmet	Moderately hazardous
Phoxim	Moderately hazardous
Profenofos	Moderately hazardous
Prothiofos	Moderately hazardous
Pyraclofos	Moderately hazardous
Quinalphos	Moderately hazardous
Sulprofos	Moderately hazardous
Acephate	Slightly hazardous
Azamethiphos	Slightly hazardous
Malathion	Slightly hazardous
Pirimiphos–methyl	Slightly hazardous
Pyridaphenthion	Slightly hazardous
Trichlorfon	Slightly hazardous
Chlorpyrifos–methyl	Unlikely to present acute hazard
Temephos	Unlikely to present acute hazard
Tetrachlorvinphos	Unlikely to present acute hazard

Appendix v Occupational Exposure Standards

Organophosphate	Long-term exposure limit	Short-term exposure limit
Azinphos–methyl	$0.2\ mg.m^{-3}$	$0.6\ mg.m^{-3}$
Chlorpyrifos	$0.2\ mg.m^{-3}$	$0.6\ mg.m^{-3}$
Diazinon	$0.1\ mg.m^{-3}$	$0.3\ mg.m^{-3}$
Dichlorvos	$0.92\ mg.m^{-3}$ (0.1 ppm)	$2.8\ mg.m^{-3}$ (0.3 ppm)
Dioxathion	$0.2\ mg.m^{-3}$	no data available
Disulfoton	$0.1\ mg.m^{-3}$	$0.3\ mg.m^{-3}$
Fenchlorphos	$10\ mg.m^{-3}$	no data available
Mevinphos	$0.09\ mg.m^{-3}$ (0.01 ppm)	$0.28\ mg.m^{-3}$ (0.03 ppm)
Naled	$3\ mg.m^{-3}$	$6\ mg.m^{-3}$
Parathion	$0.1\ mg.m^{-3}$	$0.3\ mg.m^{-3}$
Parathion–methyl	$0.2\ mg.m^{-3}$	$0.6\ mg.m^{-3}$
Phorate	$0.05\ mg.m^{-3}$	$0.2\ mg.m^{-3}$
Sulfotep	$0.2\ mg.m^{-3}$	no data available

PARAQUAT

Key Points

- Paraquat is available as colourless crystals or a yellow solid, or as a commercially prepared solution
- Paraquat is potentially extremely toxic by ingestion; dermal exposure to a concentrated solution or prolonged contact with the dilute solution may cause systemic effects; negligible absorption by inhalation would be expected
- It is an irritant to skin and mucous membranes
- If a large amount has been absorbed death may occur within the first few hours or days from multi-organ failure
- Smaller doses may cause progressive lung damage
- *In the event of a large spill, stay upwind and out of low areas. Ventilate closed spaces. Protective clothing, eye protection and breathing apparatus should be worn*

First Aid

- Terminate exposure and support vital functions
- The casualty should be moved to an uncontaminated area
- If the casualty is unconscious a clear airway should be established and maintained; **avoid giving oxygen** as this may enhance toxicity
- The casualty should be transferred to hospital as soon as possible
- **Inhalation Exposure:** If the patient stops breathing, expired air resuscitation should be started immediately using a pocket mask with a one way valve, if available. It is important where the face is contaminated that expired air resuscitation is NOT attempted unless an airway with rescuer protection is used; **avoid giving oxygen** as this may enhance toxicity
- **Dermal Exposure:** Remove contaminated clothing, if possible under a shower, and place in double, sealed, clear bags and label; store the bags in a secure area away from patients and staff
- Wash the skin thoroughly with copious amounts of water and then with soap and water, taking care to avoid abrasion
- If exposure is to agricultural concentrate or there has been prolonged exposure to dilute product, transfer to hospital
- **Eye Exposure:** Irrigate thoroughly with water or saline for 15 minutes
- **Oral exposure:** transfer to hospital *as soon as possible*

Presentation

Formulations: Aqueous concentrates or soluble granules. Purchase of concentrates (>5% paraquat ion) is restricted to farmers/horticulturalists. Products available to the general public are *granular*

Agricultural concentrates: Contain paraquat only or mixtures with one or more of diquat, simazine, terbuthylazine, terbutryn, diuron, monolinuron – assume 20% (200 mg . ml^{-1}) paraquat if an unknown concentrate; plus *emetic* and a stenching agent; a dark blue dye is added

Granular products: Agricultural, up to 8% w/w (8g . 100 g^{-1}); domestic, 2.5% w/w; packaged in 57 g sachets, 1.425 g paraquat ion per sachet.

These products may also contain diquat and other herbicides with an *emetic* added.

Concentration for agricultural/domestic use should not exceed **0.5% (5 mg . ml^{-1})**; usually much weaker, pH 6.6–7.5.

Emetic added to all paraquat-containing products since 1977. This is a potent, centrally-acting triazolo-pyridine code-named PP796

Detailed Information

- Paraquat is available as colourless, hygroscopic crystals (dichloride salts) or a yellow solid (bis(methyl sulphate)salt), or as a commercially prepared solution; it is soluble in water and has a faint ammonia-like odour

- *Common synonyms* methyl viologen, the term *paraquat* applies to two technical products: 1,1'-dimethyl-4,4'-bipyridylium dichloride (the salt most used for herbicide formulations) and 1,1'-dimethyl-4,4'-bipyridylium dimethylsulphate

Identification Numbers for Paraquat

Chemical	CAS	NIOSH/RTECS
Cation	4685-14-7	DW 1960000
Dichloride	1910-42-5	DW 2275000
Dimethyl sulphate	2074-50-2	DW 2010000

- UN 2781 bipyridilium pesticide, solid toxic

- UN 2782 or 3015 bipyridilium pesticide, liquid, flammable, toxic

- UN 3016 bipyridilium pesticide, liquid, toxic

- Molecular formula $C_{12}H_{14}N_2$ (paraquat ion)
  $C_{12}H_{14}N_2Cl_2$ (paraquat dichloride)
  $C_{12}H_{14}N_2[CH_3SO_4]_2$ (paraquat dimethyl sulphate)

- Molecular weight 186.28 (paraquat ion)
  257.2 (paraquat dichloride)
  408.5 (paraquat dimethyl sulphate)

- Paraquat is a non-selective foliage-applied bipyridyl contact herbicide

- It acts rapidly by killing tissues of green plants by contact action with foliage and by some translocation of xylem[1]

- It decomposes in the presence of UV light and is hydrolysed by alkaline solutions

- Paraquat is very soluble in water; it is slightly soluble in lower alcohols and insoluble in hydrocarbons

- When heated to decomposition (approximately 300 °C) it produces toxic gases and such vapours as nitrogen oxides, hydrogen chloride and carbon monoxide

Summary of Human Toxicity

- Paraquat is potentially extremely toxic by ingestion; the more concentrated the product the greater the risk of severe effects

- Dermal exposure to a concentrated solution or prolonged contact with the dilute solution may cause systemic effects

- It is irritant to skin and mucous membranes

- Negligible absorption by inhalation would be expected; droplets produced in spray application are too large to enter the lungs, and will not cause systemic toxicity

- Paraquat is non-volatile but commercial products contain an unpleasant 'stenching agent' which may cause nausea

- The toxicity of paraquat is due to its capacity for continual redox cycling

- Exact mechanisms of toxicity are not fully understood, probably a combination of NADPH depletion and free radical formation

- Paraquat undergoes NADPH-dependent reduction to the free radical which reacts with molecular oxygen to reform the cation and produce a superoxide free radical; this continues indefinitely in the presence of NADPH and oxygen

- Superoxide free radicals disrupt cell function and structure, causing damage through lipid peroxidation which may cause cell death

- Cellular NADPH becomes depleted, partly from continual redox cycling of paraquat but also from its requirement in detoxification of lipid hydroperoxides and of hydrogen peroxide which is formed in the presence of superoxide dismutase[2,3]

- Ingestion of massive amounts of paraquat can result in death within hours of ingestion from multi-organ failure; lower doses can cause lung damage, with death resulting days or weeks later

- Three degrees of oral toxicity are recognised:[4]

  - **Mild:** (<20 mg of paraquat ion . kg^{-1}) asymptomatic or vomiting and diarrhoea, complete recovery

  - **Moderate to severe:** (20–40 mg paraquat ion . kg^{-1}) vomiting (immediate); diarrhoea (within hours), renal failure, hepatic dysfunction (approx. 1–4 days), pulmonary fibrosis (1–4 weeks)

  - Survival possible but death 2 to 3 weeks later in majority of cases and may be delayed up to six weeks

  - Survivors may experience residual pulmonary insufficiency but generally improve within a few months

  - It is suggested that diagnosis of irreversible fibrosis should be made only after a follow-up period of one year[5]

  - **Very severe:** (>40 mg paraquat ion . kg^{-1}) death from multi-organ failure may occur within 24 hours (never more than seven days)

Acute Clinical Effects

Inhalation Effects

- Nausea may occur following exposure to commercial products

- Spray exposure may lead to local irritation of nose, throat and may cause epistaxis

Dermal Effects

- Mild exposures to a dilute form may cause irritation and erythema

- Concentrated solutions can cause severe burns

- Concentrated solutions or prolonged contact with dilute solutions may cause systemic toxicity; see systemic effects below

Eye Effects

- Exposure to diluted product may result in transient stinging sensation

- Concentrated solutions may lead to severe inflammation of the cornea and conjunctiva with loss of corneal and conjunctival epithelium; mild iritis may develop over 24 hours

- Ulceration can occur with the risk of secondary infection and residual corneal scarring

Oral Effects

- Vomiting (paraquat will cause vomiting in itself and all paraquat products should contain a potent, centally-acting emetic), diarrhoea, abdominal pain and ulceration of mouth and throat may occur

- See systemic effects below

Systemic Effects

- Headache, lethargy, myalgia and renal failure (usually due to acute tubular necrosis) hypotension and tachycardia may occur

- Hepatic damage, jaundice, pancreatitis, ECG changes, coma and convulsions (rare, probably due to cerebral oedema secondary to fluid overload)

- Pulmonary effects include productive or non-productive cough, haemoptysis, pneumothorax (rare), pleural effusion and gradual onset of pulmonary fibrosis with deteriorating lung function

- Death may occur within the first few hours or days from multi-organ failure, or be delayed for several weeks from pulmonary fibrosis

Chronic Clinical Effects

Skin rashes, nail damage and epistaxis are the only likely effects in those occupationally involved with paraquat production or application.[6,7]

Management

Urine spot test is a good indicator of exposure. Blood paraquat concentration is a good indicator of toxicity. **Note:** samples must be collected/stored in **plastic,** not glass, as paraquat is inactivated by glass

Inhalation Management

- Symptomatic and supportive care

Dermal Management

- Remove any remaining contaminated clothing, place in double, sealed, clear bags and label; store in a secure area away from patients and staff

- Wash the skin with copious amounts of soap and water but with care to avoid abrasions

- *Diluted product:* symptomatic treatment of mild irritation and erythema may be needed but no further action is required

- If skin is broken or exposure prolonged, test urine (see below) to exclude systemic absorption; severe toxicity unlikely

- *Concentrate:* treat all exposures as potentially serious; urine test (see below) for all cases

- Blistering and chemical burns may develop over 1 to 3 days and should be treated as thermal burns

- If urine test remains negative for 24 hours, systemic toxicity can be discounted

- If urine test is positive obtain blood concentration for prognostic purposes; sample >4 hours post exposure

- **Note:** Samples must be collected/stored in **plastic** not glass as paraquat is inactivated by glass

Eye Management

- Irrigate thoroughly with running water or saline for 15 minutes

- Stain with fluorescein and refer to an ophthalmologist if there is any uptake of the stain

- If splashed with the concentrate all patients should be reviewed in 24 hours

- Treatment is symptomatic and supportive

Oral Management

- Urine test (see below) can confirm whether ingestion has occurred

- Violent, protracted vomiting suggests a significant amount may have been ingested but treatment should be instituted in all cases

- Give activated charcoal 2 hourly up to 6 hours post ingestion (Fuller's earth at the same dose in a 15% aqueous solution may be used as an alternative to charcoal)

- Control vomiting, preferably with a $5HT_3$ serotonin antagonist (e.g. ondansetron); prochlorperazine if these are unavailable

- Urgent rehydration is essential

- Supplemental oxygen should be avoided as this may enhance toxicity

- Monitor renal, pulmonary and hepatic function

- If urine test remains negative for 6 hours post ingestion, treatment can be discontinued

- If urine test is positive obtain blood level for prognostic purposes; sample >4 hours post ingestion

- Samples must be collected/stored in **plastic** not glass

- Contact a poisons information service for further guidance

Urine Spot Test

- Add 100 mg sodium dithionite to 10 ml of 1M sodium hydroxide solution; add 1 ml of this to 1 ml of urine

- A blue-green colour indicates a positive result

- **Note:** Use **plastic** containers, not glass, as paraquat is inactivated by glass

Summary of Environment Hazards

- In the air paraquat persists predominantly in the particulate phase; it is removed by dry and wet deposition

- In water it will be completely removed from water within 8–12 days due to adsorption on to suspended soils and sediments

- In soil paraquat is rapidly and strongly adsorbed to clay minerals and becomes biologically inactive. It is slowly degraded, probably by soil micro-organisms; photolysis and volatilisation from soil is not important

- It is expected to be almost immobile in soil; estimated half-life in field conditions is 1,000 days[1]

- Bioconcentration in the food chain is negligible

- Drinking Water Standards:
  pesticide: $0.1 \, \mu g.l^{-1}$ (UK max)

- Soil Guidelines: no data available

- Air Quality Standards: no data available

REFERENCES

1. Hall AH & Rumack BH (eds.) *TOMES System ®* Micromedex, Englewood, Colorado. CD ROM. vol.41 (exp. 31 July 1999)

2. Smith LL, 1987. Mechanism of paraquat toxicity in lung and the relevance to treatment. *Hum Tox*; 6: 41–7

3. Bismuth C, Garnier R, Baud FJ, Muszynski J & Keyes C, 1990. Paraquat poisoning: an overview of the current status. *Drug Saf*; 5: 243–51

4. Vale JA, Meredith TJ & Buckley BM, 1987. Paraquat poisoning: clinical features and immediate general management. *Hum Tox*; 6: 41–7

5. Talbot AR & Barnes MR, 1988. Radiotherapy for the treatment of pulmonary complications of paraquat poisoning. *Hum Toxicol*; 7: 325–32

6. Howard JK, 1979. A clinical survey of paraquat formulation workers. *Brit J Ind Med*; 36: 220–3

7. Senanayake N, Gurunathan G, Hart TB, Amerasinghe P, Babapulle M, Ellapola SB, Udupihille M & Basanayake V, 1993. An epidemiological study of the health of Sri Lankan tea plantation workers associated with long term exposure to paraquat. *Brit J Ind Med*; 50: 257–63

Frances Northall

PETROLEUM SPIRIT

Key Points

- Petroleum spirit (petrol) is a clear flammable liquid with a distinctive odour
- It is irritating to both the respiratory system and gastrointestinal tract
- Petrol causes CNS depression and can sensitise the myocardium to endogenous catecholamines in large exposures
- Petrol vapours are heavier than air, can cause an explosive mixture with air and may travel great distances to an ignition source to flash back and cause fire or explosion
- *In the event of a large spill, stay up wind and out of low areas. Ventilate closed spaces. Protective clothing, eye protection and breathing apparatus should be worn*

First Aid

- Terminate exposure and support vital functions
- The casualty should be moved to an uncontaminated area
- Rescuers should, ideally, be trained personnel and must be careful *not to put themselves at risk* and *so wear appropriate protective clothing and, if available, breathing apparatus*
- If the casualty is unconscious a clear airway should be established and maintained; give 100% oxygen if available
- **Inhalation Exposure:** If the patient stops breathing, expired air resuscitation should be started immediately using a pocket mask with a one way valve, if available. It is important where the face is contaminated that expired air resuscitation is NOT attempted unless an airway with rescuer protection is used
- **Dermal Exposure:** Remove contaminated clothing, if possible under a shower and place in double, sealed, clear bags and label; store the bags in a secure area away from patients and staff
- Wash the skin thoroughly with copious amounts of water with a few drops of detergent mixed in
- **Eye Exposure:** Irrigate thoroughly with water or saline for 15 minutes
- **Oral Exposure:** Encourage small quantities of oral fluids (no more than 50–100 ml in total)

Detailed Information

- Petrol is a clear to amber volatile liquid that has a distinctive odour
- *Common synonyms* gasoline, motor fuel, petroleum, petrol
- CAS 8006-61-9
- CAS 86290-81-5 (leaded)
- UN 1203
- UN 1257 (leaded)
- NIOSH/RTECS LX 3300000
- Molecular weight 100 (approx)

- Petrol is a complex mixture of at least 150 hydrocarbons with about 60–70% alkanes, 25–30% aromatics and ether and 6–9% alkenes[1]
- It is used as a fuel for sparking ignition for internal combustion engines
- Petrol vapours may form an explosive mixture with air
- Vapours may be heavier than air and collect in low or confined areas; vapours may travel to source of ignition and flash back
- It is insoluble in water but freely soluble in absolute alcohol, chloroform and benzene
- Petrol dissolves fats, oils and natural resins
- Petrol can react with peroxides, nitric acid and perchlorates
- Products of combustion include carbon monoxide or dioxide, hydrocarbons, aromatics, 1,3-butadiene, oxides of nitrogen and sulphur, lead and other trace elements, phenols and polyaromatic hydrocarbons (PAHs)
- The concentration of aromatics, the most toxic component of petrol, is depleted to about 2% in the vapour phase; benzene is also present and represents a major concern[1]

Summary of Human Toxicity

- Petrol is absorbed through the lungs, gastrointestinal tract and skin with large exposures
- Petrol causes myocardium depression and sensitises the myocardium to endogenous catecholamines via inhalation and when ingested in large amounts; this cardiac sensitisation is unpredictable in terms of exposure duration and quantity
- Petrol may be abused by inhalation
- Tetraethyl lead in leaded petrol is not of toxic significance in acute exposures via ingestion or inhalation, although repeated, large exposure or massive skin contact may cause lead poisoning
- Ingestion of 10–15 g has been reported as fatal in children, 20–50 g causes severe intoxication in adults[2]

Toxicity of Petrol[2]

Concentration	Clinical effects
550 ppm for 1 h	No effects
900–2,000 ppm for 1 h	Dizziness, irritation of eyes, nose and throat and drowsiness
10,000 ppm	Nose and throat irritation (in 2 min) Dizziness (in 4 min) Signs of intoxication (in 4–10 min)

- Occupational Exposure Standard:
  Long-term exposure limit: no data available
  Short-term exposure limit: no data available

Acute Clinical Effects

Inhalation Effects

- Petrol is irritating to the respiratory tract leading to coughing, wheezing, dizziness

- Respiratory depression, dyspnoea, cyanosis and pulmonary oedema can also occur, these may be delayed for up to 12 hours

- Inhalation may cause CNS depression and systemic effects; see systemic effects below

Dermal Effects

- Petrol is a strong irritant causing defatting of the skin

- Prolonged exposure may lead to severe irritation and chemical burns

- Petrol is absorbed through the skin with large exposures and may cause systemic effects; see systemic effects below

- Massive exposure to leaded petrol may lead to Lead poisoning (see Lead, page 139)

Eye Effects

- Petrol vapour may cause lacrimation, irritation and pain in the eye

Oral Effects

- Petrol is a gastric irritant leading to nausea and vomiting

- Aspiration may lead to cough, wheeze, frothy sputum production, cyanosis, chemical pneumonitis and pulmonary oedema

- Ingestion of large quantities may cause CNS depression and systemic effects; see systemic effects below

Systemic Effects

- CNS effects include nausea, vomiting, diarrhoea, headache, confusion, ataxia, dizziness, euphoria, loss of judgement, coma and death

- Death may occur from ventricular arrhythmias as at high concentrations the solvent sensitises the myocardium to adrenaline and other catecholamines

Chronic Clinical Effects

Irreversible encephalopathy may occur from chronic inhalation abuse with symptoms such as ataxia, tremor and dementia; symptoms may also be due to tetraethyl lead in leaded petrol. Chronic exposure causes blurred vision, colour and peripheral visual field disturbance, altered mental status, delirium, drowsiness, convulsions and sudden death. Peripheral neuropathy involving the motor nerves may be seen. Asthma-like bronchospasm has been observed in chronic exposure. Chronic inhalation studies in rats and mice indicate possible hepatic and renal damage.[2] Repeated or chronic dermal contact may result in drying of the skin, lesions and dermatitis.

IARC has determined that petrol is a potential carcinogenic risk to humans due to the benzene content.

Management

Measurement of blood concentrations is of no clinical relevance other than to confirm exposure when this is in doubt

Inhalation Management

- Maintain a clear airway, give humidified oxygen and ventilate if necessary

- If respiratory irritation occurs assess respiratory function and if necessary perform chest X-rays to check for chemical pneumonitis

- See systemic management below

Dermal Management

- Remove any remaining contaminated clothing, place in double, sealed, clear bags and label; store in a secure area away from patients and staff

- Irrigate with copious amounts of water with a few drops of detergent added

- An emollient may be required

- Treat irritation and burns symptomatically

- For large or prolonged exposure see systemic management below

Eye Management

- Irrigate thoroughly with running water or saline for 15 minutes

- Stain with fluorescein and refer to an ophthalmologist if there is any uptake of the stain

Oral Management

- Encourage oral fluids for small ingestion

- DO NOT INDUCE EMESIS due to risk of aspiration

- Consider gastric lavage within 1 hour of a substantial ingestion, ensuring airway is protected

- Contact a poisons information service for further guidance on gut decontamination

- Symptomatic and supportive care; see systemic management below

Systemic Management

- Patients should be kept at rest

- With severe exposures the patient should be kept on a cardiac monitor for 6 hours, avoiding the use of all stimulants, except for resuscitation[3]

- Treat pulmonary oedema with PEEP or CPAP ventilation

- Symptomatic and supportive care

- For management of organo-lead toxicity see page 139

Summary of Environmental Hazards

- Run off from fire or spills to sewers may create a fire or explosion hazard

- Petrol is harmful to aquatic life at low concentrations

- It is not expected to bioconcentrate or bioaccumulate in the food chain unless lead is present

- Drinking Water Guidelines:
    hydrocarbon total: $10\,\mu g.l^{-1}$ (UK max)

- Soil Standards: Dutch Criteria:
    mineral oil: $50\,mg.kg^{-1}$ (target)
    $5000\,mg.kg^{-1}$ (intervention)

- Air Quality Standards: no data available

REFERENCES

1. Hall AH & Rumack BH (eds.) *TOMES System®* Micromedex, Englewood, Colorado. CD ROM. vol.41 (exp. 31 July 1999)

2. Clayton GD & Clayton FE (eds.) *Patty's Industrial hygiene and toxicology*, 4th edn. John Wiley & Sons, Inc., New York, 1994

3. Ellenhorn MJ, Schonwalds S, Ordog G & Wasserberger J. *Ellenhorn's Medical Toxicology – Diagnosis and Treatment of Human Poisoning*, 2nd edn. Williams & Wilkins, London, 1997

Henrietta Wheeler

PHENOL

Key Points

- Phenol crystals are colourless and hydroscopic with a characteristic odour, turning pink on exposure to air
- It may be fatal if inhaled, swallowed or through dermal exposure
- Contact to skin and eyes may cause severe burns
- Concentrated phenol is extremely corrosive and may cause oral and oesophageal burns following ingestion
- Systemic symptoms may include methaemoglobinaemia, arrythmias, pulmonary oedema, convulsions and coma; liver and renal damage may occur
- Fire may produce irritating or poisonous gases
- *In the event of a large spill, stay upwind and out of low areas. Ventilate closed spaces. Protective clothing and breathing apparatus should be worn*
- *Phenol vapours may evaporate from heavily contaminated casualties and so there is a risk that emergency personnel may become contaminated*
- *Casualties should be transported in such a way that there is no risk of the drivers of the emergency vehicles becoming contaminated by the fumes*

First Aid

- Terminate exposure and support vital functions
- The casualty should be moved to an uncontaminated area
- Rescuers should, ideally, be trained personnel and must be careful *not to put themselves at risk* and *so wear appropriate protective clothing and, if available, breathing apparatus*
- If the casualty is unconscious a clear airway should be established and maintained; give 100% oxygen if available
- **Inhalation Exposure:** If the patient stops breathing, expired air resuscitation should be started immediately using a pocket mask with a one way valve, if available. It is important where the face is contaminated that expired air resuscitation is NOT attempted unless an airway with rescuer protection is used
- **Dermal Exposure:** Rapid decontamination is vital; immediate action can make a major difference in odds of survival
- Removed contaminated clothing, if possible under a shower and place in double, sealed, clear bags and label; store the bags in a secure area away from patients and staff
- Wash the skin thoroughly with copious amounts of water; if available, the affected area should be repeatedly swabbed with fresh bandages soaked in polyethylene glycol 300 or 400
- **Eye Exposure:** Irrigate thoroughly with water or saline for 15 minutes
- **Oral Exposure:** Encourage small quantities of oral fluids (no more than 50–100 ml in total)

Detailed Information

- Phenol appears as colourless pointed crystals or as a white, crystalline mass with a characteristic odour

- *Common synonyms* carbolic acid, hydroxybenzene, phenyl alcohol, phenylic acid
- CAS 108-95-2
- UN 1671 (solid)
- UN 2312 (molten)
- UN 2821 (solution)
- NIOSH/RTECS SJ 3325000
- Molecular formula C_6H_6O
- Molecular weight 94.11
- Phenol is used as a disinfectant and as a chemical intermediate
- Phenol crystals turn pink on exposure to air; vapours mix readily with air
- Its solubility in water is limited at room temperature
- Soluble in alcohol, ether, chloroform, glycerol, carbon disulphide, petrolatum, fixed or volatile oils, and alkalies[1]
- Phenol reacts with oxidants and attacks aluminium, lead and zinc

Summary of Human Toxicity

- Phenol is readily absorbed through the skin and gastrointestinal tract
- It has a low vapour pressure and inhalation is usually not a major route of exposure although it is well absorbed by the lungs; the risk of toxicity via inhalation is greater in an industrial environment where large volumes may be involved
- The mechanism of action of phenolic compounds is poorly understood; it denatures proteins and is a protoplasmic poison toxic to all cells
- Phenol burns are often associated with no pain because phenol destroys nerve fibres either by demyelination or an unknown mechanism
- Phenol is excreted by the kidneys but some is lost via lungs and this can produce a phenolic odour on the breath of exposed patients
- Ingestion of 0.6 g has produced no symptoms[2]
- Lethal oral dose is estimated to be 15 g, although deaths from smaller doses have been reported[3]
- Fatalities have occurred after ingestion of as little as 2 g and 4.8 g of pure phenol[2]
- Occupational Exposure Standards:
  Long-term exposure limit: 5 ppm (20 mg.m^{-3})
  Short-term exposure limit: 10 ppm (39 mg.m^{-3})

Acute Clinical Effects

Inhalation Effects

- Not the usual route of exposure

- Symptoms expected would be headache, nausea, vertigo and salivation

- Systemic effects are possible following large or prolonged exposure; see systemic effects below

Dermal Effects

- Burns which are often painless with white or brown necrotic lesions

- The brown discoloration may remain after healing

- Phenol burns are generally less severe than those produced by acids or alkalis

- Even dilute solutions of 1% may cause dermal burns if contact is prolonged

- Severe effects have followed skin exposure to 5% phenol

- Systemic effects from dermal exposure occur quickly, usually within a few hours; see systemic effects below

Eye Effects

- Phenol vapour and liquid in the eye may result in severe pain and photophobia

- Concentrated phenol is extremely corrosive and injury may include epithelial ulceration and stromal opacity; partial or total loss of vision may occur

- Corneal denudation has also been reported[4]

Ingestion Effects

- Exposure to small amounts would lead to irritation of the mucous membranes and gastrointestinal tract

- Large amounts cause burning pain in the mouth, white or brown necrotic lesions in the mouth and oesophagus, vomiting and abdominal pain

- Systemic effects may develop; see systemic effects below

- Laryngeal oedema may develop

- Oesophageal stricture may subsequently occur

Systemic Effects

- Brief CNS stimulation with excitation and confusion, may be observed

- CNS depression with coma, hypothermia and respiratory depression with cyanosis and strenuous breathing

- There may be pallor, sweating and shock

- Pulmonary oedema may occur rapidly, the mechanism is unclear

- Cardiac depression may result in hypotension and ventricular tachycardia

- Convulsions may occur and are often delayed

- Metabolic acidosis and methaemoglobinaemia may occur

- The urine may be dark due to the presence of phenol metabolites

- Renal complications are possible and renal failure occasionally occurs

- Liver damage may occur

- Death is usually due to respiratory, cardiac or circulatory failure

Chronic Clinical Effects

Chronic phenol poisoning is rare, although it used to be more common when phenol was used in medicine and surgery.[5] Daily exposure can lead to tolerance. Repeated exposure to high levels can cause vomiting, difficulty swallowing, salivation, diarrhoea, weakness, headache, fainting, dizziness and mental disturbances.[6] Kidney and liver failure, muscle pain, loss of appetite and weight, dark urine, skin eruptions and changes in pigmentation have also been reported.

Management

Inhalation Management

- Not the normal route of exposure

Dermal Management

- *Health carers must wear protective clothing*

- Remove any remaining contaminated clothing, place in double, sealed, clear bags and label; store in a secure area away from patients and staff

- Any remaining liquid should be blotted off the skin

- The removal of phenolic compounds from the skin is controversial, most of these compounds are poorly soluble in water; polyethylene glycol has been used but has not been shown to be any more effective than copious amounts of water

- Decontamination of the skin should not be delayed and irrigation with copious amounts of water is probably the easiest method

- Medical observation may be required due to the risk of systemic effects; see below

Eye Management

- Irrigate thoroughly with running water or saline for 15 minutes

- Stain with fluorescein and refer to an ophthalmologist if there is any uptake of the stain

Oral Management

- DO NOT INDUCE EMESIS

- Consider gastric lavage within 1 hour of a substantial ingestion if the patient has no evidence of burns to the mouth or oesophagus

- Give activated charcoal if possible either orally or by nasogastric tube if necessary

- Contact a poisons information service for further guidance on gut decontamination

- Observation is advisable for a minimum of four hours post ingestion, with cardiac monitoring in patients showing systemic effects; see below

Systemic Management

- Management of systemic effects is supportive with IV fluids for dehydration, diazepam for convulsions and anti-arrhythmic therapy if indicated

- Monitor pH, ECG, liver and renal function

- Sodium bicarbonate may resolve some effects of systemic poisoning even in the absence of metabolic acidosis

- Patients with systemic effects should be observed in hospital for at least 24 hours

Summary of Environmental Hazards

- Phenol in the atmosphere exists mainly as vapour with an estimated half-life by reaction with hydroxyl radicals in air of 0.61 days[4]

- Biodegradation in water is rapid, in order of hours to days in freshwater and a few weeks in anaerobic conditions[4]

- Biodegradation in soil is rapid, 2 to 5 days; due to this ground water is generally not contaminated

- Phenol is very toxic to fish in low concentrations, it taints the taste of the fish if present in the marine environment at 0.1 to 1.0 ppm[4]

- Phenol is not expected to bioaccumulate in fish

- Drinking Water Standards:
  phenol: $0.5\,\mu g.l^{-1}$ (UK max)
  hydrocarbon total: $10\,\mu g.l^{-1}$ (UK max)

- Soil Guidelines: Dutch Criteria:
  $0.5\,mg.kg^{-1}$ (detection limit) (target)
  $40\,mg.kg^{-1}$ (intervention)

- Air Quality Standards: no data available

REFERENCES

1. Sax NI & Lewis RJ. *Hawley's Condensed Chemical Dictionary*, 11th edn. Van Nostrand Reinhold Company, New York, 1987

2. Ellenhorn MJ, Schonwalds S, Ordog G & Wasserberger J. *Ellenhorn's Medical Toxicology - Diagnosis and Treatment of Human Poisoning*, 2nd edn. Williams & Wilkins, London, 1997

3. Haddad LM & Winchester JF (eds.) *Clinical Management of Poisoning and Drug Overdose*, 2nd edn. WB Saunders Co., Philadelphia, 1990

4. Hall AH & Rumack BH (eds.) *TOMES System* ® Micromedex, Englewood, Colorado. CD ROM. vol.41 (exp. 31 July 1999)

5. Clayton GD & Clayton FE (eds.) *Patty's Industrial hygiene and toxicology*, 4th edn. John Wiley & Sons, Inc., New York, 1994

6. Sax NI. *Dangerous Properties of Industrial Materials*, 6th edn. Van Nostrand Reinhold, New York, 1984

Henrietta Wheeler

PHOSGENE

Key Points

- Phosgene is a colourless gas with a suffocating odour like musty hay
- Phosgene may be fatal on inhalation
- Exposure may lead to severe pulmonary, eye and dermal irritation
- Dermal and eye contact may cause burns
- Signs of toxicity from inhalation may be delayed for 24 to 72 hours
- Contact with liquid phosgene may cause frostbite
- *Phosgene is heavier than air and may accumulate in low or confined areas; in the event of a large spill, stay upwind and out of low areas. Ventilate closed spaces. Protective clothing, eye protection and breathing apparatus should be worn*

FIRST AID

- Terminate exposure and support vital functions
- The casualty should be moved to an uncontaminated area
- Rescuers should, ideally, be trained personnel and must be careful *not to put themselves at risk* and *so wear appropriate protective clothing and, if available, breathing apparatus*
- If the casualty is unconscious a clear airway should be established and maintained; give 100% oxygen if available
- **Inhalation Exposure:** If the patient stops breathing, expired air resuscitation should be started immediately using a pocket mask with a one way valve, if available. It is important where the face is contaminated that expired air resuscitation is NOT attempted unless an airway with rescuer protection is used
- **Dermal Exposure:** If frostbite occurs *do not* remove clothing, flush skin with water
- If frostbite is not present remove contaminated clothing, if possible under a shower and place in double, sealed, clear bags and label; store the bags in a secure area away from patients and staff
- Wash the skin thoroughly with copious amounts of water
- **Eye Exposure:** If the eye tissue is frozen seek medical advice as soon as possible
- If tissue is not frozen, irrigate thoroughly with water or saline for 15 minutes

Detailed Information

- Phosgene is a colourless gas which is easily liquefied to a colourless or pale yellow liquid, at low concentrations it has an aroma of musty hay
- *Common synonyms* carbonyl chloride, carbonic dichloride, chloroformyl chloride
- CAS 75-44-5
- UN 1076
- NIOSH/RTECS SY 5600000
- Molecular formula CCl_2O

- Molecular weight 98.92
- Phosgene is used in organic synthesis, especially of isocyanates, polyurethane and polycarbonate resins, carbamates, pesticides, herbicides and dye manufacture[1]
- Phosgene was developed as a chemical weapon by Germany during WW1 and was first used against British troops in December 1915; 88 tons were released, causing 1069 casualties and 120 deaths[2]
- Phosgene is heavier than air and does not photolyse or react with common reactive species (e.g. hydroxyl radicals or ozone); it is extremely persistent in the atmosphere and can therefore be transported in plumes a relatively long way from a spill
- It is slightly soluble in water and is freely soluble in benzene, chloroform, glacial acetic acid, toluene and most liquid hydrocarbons[3,4]
- Phosgene has no specific pH value, however, it is considered that solutions of phosgene in water are slightly acidic
- Reacts violently with anhydrous ammonia, strong oxidants and many other substances; reacts slowly with water forming hydrochloric acid and carbon dioxide
- Attacks many metals in moist environments
- When phosgene is heated to decomposition it gives off oxides of carbon, chloride and hydrochloric acid

Summary of Human Toxicity

- Phosgene is absorbed by inhalation, it is not absorbed by the skin
- Phosgene toxicity is divided into three phases: initial reflex phase, latent phase and oedema phase
- During the latent phase, phosgene reaches the terminal spaces of the lung where hydrolysis occurs, yielding hydrochloric acid
- Although hydrochloric acid may cause some toxic effects of phosgene, acylation of macromolecules may be the initiating event in phosgene toxicity
- Membrane function breakdown, fluid leaks from the capillaries into the interstitial space and gradual increasing pulmonary oedema occurs
- In time the air spaces are diminished, leading to insufficient oxygen
- Concentrations greater than 3 ppm may not be immediately accompanied by symptoms which may be delayed for 24 hours
- Olfactory fatigue occurs with repeated or prolonged exposure, hence phosgene is considered to have poor odour warning properties

Toxicity of Phosgene[4]

Concentration	Clinical effects
0.125 ppm	Workers exposed for a long time (time not specified), no effects
0.5–1 ppm	Odour detected
3 ppm	Throat irritation
3 ppm (for 170 min)	May be fatal
4 ppm	Immediate eye irritation
4.8 ppm	Cough
25 ppm	Short exposures may cause lower respiratory tract irritation
50 ppm	Brief exposures may be rapidly fatal

- Occupational Exposure Standards:
  Long-term exposure limit: 0.02 ppm (0.08 mg.m^{-3})
  Short-term exposure limit: 0.06 ppm (0.25 mg.m^{-3})

Acute Clinical Effects

Inhalation Effects

In many cases the course of phosgene toxicity following inhalation can be divided into three phases:

Initial reflex phase:

- **Note:** *the severity of the initial phase is no indication of the severity of the poisoning*

- Initially irritation to the eyes, lacrimation and blepharospasm may occur

- Followed by coughing, wheezing, retrosternal and epigastric pain, nausea and vomiting

- Shortness of breath, tachypnoea, hypoventilation, cyanosis and respiratory acidosis may develop

- Hypotension, bradycardia, rarely sinus arrhythmias can occur

- Severe cases (exposure level > 50 ppm) ARDS may develop with apnoea, bronchoconstriction, bronchial epithelium desquamation and inflammation

- Haemolysis in pulmonary capillaries and haematin formation; congestion by erythrocyte fragments may lead to blockage of capillary circulation

- Death may occur as a result of acute right heart failure or asphyxiation

Latent Phase

- Duration 1 to 24 hours (dose dependent)

- Clinically the patient may appear well, these symptoms may be precipitated by exercise, collapse of patients exposed to phosgene on subsequently taking exercise was often reported in WW1[2]

- Blood plasma may be present in the alveolar space leading to basal crepitations with oedema fluid visible on chest X-ray

- Raised lactate dehydrogenase and respiratory or metabolic acidosis

Oedema Phase

- Dyspnoea, hypoxia, production of frothy sputum which may be white or yellowish and then later become pink-tinged[3]

- Necrosis of bronchial mucous membranes has been reported

- Severe hypotension and tachycardia, and methaemoglobinaemia may occur

- Pulmonary arterial pressure may remain normal until the terminal phase

- Death, when it occurs, is due to respiratory arrest as a result of hypoxia

Long term sequelae

- Following recovery from an acute exposure blood gases may remain abnormal for a month, dyspnoea and increased bronchial resistance may be present for several months[5]

Dermal Effects

- On contact with liquid phosgene, burns and frostbite may occur which may lead to necrosis

Eye Effects

- *Gas:* Smarting of eyes, irritation, lacrimation, swelling and pain

- *Liquid:* Severe irritation, photophobia, corneal opacification and perforation have been reported[5]

Oral Effects

- Phosgene as a gas is unlikely to be ingested

- There is no information on ingestion of phosgene in liquid form

Chronic Clinical Effects

Chronic bronchitis and emphysema may occur following phosgene exposure.[2]

Management

Plasma phosgene concentrations are not clinically significant.

Inhalation Management

- **Note:** *The severity or duration of the initial reflex phase does not reliably indicate the potential for severe clinical effects*

- The patient should undertake no exercise and should be on strict bed rest as soon as possible

- Symptomatic and supportive care as required

- Replace fluids to maintain blood volume, but monitor pulmonary wedge pressure closely to avoid fluid overload

- Patients with cyanosis require additional oxygen via a face mask or if severe via an ET tube

- Consider early intubation of symptomatic patients using CPAP or PEEP ventilation to reduce surfactant loss

- Monitor blood gases, correct a metabolic acidosis (if present) with bicarbonate to 7.2 (slight acidosis aids oxygen delivery)

- Theophylline may be used for bronchospasm (may also alleviate bradycardia and hypotension)

- Administer antibiotics if indicated

- Early administration of corticosteroids (e.g. methylprednisolone 1g per day for 5 days) may be considered to prevent pulmonary oedema

- All patients should be admitted for 24 hours

- Take a baseline chest X-ray on admission, and repeat at 3 and 8 hours post exposure, then according to symptoms

- Animal studies have suggested that nebulised surfactants (e.g. Exosurf) may improve outcome

- **Review all symptomatic patients 2 to 3 months post exposure for lung function studies and chest X-ray**

Dermal Management

- *If frostbite has occurred:* remove clothing carefully, these may need to be soaked off with tepid water; irrigate the area

- Surgical referral may be necessary

- *If frostbite has not occurred:* remove any remaining contaminated clothing

- Irrigate with copious amounts of water

- Treat burns symptomatically

- Place any contaminated clothes in double, sealed, clear bags and label; store in a secure area away from patients and staff

Eye Management

- Irrigate thoroughly with running water or saline for 15 minutes

- Stain with fluorescein and refer to an ophthalmologist if there is any uptake of the stain

Oral Management

- Not applicable

Summary of Environmental Hazards

- Phosgene is produced in the atmosphere by photoxidation of chlorinated molecules and by the thermal and photodecomposition of several chlorinated solvents in the workplace

- It is persistent in the atmosphere and would be transported long distances from source

- In water phosgene is rapidly lost by volatilisation and slowly hydrolised

- In soil its fate is unknown; while it adsorbs relatively strongly to dry soil, it is likely to volatilise and hydrolise when released in soil with a high moisture content

- Drinking Water Standards: no data available

- Soil Guidelines: no data available

- Air Quality Standards: no data available

REFERENCES

1. Sax NI & Lewis RJ. *Hawley's Condensed Chemical Dictionary*, 11th edn. Van Nostrand Reinhold Company, New York, 1987

2. Marrs TC, Maynard RL & Sidell FR. *Chemical Warfare Agents*. John Wiley & Sons, Chichester, 1996

3. Budavari S, O'Neil MJ, Smith A, Heckelman PE & Kinneary JF (eds.) *The Merck Index*, 12th edn. Merck & Co., Inc., Whitehouse Station, 1996

4. Clayton GD & Clayton FE (eds.) *Patty's Industrial Hygiene and Toxicology*, 4th edn. John Wiley & Sons, Inc., New York, 1994

5. Wyatt JP & Allister CA, 1995. Occupational phosgene poisoning: a case report and review . *J Accid Emerg Med*; 12(3): 212–3

Henrietta Wheeler

PHOSPHINE

Key Points

- Phosphine is a colourless, flammable gas with a characteristic odour described as garlicky or of decaying fish; the odour threshold does not provide sufficient warning of dangerous concentrations
- Aluminium phosphide reacts with water to release phosphine
- Exposure to high concentrations of phosphine leads to profound hypotension, collapse and death
- Inhalation leads to severe respiratory tract irritation
- Following inhalation death may be sudden, usually occurring within four days, but may be delayed for one to two weeks; following ingestion severe clinical effects may be delayed for up to a week
- *In the event of a large spill, stay upwind and out of low areas. Ventilate closed spaces. Protective clothing, eye protection and breathing apparatus should be worn*

FIRST AID

- Terminate exposure and support vital functions
- The casualty should be moved to an uncontaminated area
- Rescuers should, ideally, be trained personnel and must be careful *not to put themselves at risk* and *so wear appropriate protective clothing and, if available, breathing apparatus*
- If the casualty is unconscious a clear airway should be established and maintained; give 100% oxygen if available
- **Inhalation Exposure:** If the patient stops breathing, expired air resuscitation should be started immediately using a pocket mask with a one way valve, if available. It is important where the face is contaminated that expired air resuscitation is NOT attempted unless an airway with rescuer protection is used
- **Dermal Exposure:** Remove contaminated clothing, if possible under a shower and place in double, sealed, clear bags and label; store the bags in a secure area away from patients and staff
- Wash the skin thoroughly with copious amounts of water
- **Eye Exposure:** Irrigate thoroughly with water or saline for 15 minutes
- **Oral Exposure:** Encourage small quantities of oral fluids (no more than 50–100 ml in total) unless perforation is suspected

Detailed Information

- Phosphine is a colourless, flammable gas with a characteristic odour described as garlicky or of decaying fish; the odour threshold does not provide sufficient warning of dangerous concentrations
- *Common synonyms* hydrogen phosphide, phosphorous trihydride
- CAS 7803-51-2
- UN 2199
- NIOSH/RTECS SY 7525000

- Molecular formula H_3P
- Molecular weight 34.00
- Phosphine is used in organic preparations, phosphonium halides, doping agent for n-type semi-conductors, polymerisation initiator and condensation catalyst; fumigant for grain and other stored products and a rodenticide[1]
- A Part 1 Schedule 1 Poison under the Poisons Rules, 1992 (UK); phosphine is only to be used by farmers, growers, foresters or other qualified users; original containers must be stored in a safe place, under lock and key
- **Note:** Aluminium phosphide, which is used as a rodenticide and fumigant, reacts with water to release phosphine; this reaction may be incomplete due to the formation of a protective layer of aluminium hydroxide on the surface[2]
- Phosphine gas is heavier than air
- It is slightly soluble in water to form a neutral solution; soluble in most organic solvents
- Phosphine reacts with metals, especially copper or copper-containing alloys, which causes severe corrosion; this reaction is enhanced in the presence of ammonia (which is given off with phosphine during the decomposition of some fumigation tablets or pellets) and in the presence of moisture and salt[2]
- Phosphine is flammable and explosive in air and can autoignite at ambient temperatures[2]
- Decomposes when heated forming phosphorous and hydrogen; phosphoric acid vapours are a product of combustion

Summary of Human Toxicity

- Inhaled phosphine is freely absorbed by the lungs and gastrointestinal tract; it is excreted via the lungs
- Phosphine inhalation is the usual route of exposure; onset of symptoms is generally within the first few hours post exposure; death may be sudden, usually occurring within four days, but may be delayed for one to two weeks
- Following ingestion severe clinical effects may be delayed for up to a week
- The mechanism of toxicity has not been determined; poisoning has become rare in recent years
- The organs with the greatest oxygen requirements appear to be especially sensitive to damage and include the brain, kidneys, heart and liver
- Exposure to high concentrations of phosphine leads to profound hypotension, collapse and death
- There are three groups of clinical effects:

  1. nervous system: headache, vertigo, tremors, ataxia, convulsions, coma and death

  2. gastrointestinal: loss of appetite, thirst, nausea, vomiting, diarrhoea and severe epigastric pain

3. respiratory symptoms: pressure and pain in chest, shortness of breath and pulmonary oedema

- Inhalation of 11 mg.m^{-3} for 1–2 hours daily for 6 weeks was fatal[3]

- Lethal oral dose of aluminium phosphide for a 70 kg male is <500 mg[4]

- **Note:** Odour of decaying fish or garlic at 2 ppm, COSHH short-term occupational exposure standard is 0.3 ppm, therefore odour threshold does not provide sufficient warning of dangerous concentrations

Toxicity of Phosphine[2,5]

Concentration	Clinical effects
2 ppm	Odour detectable
7 ppm for 0.5–1 h	No serious effects
100–190 ppm for 0.5–1 h	Serious effects
290–430 ppm for 0.5–1 h	Dangerous to life
400–600 ppm for 0.5–1 h	Death
2000 ppm	Rapidly fatal

- Occupational Exposure Standards:
    Long-term exposure limit: no data available
    Short-term exposure limit: 0.3 ppm (0.42 mg.m^{-3})

Acute Clinical Effects

Inhalation Effects

- Initially there may be a dry mouth, thirst, muscle pain, chills, cough, tight chest, headache, fatigue, dizziness, ataxia, and confusion[5]

- Severe gastrointestinal effects have been reported including nausea, vomiting and diarrhoea

- Systemic effects occur; see below

Dermal Effects

- Sweating and cyanosis have been reported

- Skin irritation may occur

Eye Effects

- Little data available but it is thought that phosphine may cause irritation, lacrimation and conjunctivitis

Oral Effects

- Ingestion may lead to local mucosal irritation, epigastric pain, nausea and vomiting, swelling of the lips, mouth and larynx and abdominal pain[4]

- Tightness of the chest, coughing, headache and dizziness may occur

- Restlessness and anxiety are common following ingestion of aluminium phosphide[6]

- Systemic effects occur; see below

Systemic Effects

- Tremor, paraesthesia, diplopia, and coma have been reported[7]

- Metabolic acidosis, hypermagnesaemia and hypokalaemia may occur[6]

- In severe cases this may progress to shock, severe hypotension, tachycardia, cardiac arrhythmias and convulsions

- Sweating, cyanosis and respiratory failure may occur; ARDS has been reported after 6–24 hours[4]

- Delayed pulmonary oedema may occur[6]

- Pericarditis, renal failure and hepatic damage, including jaundice and leukopaenia may develop

- Post-mortem examinations have revealed focal myocardial infiltration and necrosis, pulmonary oedema and widespread small vessel injury[2]

Chronic Clinical Effects

Prolonged low concentrations of phosphine may lead to toothache, anaemia, bronchitis, gastrointestinal disturbances, spontaneous fractures and visual, speech and motor disturbances.

Chronic exposure may produce jaundice or hepatic damage; additional effects included astheno–vegative syndrome, renal damage, and respiratory disease.[2]

Workers chronically exposed to phosphine and other fumigants had a significant increase in cancers of the haematopoietic system.

Management

Plasma phosphine concentrations are not clinically useful.

Inhalation Management

- Maintain a clear airway, give humidified oxygen and ventilate if necessary

- If respiratory irritation occurs assess respiratory function and if necessary perform chest X-rays to check for chemical pneumonitis

- Consider the use of steroids to reduce the inflammatory response

- See systemic management below

Dermal Management

- Remove any remaining contaminated clothing, place in double, sealed, clear bags and label; store in a secure area away from patients and staff

- Irrigate with copious amounts of water

- Symptomatic and supportive care

Eye Management

- Irrigate thoroughly with running water or saline for 15 minutes

- Stain with fluorescein and refer to an ophthalmologist if there is any uptake of the stain

Oral Management

- NO EMETIC

- Encourage oral fluids

- Consider gastric lavage followed by 50 g activated charcoal within 1 hour of a substantial ingestion, ensuring the airway is protected

- Contact a poisons information service for further guidance on gut contamination

- Give plasma expanders/blood or IV fluids for shock

- Consider the use of steroids to reduce the inflammatory response

- See systemic management below

Systemic Management

- Treat pulmonary oedema with PEEP or CPAP ventilation

- Any patient who has inhaled or ingested any quantity of phosphine should be admitted for observation for 72 hours

- Monitor cardiac, renal and liver function

- Monitor and correct electrolyte imbalance, blood gases

- Control convulsions with diazepam

- Symptomatic and supportive care

Summary of Environmental Hazards

- In the atmosphere phosphine reacts with photochemically produced hydroxyl radicals or ozone with a half-life of approximately 28 hours; sunshine increases the hydroxyl radicals and may reduce the half-life to 5 hours[2]

- In soil phosphine disappeared within 18 days in dry soil and 40 days in wet soil[2]

- Drinking Water Standards:
      Phosphorus: 2200 µg.l^{-1} (UK max)

- Soil Guidelines: no data available

- Air Quality Standards: no data available

REFERENCES

1. Sax NI & Lewis RJ. *Hawley's Condensed Chemical Dictionary*, 11th edn. Van Nostrand Reinhold Company, New York, 1987

2. International Programme on Chemical Safety. *Environmental Health Criteria 73: Phosphine and Selected Metal Phosphides*. WHO, Geneva, 1988

3. Harger & Spolyar LW, 1958. Toxicity of phosphine, with a possible fatality from this poison. *Archives of Ind Health*; 18: 497–503

4. Banjay R & Wasir HS, 1988. Epidemic aluminium phosphide poisoning in Northern India. *Lancet*; 1(8589): 820–1

5. Hathaway GJ, Proctor NH & Hughes JP. *Proctor and Hughes' Chemical Hazards of the Workplace*, 4th edn. Van Nostrand Reinhold, New York, 1996

6. Singh S, Dilawari JB, Vashist R, Malhotra HS & Sharma BK, 1985. Aluminium phosphide ingestion. *Brit Med J*; 290(6475): 1110–1

7. Misra UK, Tripathi AK, Pandey R & Bhargwa B, 1988. Acute phosphine poisoning following ingestion of aluminium phosphide. *Human Toxicol*; 7(4): 343–5

Henrietta Wheeler

PHOSPHORIC ACID

Key Points

- Phosphoric acid can appear as unstable orthorhombic crystals or as a clear syrupy liquid
- It is irritant to the skin, respiratory tract, eyes, skin, mucous membranes and gastrointestinal tract, or any tissue with which it comes into contact
- Ingestion may cause corrosive damage to the gastrointestinal tract with a risk of haemorrhage and perforation
- *In the event of a large spill, stay upwind and out of low areas. Ventilate closed spaces. Protective clothing, eye protection and breathing apparatus should be worn*

FIRST AID

- Terminate exposure and support vital functions
- The casualty should be moved to an uncontaminated area
- Rescuers should, ideally, be trained personnel and must be careful *not to put themselves at risk* and *so wear appropriate protective clothing and, if available, breathing apparatus*
- If the casualty is unconscious a clear airway should be established and maintained; give 100% oxygen if available
- **Inhalation Exposure:** If the patient stops breathing, expired air resuscitation should be started immediately using a pocket mask with a one way valve, if available. It is important where the face is contaminated that expired air resuscitation is NOT attempted unless an airway with rescuer protection is used
- **Dermal Exposure:** Remove contaminated clothing, if possible under a shower and place in double, sealed, clear bags and label; store the bags in a secure area away from patients and staff
- Wash the skin thoroughly with copious amounts of water
- **Eye Exposure:** Irrigate thoroughly with water or saline for 15 minutes
- *Oral Exposure:* Encourage small quantities of oral fluids (no more than 50–100 ml in total) unless perforation is suspected

Detailed Information

- Phosphoric acid can appear as unstable orthorhombic crystals or as a clear syrupy liquid
- *Common synonyms* orthophosphoric acid, white phosphoric acid
- CAS 7664-38-2
- UN 1805
- NIOSH/RTECS TB 6300000
- Molecular formula H_3PO_4
- Molecular weight 98.00
- It is used in the manufacture of superphosphate fertilizers, other phosphate salts, polyphosphates and detergents; used as an acid catalyst in making ethylene and purifying hydrogen peroxide; as an acidulant, flavour and sequestrant in food; in dental cements; process engraving; rustproofing of metals; coagulating rubber latex and as an analytical reagent
- A 1% solution has a pH of 2.0–2.2; soluble in water and ethanol
- When mixing with water caution must be taken due to the heat produced which can cause explosive splattering; acid must always be added to water and never the reverse
- Phosphoric acid is non-flammable; when heated to decomposition, oxides of phosphorous (POx) are emitted

Summary of Human Toxicity

- Phosphoric acid exerts its effects by virtue of its acidity; it is considered less harmful than sulphuric, hydrochloric or nitric acid[1]
- Ingestion of 90–120 ml of a cleaner containing 20% phosphoric acid caused only mild corrosive damage but marked hyperphosphataemia, hypocalcaemia and acidosis[2]
- Phosphoric acid has a low vapour pressure at room temperature and does not produce respiratory or eye irritation unless it is present as a mist or spray[1]
- Occupational Exposure Standard:
  Long-term exposure limit: no data available
  Short-term exposure limit: $2 \, mg.m^{-3}$

Acute Clinical Effects

Inhalation Effects

- Inhalation of phosphoric acid mists may cause irritation to the upper respiratory tract
- Pulmonary oedema is unlikely[1]

Dermal Effects

- Irritation and erythema
- Solutions of 75% or greater may cause severe burns[3]

Eye Effects

- Within the permissible exposure limits phosphoric acid mist or spray is irritating to the eyes but acclimatised workers easily tolerate such exposures
- Splash contact with a concentrated solution may cause more severe irritation and possibly corrosive damage

Oral Effects

- Ingestion of small quantities may lead to local mucosal irritation, epigastric pain, nausea and vomiting, which may be mucoid and coffee-ground in nature
- Ingestion of a larger quantity, or a concentrated solution, may lead to oesophageal corrosion, stricture, necrosis and perforation of the gastrointestinal tract, typically more severe in the stomach and intestinal tract rather than the oesophagus
- Hyperphosphataemia, hypocalcaemia and severe metabolic acidosis
- Shock may occur in severe cases

- Pyloric stenosis often follows several weeks to years later in patients who survive an acute episode of ingestion[4]

Chronic Clinical Effects

Dermatitis may occur from chronic exposure. There is no evidence that exposure to phosphoric acid results in phosphorous toxicity.

Management

Inhalation Management

- Maintain a clear airway, give humidified oxygen and ventilate if necessary

- If respiratory irritation occurs assess respiratory function and if necessary perform chest X-rays to check for chemical pneumonitis

- Consider the use of steroids to reduce the inflammatory response

- Treat pulmonary oedema with PEEP or CPAP ventilation

- Symptomatic and supportive care

Dermal Management

- Remove any remaining contaminated clothing, place in double, sealed, clear bags and label; store in a secure area away from patients and staff

- Irrigate with copious amounts of water

- Treat burns symptomatically

Eye Management

- Irrigate thoroughly with running water or saline for 15 minutes

- Stain with fluorescein and refer to an ophthalmologist if there is any uptake of the stain

Oral Management

- NO GASTRIC LAVAGE OR EMETIC

- Encourage oral fluids, unless perforation is suspected

- Monitor ECG, electrolytes and blood gases

- Give plasma expanders/blood or IV fluids for shock and analgesics for pain

- Consider the use of steroids to reduce the inflammatory response

- Take abdominal X-ray to check for perforation

- Symptomatic and supportive care

- If facilities are available, early gastro-oesophagoscopy should be undertaken within 12 hours of the event to assess the extent and severity of the injury

- Oral administration of phosphate binders such as aluminium hydroxide, may be given to decrease the absorption of phosphate[1]

Summary of Environmental Hazards

- Phosphoric acid, a natural constituent of many fruits and their juices, is a common air contaminant

- Acidity of phosphoric acid may be readily reduced by natural water hardness minerals; although phosphate may persist indefinitely

- On soil, phosphoric acid will infiltrate downward, the rate being greater with lower concentration due to reduced viscosity

- During transport through the soil it will dissolve some of the soil material (particularly carbonate-based materials)

- During transport down towards the groundwater the acid is neutralised to some degree, although a significant amount will remain in transport

- Phosphoric acid is dangerous to aquatic life at high concentrations; it is lethal to fish in most natural waters only when the pH is 5 or less

- Drinking Water Standards:
    phosphorus: $2200\,\mu g.l^{-1}$ (UK max)

- Soil Guidelines: no data available

- Air Quality Standards: no data available

REFERENCES

1. Von Burg R, 1992. Phosphoric acid/phosphates. *J Appl Toxicol;* 12(4):301–3

2. Caravati EM, 1987. Metabolic abnormalities associated with phosphoric acid ingestion. *Ann Emerg Med;* 16(8):904–7

3. Hall AH & Rumack BH (eds.) *TOMES System ®* Micromedex, Englewood, Colorado. CD ROM. vol.41 (exp. 31 July 1999)

4. Gosselin RE, Smith RP & Hodge HC. *Clinical Toxicology of Commercial Products,* 5th edn. Williams & Wilkins, Baltimore, 1984

Nicola Bates

POTASSIUM PERMANGANATE

Key Points

- Potassium permanganate is a highly corrosive, odourless, dark purple, crystalline solid
- Dry crystals and concentrated solutions are highly corrosive; dilute solutions are only mildly irritant
- Systemic effects are not of primary importance due to poor absorption
- Evaporation of potassium permanganate is negligible at 20°C, but harmful concentrations of airborne particles (powder) can build up rapidly
- *In the event of a large spill, stay upwind and out of low areas. Ventilate closed spaces. Protective clothing, eye protection and breathing apparatus should be worn*

FIRST AID

- Terminate exposure and support vital functions
- The casualty should be moved to an uncontaminated area
- Rescuers should, ideally, be trained personnel and must be careful *not to put themselves at risk* and *so wear appropriate protective clothing and, if available, breathing apparatus*
- If the casualty is unconscious a clear airway should be established and maintained; give 100% oxygen if available
- **Inhalation Exposure:** If the patient stops breathing, expired air resuscitation should be started immediately using a pocket mask with a one way valve, if available. It is important where the face is contaminated that expired air resuscitation is NOT attempted unless an airway with rescuer protection is used
- **Dermal Exposure:** Remove contaminated clothing, if possible under a shower and place in double, sealed, clear bags and label; store the bags in a secure area away from patients and staff
- Wash the skin thoroughly with copious amounts of water
- **Eye Exposure:** Irrigate thoroughly with water or saline for 15 minutes
- **Oral Exposure:** Encourage small quantities of oral fluids (no more than 50–100 ml in total) unless perforation is suspected

Detailed Information

- Potassium permanganate is a highly corrosive, odourless, dark purple, crystalline solid
- *Common synonyms* permanganate of potash, potassium salt of permanganic acid, potassium manganateVII
- CAS 7722-64-7
- UN 1490
- NIOSH/RTECS SD 6475000
- Molecular formula $KMnO_4$
- Molecular weight 158.03
- Potassium permanganate is widely used as a strong oxidising agent in the photographic, pharmaceutical and chemical industries

- It decomposes at 240°C, giving off oxygen, with increased risk of fire; its decomposition is accelerated by light
- Potassium permanganate is soluble in water, dissociating to potassium and permanganate ions, forming a deep purple solution
- Potassium permanganate oxidises most compounds on contact including metals, organic compounds, combustible substances and reducing agents, eventually liberating manganese dioxide then manganese (II) ions; reactions may be violent or explosive
- It reacts with hydrochloric and nitric acids giving off toxic chlorine gas and nitrous vapours respectively
- Potassium permanganate is not combustible but increases the combustibility of other substances

Summary of Human Toxicity

- Dry crystals and concentrated solutions of potassium permanganate are highly corrosive to skin, eyes, upper respiratory tract and mucous membranes, rapidly causing coagulation necrosis[1,2]
- When in contact with organic matter it is reduced to manganese dioxide and the very corrosive potassium hydroxide; the reaction is exothermic, increasing injury[3,4]
- Systemic effects are not usually of primary importance due to poor absorption; the mechanism for those reported is unclear, but has been postulated to be due to free radical generation similar to paracetamol overdose[2]
- Potassium permanganate solutions are dark purple but leave a deep brown stain of manganese dioxide
- Oral or inhalation exposures can be fatal; the lethal dose for an adult is estimated as 10 g or approximately 1.5 teaspoons of the crystals[5,6]
- Solutions over 1:5000 concentration are irritating to tissues; dilute solutions are only mildly irritant[3]
- Occupational Exposure Standards: no data available

Acute Clinical Effects

Inhalation Effects

- Exposure may lead to sore throat, coughing, shortness of breath and severe breathing difficulties
- Death may occur following bronchospasm, inflammation, airway obstruction, circulatory collapse and pulmonary oedema[2]

Dermal Effects

- Exposure may result in brown staining of the skin, redness, pain and serious burns

Eye Effects

- Redness, pain and impaired vision, brown staining and hardened eroded lesions of the conjuctiva may occur

- Contact may lead to swelling of the lids and conjunctivae and subconjunctival haemorrhages

- With prolonged contact, turbidity and brown discoloration have developed

- The cornea is more likely to be involved if its epithelium has been injured mechanically

- Recovery is usually spontaneous and complete, although total leukoma (dense, white corneal opacity) has been reported following application of a strong solution of potassium permanganate[7]

Oral Exposure

- Brown staining in the mouth and throat, sore throat, abdominal pain, vomiting, moderate gastroenteritis and some dysphagia may occur[2,5]

- Swelling and bleeding of the lips and tongue, pharyngeal oedema and swelling of the larynx; long-term sequelae including oesophageal stricture and pyloric stenosis have been reported[2,4]

- Systemic effects may include tachycardia, hypotension, methaemoglobinaemia and cyanosis, haemolysis, neurological dysfunction, disseminated intravascular coagulation, acute hepatic necrosis and acute tubular necrosis, pancreatitis and coma[2,5,8]

- Manganese may be absorbed via the injured oesophageal mucosae causing elevated whole blood and tissue concentrations;[8] plasma concentrations correlate poorly with manganese toxicity, and there is very little information about acute manganese toxicity

- Death usually results from glottal oedema and cardiovascular collapse, but multi-organ failure has been reported[2]

Chronic Clinical Effects

No data available.

Management

Whole blood manganese concentration can be used as an indicator of oral exposure in severe cases as the concentration remains elevated for more than five days post ingestion.

Normal whole blood concentration:
$7-18\,\mu g.l^{-1}$ $(120-325\,nmol.l^{-1})$.

Normal plasma/serum concentrations:
$0.5-1.3\,\mu g.l^{-1}$ $(9-24\,nmol.l^{-1})$.

Whole blood manganese concentration of $> 30\,\mu g.l^{-1}$ $(> 540\,nmol.l^{-1})$ is associated with a risk of neurological toxicity.

Inhalation Management

- Maintain a clear airway, give humidified oxygen and ventilate if necessary

- If respiratory irritation occurs assess respiratory function and if necessary perform chest X-rays to check for chemical pneumonitis

- Consider the use of steroids to reduce the inflammatory response

- Treat pulmonary oedema with PEEP or CPAP ventilation

Dermal Management

- Remove any remaining contaminated clothing, place in double, sealed, clear bags and label; store in a secure area away from patients and staff

- Irrigate with copious amounts of water

- Skin burns should be treated as a thermal injury

Eye Management

- Irrigate thoroughly with running water or saline for 15 minutes

- Stain with fluorescein and refer to an ophthalmologist if there is any uptake of the stain

Oral Management

- NO GASTRIC LAVAGE OR EMETIC

- Encourage oral fluids, unless perforation is suspected

- Emergency tracheostomy may be necessary[5]

- Give plasma expanders/blood or IV fluids for shock and analgesics for pain

- Consider the use of steroids to reduce the inflammatory response

- Take abdominal X-ray to check for perforation

- If facilities are available, early gastro-oesophagoscopy should be undertaken within 12 hours of the event to assess the extent and severity of the injury

- Symptomatic and supportive care

- Early administration of N-acetylcysteine in the same dose as used in paracetamol poisoning has been postulated as an antidote for oral exposure, but its efficacy has not been proven[2]

- Long-term follow-up is necessary following severe oral exposure due to sequelae

Summary of Environmental Hazards

- Decomposition of potassium permanganate is accelerated by light

- Potassium permanganate is likely to be persistent in water and soil

- In water it dissociates to potassium and permanganate ions

- It is highly toxic to fish

- Drinking Water Standards:
  potassium: $12\,mg.l^{-1}$ (UK max)
  manganese: $50\,\mu g.l^{-1}$ (UK max)
  $500\,\mu g.l^{-1}$ (WHO provisional guideline)

- Soil Guidelines: no data available

- Air Quality Standards:
  manganese: $1\,\mu g.m^{-3}$ (WHO guideline)

REFERENCES

1. Gosselin RE, Smith RP & Hodge HC. *Clinical Toxicology of Commercial Products*, 5th edn. Williams & Wilkins, Baltimore, 1984

2. Young RJ, Critchley JAJH, Young KK, Freebairn RC, Reynolds AP & Lolin YI, 1996. Fatal acute hepatorenal failure following potassium permanganate ingestion. *Hum Exp Toxicol*; 15: 259–61

3. Dagli AJ, Golden D, Finkel M & Austin E, 1973. Pyloric stenosis following ingestion of potassium permanganate. *Dig Dis*; 18(12): 1091–4

4. Kochhar R, Das K & Mehta SK, 1986. Potassium permanganate induced oesophageal stricture. *Hum Toxicol;* 5: 393–4

5. Huntley AC, 1984. Oral ingestion of potassium permanganate or aluminium acetate in two patients. *Arch Dermatol*; 120: 1363–5

6. Ong KL, Tan TH & Cheung WL, 1997. Potassium permanganate poisoning – a rare cause of fatal self poisoning. *J Accid Emerg Med*; 14(1): 43–5

7. Grant MW & Schuman JS. *Toxicology of the Eye*, 4th edn. Charles C Thomas, Springfield, 1993

8. Mahomedy MC, Mahomedy YH, Canham PAS, Downing JW & Jeal DE, 1975. Methaemoglobinaemia following treatment dispensed by witch doctors. Two cases of potassium permanganate poisoning. *Anaesthesia*; 30: 190–3

Catherine Farrow

PROPANE

Key Points

- Propane is a colourless, odourless gas
- Inhalation is the main route of exposure; it is a simple asphyxiant
- Propane is highly flammable and an explosion hazard
- Direct skin and mucous membrane exposure to liquid propane produces severe frostbite injury
- *In the event of a large spill, stay upwind and out of low areas. Ventilate enclosed spaces. Protective clothing, eye protection and breathing apparatus should be worn*

FIRST AID

- Terminate exposure and support vital functions
- The casualty should be moved to an uncontaminated area
- Rescuers should, ideally, be trained personnel and must be careful *not to put themselves at risk* and *so wear appropriate protective clothing and, if available, breathing apparatus*
- If the casualty is unconscious a clear airway should be established and maintained; give 100% oxygen if available
- **Inhalation Exposure:** If the patient stops breathing, expired air resuscitation should be started immediately using a pocket mask with a one way valve, if available. It is important where the face is contaminated that expired air resuscitation is NOT attempted unless an airway with rescuer protection is used
- **Dermal Exposure:** If frostbite occurs *do not* remove clothing, flush skin with water
- If frostbite is not present remove contaminated clothing, if possible under a shower and place in double, sealed, clear bags and label; store the bags in a secure area away from patients and staff
- Wash the skin thoroughly with copious amounts of water
- **Eye Exposure:** If the eye tissue is frozen seek medical advice as soon as possible
- If eye tissue is not frozen, irrigate thoroughly with water or saline for 15 minutes

Detailed Information

- Propane is a colourless, odourless gas
- *Common synonyms* dimethylmethane, propyl hydride
- CAS 74-98-6
- UN 1978
- UN 1075
- NIOSH/RTECS TX 2275000
- Molecular formula C_3H_8
- Molecular weight 44.10
- Propane, a component of liquid petroleum gas, is used as a refrigerant in chemical refining and gas processing operations; also has uses as an extractant, solvent, gas enricher, aerosol propellant, and a mixture for bubble chambers[1]

- Propane vapour is heavier than air, and flammable mixtures of the gas and air may travel significant distances from a propane leak; in the event of a spill stay out of low areas[2]
- It is soluble in ether and alcohol, and slightly soluble in water
- Propane is highly flammable and explosive; it is easily ignited by heat, sparks, or flame[3]
- On combustion propane burns to carbon dioxide and water, and to carbon monoxide and water when oxygen is limited[3]

Summary of Human Toxicity

- As propane is a simple asphyxiant, the major factor in exposure is the available oxygen[4]
- Volunteers exposed to 100,000 ppm propane for a few minutes experienced dizziness but no eye or mucous membrane irritation[3]
- At concentrations less than 1000 ppm propane is physiologically inert[5]
- Accidental inhalation by an infant caused seizures and ventricular tachycardia; the patient recovered without sequelae[6]
- In dogs propane is a weak cardiac sensitiser[3]
- Occupational Exposure Standards: simple asphyxiant (i.e. when present in high concentrations in air, reducing oxygen content by dilution to such an extent that life cannot be supported. Oxygen content should be monitored and not allowed to fall below 18% to ensure safety)

Acute Clinical Effects

Inhalation Effects

- At high concentrations propane causes CNS depression and asphyxiation
- Initially dizziness, euphoria, ataxia, light-headedness are expected[4,6]
- Vagal inhibition of the heart leads to bradycardia or cardiac arrest
- Propane may cause cardiac sensitisation
- Death is due to hypoxia or vagal inhibition

Dermal Effects

- Dermal exposure to liquid propane can cause burns and frostbite injury

Eye Effects

- Direct contact with liquid propane causes burns and frostbite injury

Oral Effects

- Unlikely route of exposure

Chronic Clinical Effects

No data available, unlikely to result in adverse health effects.

Management

Measurement of blood concentrations is of no clinical relevance other than to confirm exposure when this is in doubt.

Inhalation Management

- Recovery is normally quick once exposure has ceased but support of the respiratory and cardiovascular symptoms may be required

- Maintain a clear airway, give humidified oxygen and ventilate if necessary

- Symptomatic and supportive care

Dermal Management

- *If frostbite has occurred:* remove clothing carefully, these may need to be soaked off with tepid water; irrigate the area

- Surgical referral may be necessary

- *If frostbite has not occurred:* remove any contaminated clothing

- Irrigate with copious amounts of water

- Treat burns symptomatically

- Place any contaminated clothes in double, sealed, clear bags and label; store in a secure area away from patients and staff

Eye Management

- Irrigate thoroughly with running water or saline for 15 minutes

- Stain with fluorescein and refer to an ophthalmologist if there is any uptake of the stain

Oral Management

- not applicable

Summary of Environmental Hazards

- Propane is expected to exist almost entirely in the vapour phase in ambient air; it reacts with photochemically produced hydroxyl radicals with an average half-life of 13 days[2]

- Biodegradation of propane may occur in soil and water, although volatilisation is expected to be the main fate process

- The volatilisation half-lives from a model river and a model pond (the latter considers the effect of adsorption) have been estimated to be 1.9 hours and 2.3 days, respectively[2]

- In aquatic systems, propane may partition from the water column to organic matter contained in sediments and suspended materials

- Propane is not expected to bioaccumulate

- Drinking Water Standards:
        hydrocarbon total: 10 µg.l^{-1} (UK max)

- Soil Guidelines: Dutch Criteria: no data available

- Air Quality Standards: no data available

REFERENCES

1. Sax NI. *Dangerous Properties of Industrial Materials*, 6th edn. Van Nostrand Reinhold, New York, 1984

2. Hall AH & Rumack BH (eds.) *TOMES System* ® Micromedex, Englewood, Colorado. CD ROM. vol.41 (exp. 31 July 1999)

3. Clayton GD & Clayton FE (eds.) *Patty's Industrial Hygiene and Toxicology*, 4th edn. John Wiley & Sons, Inc., New York, 1994

4. Hathaway GJ, Proctor NH & Hughes JP. *Proctor and Hughes' Chemical Hazards of the Workplace*, 4th edn. Van Nostrand Reinhold, New York, 1996

5. Arena JM & Drew RH. *Poisoning – Toxicology, Symptoms, Treatments,* 5th edn. Charles C Thomas, Spingfield, Illinois, 1985

6. Wheeler MG, Rozycki AA & Smith RP, 1992. Recreational propane inhalation in an adolescent male. *Clin Toxicol*; 30(1): 135–9

Jennifer Butler

PYRETHROID AND PYRETHRIN INSECTICIDES

Key Points

- Pyrethroids and pyrethrins have low acute toxicity; the carrier vehicle may be talc-based powder or hydrocarbon solvent and its toxicity must be considered
- Clinical effects reported with the above include muscle tremor and spasm, opisthotonus and convulsions
- Dermal reactions are usually mild, but sensitivity reactions may occur
- On inhalation, local irritation to the upper respiratory tract may occur
- Treatment likely to be necessary only for ingestion of deltamethrin, cypermethrin, alphacypermethrin or fenvalerate
- *In the event of a large spill, stay upwind and out of low areas. Ventilate closed spaces. Protective clothing, eye protection and breathing apparatus should be worn*

FIRST AID

- Terminate exposure and support vital functions
- The casualty should be moved to an uncontaminated area
- Rescuers should, ideally, be trained personnel and must be careful *not to put themselves at risk* and *so wear appropriate protective clothing and, if available, breathing apparatus*
- If the casualty is unconscious a clear airway should be established and maintained; give 100% oxygen if available
- **Inhalation Exposure:** If the patient stops breathing, expired air resuscitation should be started immediately using a pocket mask with a one way valve, if available. It is important where the face is contaminated that expired air resuscitation is NOT attempted unless an airway with rescuer protection is used
- **Dermal Exposure:** Remove contaminated clothing, if possible under a shower and place in double, sealed, clear bags and label; store the bags in a secure area away from patients and staff
- Wash the skin thoroughly with copious amounts of water
- **Eye Exposure:** Irrigate thoroughly with water or saline for 15 minutes
- **Oral Exposure:** Encourage small quantities of oral fluids (no more than 50–100 ml in total); do not give milk or fatty foods as this will enhance absorption

Detailed Information

Pyrethrins

- Pyrethrins are brown, viscous liquids or solids with the characteristic odour of the carrier vehicle
- The name pyrethrins (or pyrethrum) refers to the six naturally occurring insecticidal components of *Pyrethrum* chrysanthemums
- The compounds are: pyrethrin I, pyrethrin II (the two most insecticidally potent compounds), cinerin I, cinerin II, jasmolin I and jasmolin II

Identification for Pyrethrins

Pyrethrin	CAS	UN	NIOSH/RTECS
Pyrethrins (Pyrethrum)	8008-34-7	2902	UR 4200000
Pyrethrin I	121-21-1	n.a.	GZ 1725000
Pyrethrin II	121-29-9	n.a.	GZ 0700000
Cinerin I	2540-06-6	n.a.	n.a.
Cinerin II	1172-63-0	n.a.	n.a.
Jasmolin I	4466-14-2	n.a.	n.a.
Jasmolin II	121-20-0	n.a.	n.a.

n.a. = number not available

Molecular formula and weight of Pyrethrins

Pyrethrin	Chemical formula	Molecular weight
Pyrethrins (Pyrethrum)		
Pyrethrin I	$C_{21}H_{28}O_3$	328.46
Pyrethrin II	$C_{22}H_{28}O_5$	372.47
Cinerin I	$C_{20}H_{28}O_3$	316.4
Cinerin II	$C_{21}H_{28}O_5$	360.4
Jasmolin I	$C_{21}H_{30}O_3$	330.4
Jasmolin II	$C_{22}H_{30}O_5$	374.4

- The ratio of pyrethrin:cinerin:jasmolin is usually 71:21:7 and most commercial products contain 20–25% pyrethrins[1]
- Pyrethrins are oxidised in light and air, with loss of insecticidal activity
- Pyrethrins are insoluble in water
- Pyrethrins are incompatible with alkaline material and strong oxidisers

Pyrethroids

- Pyrethroids are synthetic substances similar to pyrethrins, modified to improve stability
- The most commonly encountered are: allethrin, cypermethrin, deltamethrin, permethrin, tetramethrin, alphacypermethrin, resmethrin, bioresmethrin, phenothrin and fenvalerate
- Piperonyl butoxide is added to many formulations as a synergist to increase stability and insecticidal effectiveness
- Available evidence suggests that the presence of piperonyl butoxide does not alter the mammalian toxicity of these compounds
- The carrier vehicle may be talc-based powder or hydrocarbon solvent

Identification for Pyrethroids

Pyrethroid	Chemical formula	Molecular weight	CAS	NIOSH/RTECS
Allethrin	$C_{19}H_{26}O_3$	302.45	584-79-2	GZ 1476000
Cypermethrin	$C_{22}H_{19}Cl_2NO_3$	416.31	52315-07-8	GZ 1250000
Deltamethrin	$C_{22}H_{19}Br_2NO_3$	505.19	52918-63-5	GZ 1233000a
Permethrin	$C_{21}H_{20}Cl_2O_3$	391.30	52645-53-1	GZ 1255000
Tetramethrin	$C_{19}H_{25}NO_4$	331.40	7696-12-0	GZ 173000
Alphacypermethrin	$C_{22}H_{19}NO_3Cl_2$	416.31	67375-30-8	not available
Resmethrin	$C_{22}H_{26}O_3$	338.48	10453-86-8	GZ 1310000
Bioresmethrin	$C_{22}H_{26}O_3$	338.48	28434-01-7	GZ 1310500
Phenothrin	$C_{23}H_{26}O_3$	350.46	26002-80-2	GZ 1975000
Fenvalerate	$C_{25}H_{22}ClNO_3$	419.91	51630-58-1	CY 1576350

Summary of Human Toxicity

- In general the acute mammalian toxicity of most pyrethrins and pyrethroids is very low due to rapid metabolism to non-active compounds

- There have been reports of serious poisoning from fenvalerate, cypermethrin, alphacypermethrin and deltamethrin via both oral and non-oral routes in adults and children; some patients developed coma after ingestion of 200–500 ml of unspecified pyrethrins; deaths have been reported following convulsions and in one case pulmonary oedema, but the exact nature of each exposure was not stated[2]

- Pyrethrins and some pyrethroids are sensitising agents

- Occupational Exposure Standards:
  Long-term exposure limit: 5 mg.m⁻³ (pyrethrins ISO)
  Short-term exposure limit: 10 mg.m⁻³(pyrethrins ISO)

Acute Clinical Effects

Note: Toxicity of the carrier must be considered

Inhalation Effects

- Many preparations will cause local irritation to the upper respiratory tract, leading to coughing, wheezing and rhinitis

Dermal Effects

- Most pyrethroids will cause local irritation and drying

- Prolonged (several hours) contact with fenvalerate has reportedly led to severe systemic effects including convulsions

Eye Effects

- Fumes or splash contact may cause a burning sensation or itching and irritation

- Carrier ingredients e.g. solvents, shampoo may be responsible for other effects

Oral Effects

- Most cases of pyrethroid ingestion result in no more than nausea, vomiting and diarrhoea, sometimes with abdominal pain

- Ingestion of deltamethrin, fenvalerate and cypermethrin has resulted in severe effects including muscle tremor and spasm, opisthotonus, convulsions, drowsiness, coma and respiratory depression

Chronic Clinical Effects

Sensitisation may also occur resulting in dermatitis upon re-exposure, particularly with pyrethrins. In sensitised patients re-exposure may lead to asthmatic attacks, this is particularly noted with pyrethrins.

Management

No common laboratory tests have diagnostic or prognostic value.

Inhalation Management

- Reassurance, symptomatic and supportive care

Dermal Management

- Wash the exposed area well with soap and water

- A simple emollient cream may help to relieve dryness

- Sensitivity reactions may respond to emollient creams or petroleum jelly

- In cases of prolonged, extensive exposure to fenvalerate the patient should be observed for at least 4 hours

- Symptomatic and supportive care

Eye Management

- Irrigate thoroughly with running water or saline for 15 minutes

- Stain with fluorescein and refer to an ophthalmologist if there is any uptake of the stain

Oral Management

- In most cases reassurance and oral fluids only

- Consider 50 g activated charcoal within 1 hour of ingestion if more than 15 mg.kg⁻¹ of deltamethrin, fenvalerate, cypermethrin or alphacypermethrin has been ingested[3]

- The effect of activated charcoal on absorption is unknown

- Contact a poisons information service for further guidance on gut decontamination

- Observe for at least 4 hours post ingestion

- Symptomatic and supportive care

Summary of Environmental Hazards

- Pyrethrins are generally unstable in the presence of light, hydrolyzed rapidly by alkali and oxidize rapidly in air

- The atmospheric half-lives for the vapour phase pyrethrin II with ozone and photochemically produced hydroxyl radicals have been estimated to be 0.5 hour and 1.3 hour, respectively

- Biological degradation is fairly rapid and agricultural dose rates are low so residues are unlikely to reach significant levels

- Permethrin disappears from ponds and streams within 6–24 hours, pond sediment within 7 days and foliage and forest soil within 58 days[4]

- Pyrethroids are highly toxic to fish; the bioaccumulation factor of cypermethrin in fish is approximately 1000 when measured experimentally, although the potential for toxicity is not reached in the field[5]

- Under aerobic conditions in soil, permethrin degrades with a half-life of 28 days

- Drinking Water Standards:
     pesticide: 0.1 μg.l^{-1} (UK max)

- Soil Guidelines: no data available

- Air Quality Standards: no data available

REFERENCES

1. Tomlin CDS (ed.) *The Pesticide Manual*, 11th edn. Crop Protection Publications, Farnham, 1997

2. He F, Wang S, Liu L, Chen S, Zhang Z & Sun J, 1989. Clinical manifestations and diagnosis of acute pyethroid poisoning. *Arch Toxicol;* 63: 54–5

3. Bates N, Edwards N, Roper J & Volans G (eds.) *Paediatric Toxicology. Handbook of Poisoning in Children*. Macmillan, London, 1997

4. International Programme on Chemical Safety. *Health and Safety Guide 33: Permethrin*. WHO, Geneva, 1989

5. International Programme on Chemical Safety. *Environmental Health Criteria 82. Cypermethrin*. WHO, Geneva, 1989

Elizabeth Schofield

PYRIDINE

Key Points

- Pyridine is a colourless or yellow liquid with a distinctive odour
- It is an irritant to the skin and mucous membranes
- Pyridine exposure may cause CNS and respiratory depression
- Ingestion of pyridine may cause liver and renal damage
- *Pyridine vapours are heavier than air and may accumulate in low or confined areas; in the event of a large spill, stay upwind and out of low areas. Ventilate closed spaces. Protective clothing, eye protection and breathing apparatus should be worn*

FIRST AID

- Terminate exposure and support vital functions
- The casualty should be moved to an uncontaminated area
- Rescuers should, ideally, be trained personnel and must be careful *not to put themselves at risk* and *so wear appropriate protective clothing and, if available, breathing apparatus*
- If the casualty is unconscious a clear airway should be established and maintained; give 100% oxygen if available
- **Inhalation Exposure:** If the patient stops breathing, expired air resuscitation should be started immediately using a pocket mask with a one way valve, if available. It is important where the face is contaminated that expired air resuscitation is NOT attempted unless an airway with rescuer protection is used
- **Dermal Exposure:** Remove contaminated clothing, if possible under a shower and place in double, sealed, clear bags and label; store the bags in a secure area and away from patients and staff
- Wash the skin thoroughly with copious amounts of water until the pH is no longer alkaline
- **Eye Exposure:** Irrigate thoroughly with water or saline until pH is no longer alkaline
- *NB:* Testing the pH of the skin and eyes immediately following irrigation may be misleading. It is recommended that 15 minutes elapse before this is undertaken. Continue the irrigation process if the skin or eyes are still alkaline
- **Oral Exposure:** Encourage small quantities of oral fluids (no more than 50–100 ml in total)

Detailed Information

- Pyridine is a colourless or yellow liquid with a nauseating odour and a burning taste
- *Common synonyms* azabenzene, azine
- CAS 110-86-1
- UN 1282
- NIOSH/RTECS UR 8400000
- Molecular formula C_5H_5N
- Molecular weight 79.10

- Pyridine is used as a solvent in drug manufacture, a reagent in industrial processes where hydrochloric acid is evolved and for making polyhydrocarbonate resins; it is the starting material for waterproofing agents and occasionally used as a dyeing assistant in the textile industry; the majority of pyridine is used in the manufacture of herbicides and insecticides[1]
- Several food items have been found to contain pyridine, either present in the raw food or formed during cooking; pyridine is found in tobacco smoke[2]
- Prior to 1950 pyridine was occasionally used as an anticonvulsant agent[2]
- It is an alkaline liquid that forms salts with strong acids and a 0.2 M solution of pyridine in water has a pH of 8.5[2]
- Vapours may form explosive mixtures with air; vapours may travel to source of ignition and flash back
- Pyridine is miscible with water, alcohol, ether, petroleum ether, oils and many other organic liquids
- Pyridine is flammable and easily ignited by heat, sparks and flames; products of combustion include carbon monoxide, carbon dioxide and oxides of nitrogen
- When heated to decomposition, cyanide fumes are released

Summary of Human Toxicity

- Pyridine is absorbed orally, dermally and via the respiratory tract
- Urinary excretion of both the parent compound and metabolites is the primary method of elimination, although some is excreted in the faeces, skin and lungs
- Odour detection is not sufficient warning of the hazard as olfactory fatigue occurs
- Ingestion of half a cup (specific amount not stated) was fatal in an adult
- Ingestion of 1.8 to 2.5 ml per day for 2 months produced liver and kidney damage in adults
- The lethal dose in man is estimated to be 0.5 to 5 $g.kg^{-1}$ [2]

Toxicity of Pyridine[2,3]

Concentration	Clinical effects
0.17 ppm	Odour threshold
1 ppm	Unpleasant odour detectable
10 ppm	Objectionable odour
6–12 ppm	CNS symptoms commence
6–12 ppm – chronic exposure	Headache, vertigo, insomnia, nausea, vomiting
125 ppm 4 h per day for 1 to 2 weeks	Nausea, headache, insomnia and nervousness

- Occupational Exposure Standards:
  Long-term exposure limit: 5 ppm (16 mg.m^{-3})
  Short-term exposure limit: 10 ppm (33 mg.m^{-3})

Acute Clinical Effects

Inhalation Effects

- Pyridine exposure may lead to respiratory, eye and nose irritation

- Peripheral weakness may occur following inhalation

- Inhalation of vapours has caused urinary frequency without renal damage; lower back pain has been reported with urinary frequency and systemic toxicity; see below

Dermal Effects

- Skin irritation may occur with prolonged or repeated contact

- Pyridine may be a photosensitising agent and prolonged exposure may lead to systemic toxicity; see below

Eye Effects

- Pyridine vapours are irritating to the eyes

Oral Effects

- Ingestion of pyridine may lead to gastrointestinal irritation and systemic toxicity; see below

- Pyrexia and dysphagia have been reported following acute ingestion[2]

Systemic Effects

- Headache, insomnia and nervousness may occur following all routes of exposure

- Nausea, vomiting, abdominal discomfort, diarrhoea and anorexia have been reported

- Severe exposure may result in respiratory depression; drowsiness leading to coma; pulmonary oedema and bronchitis have been reported from both ingestion and inhalation

- CNS depression and respiratory depression may lead to respiratory paralysis

- Theoretically pyridine exposure may induce methaemoglobinaemia

Chronic Clinical Effects

Transient symptoms of overexposure are nausea, headache, insomnia and nervousness. Chronic exposure may lead to emphysema and chronic bronchitis. Chronic oral exposure may lead to faintness, dizziness, fatigue, headaches, insomnia, nervousness and speech disorder. Chronic ingestion may lead to liver and kidney damage.

Allergic contact dermatitis may occur after prolonged exposure.

Management

If the pyridine exposure is to the decomposition product of pyridine, refer also to the cyanide entry on page 124.

Inhalation Management

- Maintain a clear airway, give humidified oxygen and ventilate if necessary

- If respiratory irritation occurs assess respiratory function and if necessary perform chest X-rays to check for chemical pneumonitis

- Consider the use of IV steroids to reduce the inflammatory response

- Treat pulmonary oedema with PEEP or CPAP ventilation

- Symptomatic and supportive care; see systemic management below

Dermal Management

- Remove any remaining contaminated clothing, place in double, sealed, clear bags and label; store in a secure area away from patients and staff

- Wash contaminated area with copious amounts of water until the skin is no longer alkaline

- Testing the pH of the skin immediately after irrigation may be misleading. It is recommended that 15 minutes elapse before this is undertaken. Continue the irrigation process if the skin is still alkaline

- Treat dermal irritation or burns symptomatically

- See systemic management below

Eye Management

- Irrigate thoroughly with running water or saline for 15 minutes

- Testing the pH of the eyes immediately following irrigation may be misleading. It is recommended that 15 minutes elapse before this is undertaken. Continue the irrigation process if the eye is still alkaline

- Stain with fluorescein and refer to an ophthalmologist if there is any uptake of the stain

Oral Management

- NO GASTRIC LAVAGE OR EMETIC

- Encourage oral fluids, unless perforation is suspected

- See systemic management below

Systemic Management

- Monitor respiratory function, arterial pH and administer oxygen if necessary

- Ventilate if respiratory function diminished

- Monitor liver and renal functions

- Symptomatic and supportive care

Summary of Environmental Hazards

- Pyridine released into the atmosphere reacts slowly with photochemically produced hydroxyl radicals with a half-life of 32 and 16 days in clean and moderately polluted atmospheres respectively and will be scavenged by rain[2]

- In polluted areas containing appreciable nitric acid vapour, reaction with nitric acid may be the major removal process

- The half-life of pyridine in water via volatilisation is estimated at 90 hours[2]

- Pyridine in soil will biodegrade in approximately 8 days

- Bioaccumulation is estimated to be low

- Drinking Water Standards:
  hydrocarbon total: $10\,\mu g.l^{-1}$ (UK max)
  pesticide: $0.1\,\mu g.l^{-1}$ (UK max)

- Soil Guidelines: Dutch Criteria:
  $0.1\,mg.kg^{-1}$ (target)
  $1\,mg.kg^{-1}$ (intervention)

- Air Quality Standards: no data available

REFERENCES

1. Clayton GD & Clayton FE (eds.) *Patty's Industrial Hygiene and Toxicology*, 4th edn. John Wiley & Sons, Inc., New York, 1994

2. Hall AH & Rumack BH (eds.) *TOMES System* ® Micromedex, Englewood, Colorado. CD ROM. vol. 41 (exp. 31 July 1999)

3. Hathaway GJ, Proctor NH & Hughes JP. *Proctor and Hughes' Chemical Hazards of the Workplace*, 4th edn. Van Nostrand Reinhold, New York, 1996

Henrietta Wheeler

SARIN

Key Points

- Sarin is a chemical warfare nerve agent that is colourless and odourless in its pure form
- It is a typical organophosphate cholinesterase inhibitor which has a rapid onset of action
- Effects from exposure can be divided into three types:

Muscarinic (parasympathetic) effects: Bradycardia, bronchospasm, bronchorrhoea, sweating, salivation, lacrimation, vomiting, diarrhoea and constricted pupils may occur

Nicotinic (sympathetic and motor) effects: Tachycardia, hypertension, muscle fasciculation and cramps, weakness and respiratory paralysis

Central effects: CNS depression, agitation, confusion, psychosis, delirium, coma and convulsions may occur; the CNS effects may be slowly reversible or irreversible

- The constricted pupils and subsequently blurred vision can persist for several months
- Death is usually due to respiratory insufficiency and paralysis
- *In the event of a large spill, stay upwind and out of low areas. Ventilate closed spaces. Protective clothing, eye protection and breathing apparatus should be worn*
- *Sarin vapours may evaporate from heavily contaminated casualties and so there is a risk that emergency personnel may become contaminated*
- *Casualties should be transported in such a way that there is no risk of the drivers of the emergency vehicles becoming contaminated by the fumes*

FIRST AID

- *Rescuers must not enter a contaminated area without full personal protective equipment and self-contained breathing apparatus*
- Terminate exposure and support vital functions
- The casualty should be moved to an uncontaminated area
- If the casualty is unconscious a clear airway should be established and maintained; give 100% oxygen if available
- **Inhalation Exposure:** If the patient stops breathing, expired air resuscitation should be started immediately using a pocket mask with a one way valve, if available. It is important where the face is contaminated that expired air resuscitation is NOT attempted unless an airway with rescuer protection is used
- **Dermal Exposure:** Rescuers must wear self-contained breathing apparatus
- Remove contaminated clothing, if possible under a shower and place in double, sealed, clear bags and label; store the bags in a secure area away from patients and staff
- Wash the skin thoroughly with copious amounts of water
- **Vapour Exposure:** Rescuers must wear full personal protective equipment
- Remove contaminated clothing, place in double, sealed, clear plastic bags and label; store the bags in a secure area away from patients and staff. Place casualty in clean clothes or gown
- **Eye Exposure:** Irrigate thoroughly with water or saline for 15 minutes

Detailed Information

- Sarin is a colourless liquid or vapour with no odour in its pure state, although impurities may alter it to be yellow to brown with a slight odour
- *Common Synonyms* GB, isopropyl methylphosphonofluoridate, MFI
- CAS 107-44-8
- UN 3018
- NIOSH/RTECS TA 8400000
- Molecular formula $C_4H_{10}FO_2P$
- Molecular weight 140.11
- Sarin is a rapid-acting military chemical warfare agent
- It is miscible in water; nerve agents are moderately soluble in water and highly soluble in lipids[1]
- It is destroyed rapidly by strong alkalis and chlorinated compounds
- It can release hydrogen fluoride when in contact with acids or possibly acid vapours
- Toxic and irritating fumes of fluoride and oxides of phosphorous are released when heated to decomposition[2]

Summary of Human Toxicology

- Sarin can be absorbed following inhalation, ingestion and dermal contact
- Death is usually due to respiratory insufficiency and paralysis
- A small drop of liquid sarin on the skin may be sufficient to cause death[1]
- Sarin is an anticholinesterase organophosphate compound, producing cholinergic overdrive at muscarinic, nicotinic and CNS sites, due to inhibition of the acetylcholinesterase enzyme with accumulation of acetylcholine and subsequent excess stimulation
- Sarin and other nerve agents differ from organophosphate insecticides in their potency and rapidity of 'ageing' of the OP-enzyme complex; the ageing half-life of sarin is estimated at 5 hours[3]

Antidotes: diazepam, atropine and pralidoxime are used for the treatment of sarin poisoning:

- *Diazepam* is effective for CNS effects especially convulsions
- *Atropine* is primarily effective for muscarinic effects, as it blocks the action of acetylcholine at the muscarinic receptors. *Hypoxia must be corrected before atropine is given*

- *Oximes* (pralidoxime) are primarily effective for nicotinic effects as they act by reactivating acetylcholinesterase; for greatest efficacy, pralidoxime should be given soon after poisoning before dealkylation (ageing) of the nerve agent occurs (for sarin this occurs within 5 hours). Oximes may also be effective for some CNS effects

- Vapour concentrations of 0.09 mg.m^{-3} lead to depressed cholinesterase concentrations and intense constricted pupils; pupillary reflexes were abolished for 11 days and normal pupillary dilatation required 30–40 days to return to normal in one reported series[4]

- Sarin has been used twice against civilian populations in Japan

- In the first incident 7 people died and there were over 200 casualties,[5] in the second attack there were a total of 11 deaths and over 5, 000 casualties;[6] in both cases casualties suffered symptoms of typical organophosphate poisoning

- Hospital staff suffered the effects of sarin toxicity due to the lack of decontamination of the casualties[7]

Toxicity of Sarin[8]

Concentration	Clinical effects
3 mg.m^{-3} for 1 min	Constricted pupils and rhinorrhoea
10 mg.m^{-3} of vapour for 10 min	Death in half those exposed
100 mg.m^{-3} of vapour for 1 min	Death in half those exposed
2 μg.kg^{-1}	Lowest toxic oral dose
90 μg.m^{-3}	Lowest toxic inhalation dose

Acute Clinical Effects

Note: the following symptoms are for general organophosphate exposures and may not have been recorded for sarin, but could potentially occur.

Inhalation Effects

- Nerve agents and their solvent carriers may cause irritation of the upper respiratory tract leading to coughing, wheezing and shortness of breath

- Systemic effects may occur; see below

Dermal Effects

- Nerve agents and their solvent carriers may cause irritation and erythema

- Systemic poisoning may occur; see below

Eye Effects

- Sarin vapour exposure may cause constricted pupils, blurred vision and lacrimation which is generally not long lasting[9]

- Dimmed vision, conjuctival hyperaemia may occur

- In severe exposures mydriasis occurs occasionally

- A sarin intermediate (difluoro) is extremely irritating to the eye and may cause permanent damage[9]

Oral Effects

- Aspiration leading to chemical pneumonitis is possible if sarin is mixed with a hydrocarbon solvent

- Ingestion may lead rapidly to severe poisoning

- See systemic effects below

Systemic Effects

- Nausea, vomiting, abdominal pain, constricted pupils, sweating, salivation, muscle weakness, pyrexia, drowsiness, tachycardia, muscle fasciculation, profuse urinary and faecal incontinence may occur

- Tracheobronchial oversecretion with bronchoconstriction leading to pulmonary oedema occurs in severe cases

- In severe cases, coma, convulsions, cyanosis and hypoxia

- The main cause of death is respiratory depression (due to muscular paralysis)

- Hyperglycaemia and glycosuria indicating acute pancreatitis may also be present; neutrophil leucocytosis and renal damage have been reported

- Cardiac effects include bradycardia, cardiac arrhythmias including prolonged QT interval, in rare cases torsade de pointes and cardiac arrest

- Prolonged neurological manifestations including peripheral neuropathy, depression, poor memory and insomnia may develop following severe exposures

Intermediate Syndrome

- Some patients, after apparent recovery, develop acute respiratory failure 24 to 96 hours after the cholinergic phase of poisoning, with paralysis of the proximal limb muscles, motor cranial muscles and respiratory muscles

- The Intermediate Syndrome is refractory to atropine and pralidoxime treatment and ventilation is required; it is thought that the Intermediate Syndrome may not occur in patients who have received adequate pralidoxime therapy during the acute cholinergic phase

Chronic Sequelae

Sarin exposure has resulted in persistent changes in the electroencephalogram (EEG) in humans.[1] Anxiety, irritability, impaired concentration and memory, confusion, slurred speech, drowsiness, depression and nightmares may be prolonged.

Management

All medical staff should wear full personal protective equipment when decontaminating patients: if they also become symptomatic they should be treated as below.

Monitor plasma and erythrocyte cholinesterase activity in every symptomatic case. Although the correlation between cholinesterase values and clinical effects is poor, generally depression in excess of 50% of red blood cell cholinesterase is associated with severe effects; for less severe exposures, correlation is very poor. Depression of cholinesterase activity is a good indicator of exposure.

Inhalation Management

- See admission criteria below

Dermal Management

- Potential for secondary contamination is HIGH before decontamination and LOW after decontamination if all sarin is removed from the skin

- Remove any remaining clothing and place in double, sealed, clear bags and label; store in a secure area away from patients and staff

- *Note:* leather goods are very hard to decontaminate and may need to be incinerated

- Wash ALL skin 3 times with copious amounts of soap and water

- The use of diluted domestic bleach (1:10 with water) or ethanol may be more effective

- See admission criteria below

Eye Management

- Irrigate eye thoroughly with copious amounts of water or normal saline for at least 15 to 20 minutes

Oral Management

- See admission criteria below

Admission Criteria[10]

All casualties must be triaged by a designated member of the medical profession

Monitor red blood cell cholinesterase concentrations in every symptomatic case 4 to 6 hourly until recovery

Mild symptoms:

- Observe for 2 hours

- Some individuals may suffer painful eyes, atropine eye drops may be considered, these individuals must not be allowed home alone as atropine may impair the visual fields; drops are not necessary for patients who are suffering constricted pupils alone

- If symptoms improve or the patient has not deteriorated within 2 hours then casualty should be discharged with information on criteria to seek further medical advice

Moderate symptoms:

- Observe for 24 hours in a 'holding facility' (i.e. a ward, chapel or other designated area with beds/mattresses)

- Medical staff must observe carefully for a deterioration in medical condition and be prepared to move to severe symptom group if necessary

- Administer antidotes as appropriate; see below

- If symptoms improve or patient has not deteriorated within 24 hours then casualty should be discharged with information on criteria to seek further medical advice

Severe symptoms:

- These casualties will be admitted to ITU or equivalent wards.

- Administer antidotes as appropriate; see below

- Supplemental oxygen may be required for respiratory distress; excess secretions may require removal by suction

- Monitor ECG and adequacy of respiration; ventilate if necessary

- Monitor red blood cell cholinesterase concentrations daily until symptoms improve; it is vital to treat the symptoms and not be lead by the cholinesterase concentration

Antidotes

Contact a poisons information service for further guidance and paediatric doses

- **Diazepam** may have an overall benefit as well as controlling twitching and convulsions

- **Atropine** *NOTE hypoxia must be corrected before atropine is given.*
  Adult: 2 mg repeatedly SC or IV until atropinisation is achieved and maintained (atropinisation is characterised by decreased bronchial secretions, heart rate >100 bpm, dry mouth, dilated pupils)

- **Pralidoxime** *NOTE pralidoxime should be given as an adjunct to, not as a replacement for, atropine and should be given in every case where atropine therapy is deemed necessary.*
  Traditional dose: 1 g (or 2 g in very severe cases) by slow IV injection over 5–10 minutes. 1–2 g 4 hourly (maximum dose 12 g in 24 hours) until clinical and analytical recovery is achieved and maintained

Summary of Environmental Hazards

- Sarin is miscible in water; nerve agents are moderately soluble in water and highly soluble in lipids[1]

- Sarin is slowly hydrolised in the environment to less toxic and non-toxic substances

- The physical properties of sarin allow evaporation and dispersion over several hours and is thus known as 'non-persistent'

- Nerve agents are rapidly detoxified by strong alkalis and chlorinated compounds

- Drinking Water Standards: no data available

- Soil Guidelines: no data available

- Air Quality Standards: no data available

REFERENCES

1. Hall AH & Rumack BH (eds.) *TOMES System* ® Micromedex, Englewood, Colorado. CD ROM. vol. 41 (exp. 31 July 1999)

2. Sax NI. *Dangerous properties of Industrial Materials*, 6th edn. Van Nostrand Reinhold, New York, 1984

3. Marrs TC, Maynard RL, Sidell FR. *Chemical Warfare Agents: Toxicology and Treatment*. John Wiley & Sons, Chichester, 1996

4. Rengstorff RH, 1985. Accidental exposure to sarin: vision effects. *Arch Toxicol*; 56(3): 201–3

5. Suzuki T, Morita H, Ono K, Maekawa K, Nagai R, Yazaki Y, 1995. Sarin poisoning in Tokyo subway. (letter). *Lancet*, 345(1): 980

6. Okumura T, Takasu N, Ishimatsu S, Miyanoki S, Mitsuhashi A, Kumada K, Tanaka K, Hirohara S, 1995. Report of 640 victims of the Tokyo subway Sarin attack. *Ann Emerg Med*; 28(2): 129–35

7. Masuda N, Takatsu M, Morinari H, Ozawa T, 1995. Sarin poisoning in Tokyo subway. *Lancet* 345 (1): 1446

8. Sidell FR, 1996. Chemical agent terrorism. *Ann Emerg Med*; 28(2): 223–4

9. Grant MW & Schuman JS. *Toxicology of the Eye*, 4th edn. Charles C Thomas, Springfield, 1993

10. Maynard R, 1997. Personal communications. Department of Health, London

Henrietta Wheeler

SODIUM CHLORATE

Key Points

- Sodium chlorate is a colourless, odourless, white crystal or granule, with a cooling, saline taste
- It is readily absorbed by ingestion; rate of absorption via the respiratory tract is unclear
- It may cause irritation to the skin, eyes and respiratory tract
- Dermal absorption associated with agricultural use of sodium chlorate is not sufficient to cause systemic poisoning
- Clinical effects include gastrointestinal irritation, convulsions, renal failure and liver damage
- Sodium chlorate induces methaemoglobinaemia which does not respond to methylene blue therapy
- Death generally occurs due to tissue hypoxia within a few hours to days post exposure
- *In the event of a large spill, stay upwind and out of low areas. Ventilate closed spaces. Protective clothing, eye protection and breathing apparatus should be worn*

FIRST AID

- Terminate exposure and support vital functions
- The casualty should be moved to an uncontaminated area
- Rescuers should, ideally, be trained personnel and must be careful *not to put themselves at risk* and *so wear appropriate protective clothing and, if available, breathing apparatus*
- If the casualty is unconscious a clear airway should be established and maintained; give 100% oxygen if available
- **Inhalation Exposure:** If the patient stops breathing, expired air resuscitation should be started immediately using a pocket mask with a one way valve, if available. It is important where the face is contaminated that expired air resuscitation is NOT attempted unless an airway with rescuer protection is used
- **Dermal Exposure:** Remove contaminated clothing, if possible under a shower and place in double, sealed, clear bags and label; store the bags in a secure area and away from patients and staff
- Wash the skin thoroughly with copious amounts of water with a few drops of detergent mixed in
- **Eye Exposure:** Irrigate thoroughly with water or saline for 15 minutes
- **Oral Exposure:** Encourage small quantities of oral fluids (no more than 50–100 ml in total)

Detailed Information

- Sodium chlorate is a colourless, odourless, white crystal or granule; it has a cooling, saline taste

- *Common synonyms* chloric acid sodium salt, chlorate of soda, chlorate salt of sodium

- CAS 7775-09-9

- UN 1495 (sodium chlorate)

- UN 2428 (sodium chlorate, aqueous solution)

- NIOSH/RTECS FO 0525000

- Molecular formula $NaClO_3$

- Molecular weight 106.44

- Sodium chlorate is used as a herbicide; as a bleaching agent for textiles and paper pulp; in the manufacture of explosives, flares and pyrotechnics; in leather tanning and finishing

- Commercial products are approximately 99% pure; in some formulations, sodium chloride or other salts are included as fire retardants

- The crystals are applied dissolved in water; aqueous solutions are neutral

- Soluble in ethanol and glycerol

- A strong oxidising agent, sodium chlorate reacts with organic materials in the presence of sunlight

- Sodium chlorate can explode if subjected to intense heat with or without sudden pressure

- If mixed with sulphur, sugar or some other oxidisable materials, it forms an explosive mixture that may be more powerful than gunpowder[1]

Summary of Human Toxicity

- Sodium chlorate is readily absorbed by ingestion; rate of absorption via the respiratory tract is unclear

- It is not readily absorbed through intact skin; dermal absorption associated with agricultural use of sodium chlorate is not sufficient to cause systemic poisoning

- After absorption haemoglobin is rapidly oxidised to methaemoglobin and intravasular haemolysis results; small amounts can produce severe effects

- The rate of methaemoglobin formation is relatively slow and dangerous concentrations can occur insidiously and without warning[2]

- Cyanosis becomes clinically detectable when the proportion of methaemoglobin exceeds 10%; values of 70% are fatal[3]

- Hypotension and irregular pulse may result from red blood cell lysis[4]

- Sodium chlorate is nephrotoxic and causes acute tubular necrosis and renal failure which may be compounded by haemoglobinuria

- Death has occurred from 4 hours to 34 days after ingestion with an average of just over 4 days[1]

- Death occurring within a few hours of ingestion is due to tissue hypoxia resulting from severe methaemoglobinaemia or hyperkalaemia resulting from massive haemolysis

- Sodium thiosulphate is the specific antidote that may inactivate the chlorate ion to form less toxic chloride and may be given orally or intravenously

- Methylene blue and ascorbic acid are usually used for the treatment of methaemoglobinaemia in an attempt to reduce methaemoglobin, its efficacy in treatment of chlorate-induced methaemoglobinaemia may be limited compared to its efficacy in treatment of other oxidant-induced methaemoglobinaemias; this may be due to inactivation of chlorates of glucose-6-phosphate dehydrogenase, an enzyme required for methylene blue's reduction of methaemoglobin[5]

- Fatal doses from ingestion of sodium chlorate range from 150 to 300 g, although doses of 150 to 200 g have been survived[3,6,7]

- Inhalation following use in a concentrated solution in an atomiser resulted in renal failure and subsequent recovery[8]

- Occupational Exposure Standard: no data available

Acute Clinical Effects

Inhalation Effects

- Respiratory tract irritation, coughing, wheezing

- Systemic effects may occur following large or prolonged exposure; see systemic effects below

Dermal Effects

- Irritation to the skin

Eye Effects

- Irritation to eyes

Oral Effects

- Irritation to mucous membranes, nausea, vomiting, diarrhoea and abdominal pain

- Pallor may be the only sign during a latent period

- Systemic effects occur; see below

Systemic Effects

- Onset may be delayed for 12 hours

- Ataxia, haematuria, dizziness and cardiovascular collapse occurs

- Anoxia, cyanosis, collapse, dyspnoea, convulsions and coma are common

- Hypotension, tachycardia and irregular pulse have been reported

- Haemolysis, methaemoglobinaemia, DIC and haemolytic anaemia and resultant haemoglobinuria

- Brown or black urine with casts, red cells, free haemoglobin and methaemoglobin

- Elevated liver function, hepatomegaly and jaundice may occur

- Acute tubular necrosis and renal failure marked by oliguria or anuria may occur[3,6,9]

- Hyperkalaemia may occur secondary to haemolysis and renal failure[7]

Chronic Clinical Effects

Chronic exposure may lead to dermatitis, sweating, skin lesions, nausea and sore throat. However, it has been suggested that these effects may have been due to borates which are often present in herbicides as fire suppressants.[2]

Management

Methaemoglobin concentration is the best indicator of toxicity.

Inhalation Management

- Maintain a clear airway, give humidified oxygen and ventilate if necessary

- If respiratory irritation occurs assess respiratory function and if necessary perform a chest X-rays to check for chemical pneumonitis

- Consider the use of IV steroids to reduce the inflammatory response

- Treat pulmonary oedema with PEEP or CPAP ventilation

- Symptomatic and supportive care; see systemic management below

Dermal Management

- Remove any remaining contaminated clothing, place in double, sealed, clear bags and label; store in a secure area away from patients and staff

- Irrigate with copious amounts of water

- An emollient may be required

- Treat irritation symptomatically

Eye Management

- Irrigate thoroughly with running water or saline for 15 minutes

- Stain with fluorescein and refer to an ophthalmologist if there is any uptake of the stain

Oral Management

- Encourage oral fluids

- DO NOT INDUCE EMESIS

- Consider gastric lavage within 1 hour of a substantial ingestion, ensuring airway is protected

- Contact a poisons information service for further guidance on gut decontamination

- See systemic management below

Systemic Management

- Monitor plasma potassium concentrations, liver and renal function

- Sodium thiosulphate: 2.5 g in 200 ml of 5% sodium bicarbonate, may be given orally or intravenously[3]

- 100% humidified oxygen should be given for hypoxia, although this may be ineffective; if required PEEP or CPAP ventilation

- Peritoneal dialysis, haemodialysis and exchange transfusion combined with haemodialysis have all been successful in the treatment of severe poisoning and renal failure

- The presence of methaemoglobin in red blood cells, but not in serum, 2 hours post ingestion suggests that early exchange transfusion may be beneficial[7]

- The use of methylene blue to treat sodium chlorate-induced methaemoglobinaemia is controversial; contact a poisons information service for further guidance

- Symptomatic and supportive care

Summary of Environmental Hazards

- Sodium chlorate is soluble in water to 75% and is most persistent in areas of low rainfall, where it may remain toxic for as long as 5 years[2]

- It does not persist in soils with a high organic content

- It is harmful to aquatic life in high concentrations

- Drinking Water Standards:
       sodium: 150 mg.l[-1] (UK max)
       pesticide: 0.1 µg.l[-1] (UK max)

- Soil Guidelines: no data available

- Air Quality Standards: no data available

REFERENCES

1. Hayes WJ & Laws ER (eds). *Handbook of Pesticide Toxicology*. Academic Press, Inc., San Diego, 1991

2. Hall AH & Rumack BH (eds.) *TOMES System* ® Micromedex, Englewood, Colorado. CD ROM. vol. 41 (exp. 31 July 1999)

3. Helliwell M & Nunn J, 1979. Mortality in sodium chlorate poisoning. *Brit Med J*; 1(6171): 1119

4. Smith EA & Oehme FW, 1991. A review of selected herbicides and their toxicities. *Vet Hum Toxicol*; 33(6): 596–608

5. Goldfrank LR, Flomenbaum NE, Lewin NA, Weisman RS, Howland MA & Hoffman RS. *Goldfrank's Toxicologic Emergencies*, 5th edn. Appleton & Lange, Norwalk, 1994

6. Steffen C & Rainer S, 1981. Severe chlorate poisoning: report of a case. *Arch Toxicol*; 48(4): 281–8

7. O'Grady J & Jarecsni E, 1971. Sodium chlorate poisoning. *Brit J Clin Pract*; 25(1): 38–9

8. Jackson RC, Elder WJ & McDonnel H, 1961. Sodium-chlorate poisoning complicated by acute renal failure. *Lancet*; 2: 1381–3

9. Steffen C & Wetzel E, 1993. Chlorate poisoning: mechanism of toxicity. *Toxicology*; 84(1–3): 217–31

Henrietta Wheeler

SODIUM DICHROMATE AND HEXAVALENT CHROMIUM SALTS

Key Points

- Sodium dichromate is odourless and orange or red when dissolved in water
- It is a severe irritant to the lungs, eyes, skin, mucous membranes and oesophagus
- Sodium dichromate may be fatal following inhalation, ingestion and dermal exposure
- On ingestion it can cause severe ulceration, burns, multi-system shock, collapse and death
- On inhalation it is a respiratory irritant and may cause asthma or pulmonary oedema
- Systemic toxicity is likely if the skin is broken or if chrome ulcers develop
- *In the event of a large spill, stay upwind and out of low areas. Ventilate closed spaces. Protective clothing, eye protection and breathing apparatus should be worn*

First Aid

- Terminate exposure and support vital functions
- The casualty should be moved to an uncontaminated area
- Rescuers should, ideally, be trained personnel and must be careful *not to put themselves at risk* and *so wear appropriate protective clothing and, if available, breathing apparatus*
- If the casualty is unconscious a clear airway should be established and maintained; give 100% oxygen if available
- **Inhalation Exposure:** If the patient stops breathing, expired air resuscitation should be started immediately using a pocket mask with a one way valve, if available. It is important where the face is contaminated that expired air resuscitation is NOT attempted unless an airway with rescuer protection is used
- **Dermal Exposure:** Remove contaminated clothing, if possible under a shower and place in double, sealed, clear bags and label; store the bags in a secure area and away from patients and staff
- Wash the skin thoroughly with copious amounts of water
- **Eye Exposure:** Irrigate thoroughly with water or saline for 15 minutes
- **Oral Exposure:** Encourage small quantities of oral fluids (no more than 50-100 ml in total) unless perforation is suspected

Detailed Information

- Sodium dichromate is odourless and orange or red when dissolved in water
- *Common synonyms* bichromate of soda, disodium dichromate, sodium bichromate, sodium chromate
- CAS 10588-01-9
- NIOSH/RTECS HX 7700000
- Molecular formula $Na_2Cr_2O_7$
- Molecular weight 261.98

- Used in copper electroengraving, as a complexing agent and an oxidising agent in dye, ink and synthetic organic chemical manufacture; important in chrome-tanning of hides, in electric batteries, as a bleaching agent, in refining petroleum and for defoliation for cotton and other plants
- Chromium exists in several oxidation states which vary in their water solubility and toxicity; sodium dichromate is a soluble hexavalent chromium salt
- Sodium dichromate is soluble in water and alcohol
- In aqueous solution it attacks many materials especially in the presence of acids
- Sodium dichromate is a strong oxidant which reacts violently with combustible and reducing agents with the risk of fire and explosion

Summary of Human Toxicity

- Sodium dichromate may be fatal following inhalation, ingestion and dermal exposure
- It is more easily absorbed than insoluble or trivalent chromium salts
- Serious toxicity resulted from ingestion of 0.5 g of hexavalent chromium[1]
- Dermal involvement of 10% of the body surface may be fatal
- Hexavalent chromate salts are strong oxidising agents which produce corrosive burns by denaturing tissue protein
- Early administration of ascorbic acid converts the toxic hexavalent ions into the trivalent form which is of lower toxicity; trivalent ions are poorly adsorbed across the gut, lungs and skin
- Maximum Exposure Limit: Chromium (VI) compounds (as Cr)
      Long-term exposure limit: $0.05 \, mg \, . \, m^{-3}$
      Short-term exposure limit: no data available

Acute Clinical Effects

Inhalation Effects

- Sodium dichromate is a respiratory irritant
- Nasal hyperaemia may result from acute exposure
- Rhinitis, nosebleed and perforated eardrums may be observed
- Asthma may occur within 4–8 hours post exposure
- Pulmonary oedema may be delayed for up to 72 hours
- Muscle cramps have been reported

Dermal Effects

- Sodium dichromate fumes may cause irritant dermatitis and ulceration
- It is extremely corrosive to the skin and mucous membranes
- Deep perforating ulcers (chrome holes) may develop, these ulcers may be relatively painless[2]

- Burns may initially resemble first and second degree burns, but extend to subcutaneous tissue within a few days

- Systemic symptoms and death have occurred after external burns, with delayed gastric intestinal symptoms of hours or days; see oral effects below[3]

Eye Effects

- Sodium dichromate is extremely irritating in the eye and may cause severe corneal injury including swelling of the corneal stoma

- Astigmatism and anaesthesia of affected surfaces may persist after acute exposure

Oral Effects

- Ingestion of sodium dichromate is corrosive to mucous membranes and may cause violent gastroenteritis with rice-water stools, yellow–green or coffee-ground vomit, corrosive burns of the mouth, oesophagus, and gastrointestinal tract

- Haemorrhage may occur shortly after oral exposure

- Acute ingestion may produce acute multi-system shock within a few hours, renal failure and, less commonly, hepatic injury

- Haematuria and oliguria, followed by acute renal failure may occur after 1–2 days

- Haemorrhagic diathesis, thrombocytopenia and anaemia have been observed up to 3–7 days post ingestion; DIC may occur within 48 hours

- Chromates are capable of oxidising free haemoglobin *in vitro* to methaemoglobin and methaemoglobinemia may occur

- Toxic or hepatic encephalopathy has been reported within 15 minutes of oral ingestion

- ARDS may be noted with significant oral ingestion

- Pulmonary oedema may be delayed up to 72 hours post exposure

- Ingestion of a small amount may result in erythematous rash with onset within 1 week

Chronic Clinical Effects

Oral ingestion within the daily range or inhalation of chromium fumes may result in exacerbation of chronic dermatitis.[4] Erosion and discolouration of teeth has been observed.

Hypersensitivity occupational dermatitis has been observed. It is thought that hexavalent salts penetrate the skin where it is reduced to the trivalent form of chromium and reacts with the skin proteins to form antigens. Ulceration of larynx and nasal septum are common in chronic inhalation, with perforation of the nasal septum observed. A purulent nasal discharge with crust formation and breathing difficulties are the primary symptoms.[5] Chronic chrome ulcers generally heal in several weeks with no specific treatment.

Chronic industrial exposure may result in bronchitis, pulmonary fibrosis, emphysema, or pulmonary cancer. Chronic keratits, conjuntival inflammation and brown bands of discoloration in the corneal surface may occur with chronic exposure.[6]

Management

Meaurement of serum chromium concentration has no clinical relevance other than confirming exposure when this is in doubt

N.B. the use of a cannula to collect blood samples is advisable to avoid contamination of the sample with chrome from a needle

Inhalation Management

- Maintain a clear airway, give humidified oxygen and ventilate if necessary

- If respiratory irritation occurs assess respiratory function and if necessary perform chest X-rays to check for chemical pneumonitis

- Consider the use of steroids to reduce the inflammatory response

- Treat pulmonary oedema with PEEP or CPAP ventilation

- See systemic management below

Dermal Management

- Remove any remaining contaminated clothing, place in double, sealed, clear bags and label; store in a secure place away from patients and staff

- Irrigate with copious amounts of water for at least 30 minutes

- Treat burns symptomatically, excision of affected skin is recommended in severe exposures

- Chronic chrome ulcers generally heal in several weeks with no specific treatment

- Barrier cream may be of use in the treatment of contact dermatitis

- See systemic management below

Eye Management

- Irrigate thoroughly with running water or saline for 15 minutes

- Stain with fluorescein and refer to an ophthalmologist if there is any uptake of the stain

Oral Management

- NO GASTRIC LAVAGE OR EMETIC, RAPID TREATMENT IS ESSENTIAL

- Treat for any amount ingested

- Give 5–15 g of ascorbic acid orally (unless perforation is suspected) or IV (orally is better up to 2 hours post ingestion)

- For serious cases give repeat doses of ascorbic acid (IV or orally) at 40, 80, 120 minutes after the original dosing, then at 6, 12 and 24 hours if necessary

- Consider the early use of steroids to reduce the inflammatory response[7]

- Monitor blood pressure, give plasma expanders, blood or IV fluids for shock and diazepam for pain or convulsions

- Haemolysis or renal damage may require treatment with peritoneal or haemodialysis

- If facilities are available, early gastro-oesophagoscopy should be undertaken within 12 hours of the event to assess the extent and severity of the injury

- See systemic management below

Systemic Management

- Monitor haematocrit, platelet count and potassium

- Monitor renal and hepatic function

- If cyanosis is present or methaemoglobinaemia is suspected, determine methaemoglobin concentration

- See nitrates (page 181) for management of methaemoglobinaemia

- Symptomatic and supportive care

Summary of Environmental Hazards

- Chromium in soil can be transported to the atmosphere by aerosol formation; chromium is also transported from soil through run-off and leaching of water

- Run-off could remove both chromium ions and bulk precipitates of chromium, with final deposition on either a different land area or a water body

- Most of the chromium in surface waters may be present in particulate form as sediment

- Chromium is removed from air through wet and dry depositions; by analogy with the residence time of atmospheric copper, the residence time of atmospheric chromium is expected to be less than 10 days

- Chromium is highly toxic to fish; chromium (VI) is expected to bioaccumulate in the food chain

- Drinking Water Standards:
    sodium: 150 mg.l^{-1} (UK max)
    chromium: 50 μg.l^{-1} (UK max)
                  50 μg.l^{-1} (WHO provisional guideline)

- Soil Guidelines: Dutch Criteria:
    100 mg.kg^{-1} (target)
    380 mg.kg^{-1} (intervention)

- Air Quality Standards: no safe level for hexavalent chromium recommended due to carcinogenic properties

REFERENCES

1. Hall AH & Rumack BH (eds.) *TOMES System* ® Micromedex, Englewood, Colorado. CD ROM. vol. 41 (exp. 31 July 1999)

2. Deng JF, Fleeger AK & Sinks T, 1990. An outbreak of chromium ulcer in a manufacturing plant. *Vet Hum Toxicol*; 32 (2): 142–6

3. Kelly WF, Ackrill P & Day JP, 1982. Cutaneous absorption of trivalent chromium: tissue levels and treatment by exchange transfusion. *Br J Ind Med*; 39(4): 397–400

4. Kaaber K & Veien NK, 1978. Chromate ingestion in chronic chromate dermatitis. *Contact Dermat*; 4(2): 119–20

5. Dixon FW, 1929. Perforations of the nasal septum in chromium worker: report of 18 cases. *J Am Med Assoc*; 93: 837–8

6. Grant MW and Schuman JS, 1993. *Toxicology of the eye,* 4th edn. Charles C Thomas, Springfield, 1993

7. Walpole IR, Johnston K and Clarkson R, 1985. Acute chromium poisoning in a 2 year old child. *Aust Paediatr J* 21(1):65–7

Henrietta Wheeler

SODIUM FLUORIDE AND FLUORIDE SALTS

Key Points

- The majority of the following information relates to sodium fluoride, unless otherwise stated
- Sodium fluoride is a white, odourless crystalline powder
- Fluoride enters the body by ingestion or inhalation
- Death from ingestion may be due to sudden cardiovascular collapse
- The most important chronic toxic effect of fluoride on human beings is skeletal fluorosis
- *In the event of a large spill, stay upwind and out of low areas. Ventilate closed spaces. Protective clothing, eye protection and breathing apparatus should be worn*

First Aid

- Terminate exposure and support vital functions
- The casualty should be moved to an uncontaminated area
- Rescuers should, ideally, be trained personnel and must be careful *not to put themselves at risk* and *so wear appropriate protective clothing and, if available, breathing apparatus*
- If the casualty is unconscious a clear airway should be established and maintained; give 100% oxygen if available
- **Inhalation Exposure:** If the patient stops breathing, expired air resuscitation should be started immediately using a pocket mask with a one way valve, if available. It is important where the face is contaminated that expired air resuscitation is NOT attempted unless an airway with rescuer protection is used
- **Dermal Exposure:** Remove contaminated clothing, if possible under a shower and place in double, sealed, clear bags and label; store the bags in a secure area and away from patients and staff
- Wash the skin thoroughly with copious amounts of water
- **Eye Exposure:** Irrigate thoroughly with water or saline for 15 minutes
- **Oral Exposure:** Encourage small quantities of oral fluids (no more than 50-100 ml in total)

Detailed Information

- Sodium fluoride is the most important of the fluoride salts, it is a white, odourless crystalline powder with a salty state
- CAS 7681-49-4 (sodium fluoride)
- UN 1690 (sodium fluoride)
- NIOSH/RTECS WB 0350000 (sodium fluoride)
- Molecular formula F (fluoride), NaF (sodium fluoride)
- Molecular weight 42.0 (NaF)
- Sodium fluoride is used in mouthwashes, drops and tablets for the prevention of dental caries; as an insecticide (particularly for cockroaches and ants); as a constituent of vitreous enamel and glass mixes; in electroplating; glass frosting; removing hydrofluoric acid from exhaust fumes; for preserving wood; in fluxes; as a steel degreasing agent; and in the manufacture of coated paper

- Sodium fluoride, sodium monophosphate and stannous fluoride are used for water fluoridation and fluoride toothpaste; this was introduced in some areas of the UK since 1970;[1] water is fluoridated in the Republic of Ireland
- Fluorides are highly soluble in water; sodium fluoride is insoluble in alcohol and non-flammable
- Sodium fluoride reacts with acids, forming hydrogen fluoride; decomposes in flames or on hot surfaces, giving off hydrogen fluoride

Summary of Human Toxicity

- Sodium fluoride is rapidly and almost completely absorbed from the gastrointestinal tract, 75–90% is absorbed within 90 minutes of ingestion; once absorbed it is stored in bones and developing teeth
- Co-ingestion of calcium decreases absorption since the calcium fluoride formed is almost 3000 times less soluble and poorly absorbed from the gastrointestinal tract
- Total fluoride daily intake generally depends upon the concentrations in ingested food and beverages, fluoride added to water contributes considerably to fluoride levels in food preparation
- Fluoride chelates calcium and excessive ingestion may cause hypocalcaemia, this effect has contributed to severe poisoning and death
- Cardiovascular collapse seen with fluoride poisoning is thought to be due to hypocalcaemia, but is probably a result of a potassium efflux; animal experiments have shown that inhibiting the potassium efflux even without correcting the hypocalcaemia was effective in preventing sudden cardiac effects[2]
- Fluoride prevents caries through a variety of actions including interfering with bacterial metabolism and influencing synthesis of extracellular polysaccharide, sugar transport system, enolase and ATPase. The main effect, however, is promotion of remineralisation of early caries and prevention of demineralisation
- Excess fluoride of 150 mg.l^{-1} in a water supply resulted in nausea, vomiting, diarrhoea, abdominal pain and paraesthesias; one patient died after 24 hours of vomiting when in an attempt to remain hydrated they had drunk an estimated 10 litres of the contaminated water[3,4]
- Occupational Exposure Standards for fluoride (as the ion):
    Long-term exposure limit: 2.5 mg.m^{-3}
    Short-term exposure limit: no data available

Acute Clinical Effects

Inhalation Effects

- Irritant, with cough occurring 1–2 hours post exposure, then pyrexia and chest tightness after 2–4 days, resolving over 10–30 days[5]

Dermal Effects

- Irritant, skin eruptions may occur

- No definitive reports of allergic skin reactions to fluoride

Eye Effects

- Irritant to the eye

Oral Effects

- Clinical effects can occur within minutes after ingestion of fluoride

- Gastrointestinal irritation with nausea, vomiting, diarrhoea and abdominal pain, salivation and a metallic or bitter taste

- Severe fluoride poisoning may cause gastrointestinal haemorrhage, dysphagia, coma, weakness, hypotension and convulsions

- Hypocalcaemia, hypomagnesaemia and hyper- or hypokalaemia may occur; the hypocalcaemia may lead to hyperreflexia and tetany

- Respiration is initially stimulated and then becomes irregular and shallow

- Cardiac effects including arrhythmias, conduction defects and decreased cardiac output

- Death is usually from cardiac failure or respiratory paralysis

- Patients can deteriorate suddenly and without warning

Chronic Clinical Effects

Skeletal and dental fluorosis are endemic to some areas of the world where the water contains a high fluoride concentration from natural sources. Dental fluorosis usually affects the permanent teeth with deciduous teeth normally only affected at very high concentrations of fluoride. Early changes are chalk-white, irregular distributed patches on the enamel surface which may be infiltrated by yellow or brown staining giving a mottled appearance to the teeth; in severe fluorosis the teeth are weakened resulting in pitting. These teeth are highly resistant to caries.

Signs of skeletal fluorosis may appear after ingestion of a high concentration (8–10 ppm or more) of fluoride for 10 years or more, although skeletal fluorosis does not always occur in such circumstances and may be due to predisposing factors such as dietary deficiencies or population differences in the metabolism of fluoride.[6] It results in bony changes and calcification of ligaments and muscles. Onset of symptoms is usually insidious with back pain and stiffness and progressive weakness of the extremities. Skeletal fluorosis has been reported following exposure at work.[7]

Renal damage may also occur from chronic fluoride exposure. It is not clear whether occupational exposure to fluoride increases the risk of cancer. Neuropsychological evaluation of a worker exposed to sodium fluoride dust revealed a number of deficits including balance, co-ordination and fine motor control.[8]

Decreased testosterone concentrations have been observed in skeletal fluorosis patients and in males drinking the same water as these patients but without clinical features of fluorosis.[9]

Management

Measurement of the plasma fluoride level is rarely indicated in acute intoxication and is only worthwhile doing within one hour of ingestion due to very rapid absorption. The normal range is $10–370 \, ng.ml^{-1}$ in plasma, it varies depending on fluoridation of the water supply and ingestion from other sources.

Inhalation Management

- Maintain a clear airway, give humidified oxygen and ventilate if necessary

- If respiratory irritation occurs assess respiratory function and if necessary perform chest X-rays to check for chemical pneumonitis

- Symptomatic and supportive care

Dermal Management

- Remove any remaining contaminated clothing, place in double, sealed, clear bags and label; store in a secure area away from patients and staff

- Irrigate with copious amounts of water

- Treat irritation symptomatically

Eye Management

- Irrigate thoroughly with running water or saline for 15 minutes

- Stain with fluorescein and refer to an ophthalmologist if there is any uptake of the stain

Oral Management

- More than $8 \, mg.kg^{-1}$ fluoride is potentially toxic;[10] consider gastric lavage within 1 hour of ingestion due to its very rapid absorption

- Calcium supplements, milk or antacids containing aluminium or magnesium will bind fluoride and reduce its absorption, therefore milk may be of use for ingestion of $< 8 \, mg.kg^{-1}$

- Contact a poisons information service for further guidance on gut decontamination

- Hypocalcaemia should be corrected with calcium gluconate $(0.1–0.2 \, ml.kg^{-1}$ of a 10% solution $[0.02–0.04 \, mmol.kg^{-1}]$); hyperkalaemia, if severe, should be corrected with sodium bicarbonate, glucose and insulin or with potassium binding ion resin

- Ventilation may be necessary in patients with severe respiratory depression

- Lignocaine, procainamide are recommended for arrhythmias, pacing may be required

- Excretion of fluoride is almost exclusively renal, therefore maintain an adequate urine output

- Haemoperfusion or dialysis can be used to decrease fluoride levels but may only be of benefit in patients with renal insufficiency; non-fluoridated water must be used

Chronic Management

Diagnosis of chronic fluorosis is based on clinical manifestations and X-rays showing increased bone density, mineral deposits in ligaments, tendons and muscles and by the presence of periosteal outgrowths.

Summary of Environmental Hazards

- Sodium fluoride dissociates in water to sodium and fluoride ions

- Drinking Water Standards:
    sodium: 150 mg.l⁻¹ (UK max)
    fluoride: 1.5 mg.l⁻¹ (UK max)
                1.5 mg.l⁻¹ (WHO guideline)

- Soil Guidelines: no data available

- Air Quality Standards: no data available

REFERENCES

1. Bowen WH, 1995. The role of fluoride toothpastes in the prevention of dental caries. *J R Soc Med*; 88: 505–7

2. McEvoy GK (ed.) *American Hospital Formulary Service Drug Information*. American Society of Health–System Pharmacists Inc., 1996

3. Flanders RA & Marques L, 1993. Fluoride overfeeds in public water supplies. *Ill Dent J*; 62(3): 165–9

4. Gessner BD, Beller M, Middaugh JP & Whitford GM, 1994. Acute fluoride poisoning from a public water system. *N Engl J Med*; 330: 95–9

5. Hall AH & Rumack BH (eds.) *TOMES System ®* Micromedex, Englewood, Colorado. CD ROM. vol. 41 (exp. 31 July 1999)

6. Whitford GM. The metabolism and toxicity of fluoride. In *Monographs in oral science,* No 13. Myers HM (ed.) Basel, 1989

7. Grandjean P, Juel K & Jewson OM, 1985. Mortality and cancer morbidity after heavy occupational fluoride exposure. *Am J Epidemiol*; 121(1): 57–64

8. Franzen MD & Golden CJ, 1984. Case study: report of a case of fluoride poisoning. *Int J Clin Neuropsychol*; 6(4): 264–9

9. Susheela AK & Jethanandani P, 1996. Circulating testosterone levels in skeletal fluorosis patients. *J Toxicol-Clin Toxicol*; 34(2): 183–9

10. Phillips S, Burkhart K, Hartman P, McKinney P, Brent J, Kulig K & Rumack B, 1992. Can dental fluoride exposure ≤ 8 mg/kg be managed at home? (abstract 72). *Vet Hum Toxicol*; 34 (4): 334

Nicola Bates

SODIUM HYDROXIDE

Key Points

- Sodium hydroxide is a white, deliquescent solid available as flakes, pellets or beads
- Sodium hydroxide, both solid and in solution, has a very corrosive action upon all body tissue causing burns and frequently deep ulceration, with ultimate scarring
- It is extremely corrosive to skin, eyes and respiratory tract
- Inhalation can cause respiratory irritation and pulmonary oedema
- Burns may not be immediately apparent
- *In the event of a large spill, stay upwind and out of low areas. Ventilate closed spaces. Protective clothing, eye protection and breathing apparatus should be worn*

FIRST AID

- Terminate exposure and support vital functions
- The casualty should be moved to an uncontaminated area
- Rescuers should, ideally, be trained personnel and must be careful *not to put themselves at risk* and *so wear appropriate protective clothing and, if available, breathing apparatus*
- If the casualty is unconscious a clear airway should be established and maintained; give 100% oxygen if available
- **Inhalation Exposure:** If the patient stops breathing, expired air resuscitation should be started immediately using a pocket mask with a one way valve, if available. It is important where the face is contaminated that expired air resuscitation is NOT attempted unless an airway with rescuer protection is used
- **Dermal Exposure:** Remove contaminated clothing, if possible under a shower and place in double, sealed, clear bags and label; store the bags in a secure area and away from patients and staff
- Wash the skin thoroughly with copious amounts of water
- **Eye Exposure:** Irrigate thoroughly with water or saline until pH is no longer alkaline
- *NB:* Testing the pH of the skin and eyes immediately following irrigation may be misleading. It is recommended that 15 minutes elapse before this is undertaken. Continue the irrigation process if the skin or eyes are still alkaline
- **Oral Exposure:** Encourage small quantities of oral fluids (no more than 50–100 ml in total) unless perforation is suspected

Detailed Information

- Sodium hydroxide is a white, deliquescent solid available as flakes, pellets or beads, also available in 50% and 73% aqueous solutions
- *Common synonyms* caustic flake, caustic soda, lye (sodium hydroxide solution, but may also refer to potassium hydroxide solution), sodium hydrate, white caustic
- CAS 1310-73-2
- UN 1823 (dry solid)

- UN 1824 (solution)
- NIOSH/RTECS WB 4900000 (dry solid)
- NIOSH/RTECS WB 4905000 (liquid)
- Molecular formula NaOH
- Molecular weight 40.01
- Used in chemical manufacture of rayon, Cellophane and as neutralising agent in petroleum refining; pulp and paper, aluminium, detergents, soap, textile processing and rubber reclaiming; peeling of fruit and vegetables in the food industry and etching and electroplating[1]
- It is a very strong alkali
- Sodium hydroxide is soluble in water, alcohol and glycol
- Generates heat when dissolving or when mixed with an acid; reacts violently with acids and corrodes aluminium, zinc and other metals
- Also reacts violently with halogenated hydrocarbons and nitrogen compounds, with risk of fire and explosion
- Sodium hydroxide is not combustible but the solid form on mixing with moisture or water may produce sufficient heat to ignite combustible material; when heated to decomposition, sodium oxide (Na_2O) is produced

Summary of Human Toxicity

- Sodium hydroxide is responsible for some of the most severe, blinding injuries to the eye
- It is hygroscopic and absorbs water from tissues, thus causing adherence and deep penetration into the tissues
- Sodium hydroxide saponifies the fat in the cell membrane, destroying the cells
- It is an irritant of the skin, eyes and mucous membranes; severe burns with deep ulceration may occur[2]
- Dermal and ocular burns have been reported from contact with the white powdery residue (talc and sodium hydroxide) of air bags in motor vehicles[3]

Toxicity of Sodium Hydroxide[4,5,6]

Exposure route	Concentration	Clinical effects
Dermal	4% for several h	Skin irritation
	25–50% for 3 min	Skin irritation
	1 N for 15–180 min	Dissolution of cells in horny layer of the skin leading through oedema to total destruction of epidermis in 60 min
Inhalation	0.005–0.7 mg.m⁻³	Burning and redness of nose, throat or eyes
	0.24–1.86 mg.m⁻³, brief exposure	Irritation of throat or eyes
	250 mg.m⁻³	Considered immediately dangerous to life
Ingestion	1.95 g	Death in man

- Occupational Exposure Standards:
    Long-term exposure limit: no data available
    Short-term exposure limit: $2\,mg\,.\,m^{-3}$

Acute Clinical Effects

Inhalation Effects

- Respiratory irritation with coughing, wheezing and shortness of breath

- Pulmonary oedema may occur following inhalation of vaporised caustics

Dermal Effects

- Severe burns; the skin may become discoloured brown or black which can make initial assessment of injury difficult

- *Note:* After exposure to low concentration of alkalis the affected area may remain painless for several hours

Eye Effects

- Intense pain with blepharospasm; visual acuity is decreased due to injury to the corneal epithelium and corneal oedema

- Mild cases may include sloughing of the corneal and conjunctival epithelium

- Severe cases may include conjunctival swelling (chemosis) and ischaemic necrosis with hazing or opacity of the cornea

- Pupils may be dilated and unreactive due to damage to the sphincter and dilator muscles; retinal damage may also occur

- There may be erythema, oedema, blistering or in severe cases ischaemic necrosis of the lids

- Assessment of the injury and a prognosis may be difficult until 48–72 hours after the exposure

- Late complications of severe ocular burns may occur

Oral Effects

- An immediate burning pain in the mouth, oesophagus and stomach (retrosternal and epigastric pain), with swelling of the lips, followed by vomiting, haematemesis, increased salivation, ulcerative mucosal burns, dyspnoea, stridor, dysphagia and shock; oesophageal and pharyngeal oedema may also occur

- *Note that oesophageal damage may occur in the absence of oral burns*

- Acute complications include gastrointestinal haemorrhage and perforation of the gut

- Late complications may include oesophageal stricture and pyloric stenosis

- Gastric necrosis and stricture may occur, usually in patients who have oesophageal injury as well

- Alkalis are known to increase the risk of oesophageal cancer, which can occur years after the initial injury

- The incidence of carcinoma following oesophageal injury from sodium hydroxide is 0.8–4%

Chronic Clinical Effects

Obstructive airway disease has been reported following chronic occupational exposure to sodium hydroxide mist. The patient developed cough, dyspnoea and tachypnoea after a 20 year exposure to sodium hydroxide. The solution was used to clean jam containers which were boiled in it for two hours. He had a barrel chest with limited movement and diffuse expiratory wheezing. A chest X-ray showed severe pulmonary hyperinflation.[7]

Management

Inhalation Management

- Maintain a clear airway, give humidified oxygen and ventilate if necessary

- If respiratory irritation occurs assess respiratory function and if necessary perform chest X-rays to check for chemical pneumonitis

- Consider the use of steroids to reduce the inflammatory response

- Treat pulmonary oedema with PEEP or CPAP ventilation

- Symptomatic and supportive care

Dermal Management

- Remove any contaminated clothing, place in double, sealed, clear bags and label; store in a secure area away from patients and staff

- Immediately irrigate with copious amounts of water

- If an extensive area has been exposed whole body irrigation should be undertaken preferably using a high-flow shower unit

- Irrigation should be continued until the pH of the skin is no longer alkaline

- Testing the pH of the skin immediately after irrigation may be misleading. It is recommended that 15 minutes elapse before this is undertaken. Continue the irrigation process if the skin is still alkaline

- Referral to a burns unit is recommended

Eye Management

- Irrigate thoroughly with running water or saline for 15 minutes

- Pain and blepharospasm may make irrigation difficult and the use of anaesthetic drops may be needed to facilitate thorough irrigation

- A lid speculum may be used if required. It is essential that the whole eye is irrigated including under the upper and lower lids

- Testing the pH of the eyes immediately following irrigation may be misleading. It is recommended that 15 minutes elapse before this is undertaken. Continue the irrigation process if the eye is still alkaline

- Stain with fluorescein and refer to an ophthalmologist if there is any uptake of the stain

Oral Management

- NO GASTRIC LAVAGE OR EMETIC

- Neutralising chemicals should never be given because heat is produced during neutralisation and this could exacerbate any injury

- Patients must be admitted for observation until the extent of the injury (if any) can be determined

- Encourage oral fluids unless there is evidence of severe injury

- Oral feeding should be maintained if the patient is able to tolerate it, otherwise tube feeding or parenteral nutrition should be provided

- Give plasma expanders/IV fluids for shock and check and correct the acid/base balance. Analgesia will almost certainly be needed

- Aggressive intervention is essential for severely affected patients. Urgent assessment of the airway is required

- Early gastro-oesophagoscopy should be undertaken within 12 hours of the event to assess the extent and severity of the injury

- A supraglottic-epiglottic burn with erythema and oedema is usually a sign that further oedema will occur which will lead to airway obstruction and is an indication for early intubation or tracheostomy[8]

- Intubation and ventilation may be necessary for patients with respiratory distress

- On discharge all patients must be advised of the possibility of late-onset sequelae and advised to return if necessary

- Strictures that prevent adequate nutritional intake and do not respond to dilatation require oesophagectomy and colonic interposition

- Oesophageal strictures which result in a lumen >10 mm do not impede normal life and should not require intervention.[9] Surgical intervention may be required for gastrointestinal perforation or haemorrhage

Summary of Environmental Hazards

- Sodium hydroxide is highly soluble in water; a small amount in water raises the pH and larger amounts raise it for extended periods

- It is persistent in water, dissociating to sodium and hydroxide ions, giving a higher pH, depending upon the buffering capacity of the water

- Sodium hydroxide does not bioaccumulate

- Drinking Water Standards:
    sodium: 150 mg.l^{-1} (UK max)

- Soil Guidelines: no data available

- Air Quality Standards: no data available

REFERENCES

1. Sax NI & Lewis RJ. *Hawley's Condensed Chemical Dictionary*, 11th edn. Van Nostrand Reinhold Company, New York, 1987

2. Hathaway GJ, Proctor NH & Hughes JP. *Proctor and Hughes' Chemical Hazards of the Workplace*, 4th edn. Van Nostrand Reinhold, New York, 1996

3. Swanson-Biearman B, Mvros R, Dean BS & Krenzelok EP, 1993. Air bags: lifesaving with toxic potential? *Am J Emerg Med*; 11: 38–9

4. Clayton GD & Clayton FE (eds.) *Patty's Industrial Hygiene and Toxicology*, 4th edn. John Wiley & Sons, Inc., New York, 1994

5. Sax NI. *Dangerous Properties of Industrial Materials*, 6th edn. Van Nostrand Reinhold, New York, 1984

6. Hall AH & Rumack BH (eds.) *TOMES System* ® Micromedex, Englewood, Colorado. CD ROM. vol. 41 (exp. 31 July 1999)

7. Rubin AE, Bentur L & Bentur Y, 1992. Obstructive airway disease associated with occupational sodium hydroxide inhalation. *Brit J Indust Med*, 49: 213–4

8. Meredith JW, Kon ND and Thompson JN, 1988. Management of injuries from liquid lye ingestion. *J Trauma*; 28(8): 1173–80

9. Sarfati E, Assens P & Celerier M, 1987. Management of caustic ingestion in adults. *Br J Surg*; 74: 146–8

Nicola Bates

SODIUM HYPOCHLORITE

Key Points

- Sodium hypochlorite is a white solid crystalline substance, usually used as a solution
- It generally causes gastrointestinal irritation with corrosive damage occurring only with concentrated solutions or when large amounts are ingested
- On mixing with an acid, chlorine is released; chloramine is released on mixing with ammonia
- *In the event of a large spill, stay upwind and out of low areas. Ventilate closed spaces. Protective clothing, eye protection and breathing apparatus should be worn*

First Aid

- Terminate exposure and support vital functions
- The casualty should be moved to an uncontaminated area
- Rescuers should, ideally, be trained personnel and must be careful *not to put themselves at risk* and *so wear appropriate protective clothing and, if available, breathing apparatus*
- If the casualty is unconscious a clear airway should be established and maintained; give 100% oxygen if available
- **Dermal Exposure:** Remove contaminated clothing, if possible under a shower and place in double, sealed, clear bags and label; store the bags in a secure area and away from patients and staff
- Wash the skin thoroughly with copious amounts of water
- **Eye Exposure:** Irrigate thoroughly with water or saline for 15 minutes
- **Oral Exposure:** Encourage small quantities of oral fluids (no more than 50–100 ml in total) unless perforation is suspected

Detailed Information

- Sodium hypochlorite is a white solid crystalline substance that is usually used as a solution; the aqueous solution is a green to yellowish watery liquid with the odour of bleach
- *Common synonyms* bleach, sodium chloride oxide, sodium oxychloride
- CAS 7681-52-9
- UN 1791
- RTECS NH 3486300
- Molecular formula NaOCl
- Molecular weight 74.44
- Used for bleaching paper pulp and textiles; as a chemical intermediate, fungicide, disinfectant and germicide and in water purification
- Household sodium hypochlorite bleaches are solutions of up to 10%, but are most commonly about 5%; industrial bleaches may be more concentrated (up to 50%)
- The aqueous solution is usually a green to yellow liquid with a bleach odour

- On mixing with an acid, chlorine is released; chloramine is released on mixing with ammonia
- Industrial solutions are strong bases, which react violently with acids and are corrosive to most metals
- Domestic sodium hypochlorite solutions often contain other agents, including sodium hydroxide to maintain a pH-dependent equilibrium between hypochlorite and chlorine
- It is non-combustible but decomposes when heated and gives off chlorine

Summary of Human Toxicity

- Sodium hypochlorite solution causes moderate mucosal irritation, the extent of which depends on the volume ingested, the viscosity and concentration of the preparation and the duration of contact
- In the majority of cases of exposure to sodium hypochlorite the effects are usually minor[1]
- Although solutions are alkaline they do not tend to cause corrosive damage except in large volumes or where concentrated solutions are ingested
- Sodium hypochlorite may release small amounts of chlorine and hypochlorous acid when acidified in the stomach, but usually in concentrations too small to cause any damage
- Severe effects from bleach ingestion are rare, most severe cases are due to large intentional ingestion in adults, particularly the elderly
- Pulmonary complications, usually resulting from aspiration, often contribute to death
- The ingestion of industrial bleaches may pose a risk because of additional chemicals or higher alkalinity
- Occupational Exposure Standards:
    Long-term exposure limit: no data available
    Short-term exposure limit: no data available

Acute Clinical Effects

Inhalation Effects

- Inhalation injury from the normal use of bleach is unlikely
- When mixed with acid (chlorine) or ammonia (chloramine) irritant gases are produced which may cause coughing, sore throat and shortness of breath and chemical pneumonitis
- If aspiration occurs following ingestion, chemical pneumonitis or pulmonary oedema may develop

Dermal Effects

- Dermal irritation
- Hypersensitivity may occur[2]

Eye Effects

- Immediate pain, irritation, lacrimation and burning sensation
- Mild transient corneal injury may result from most household bleaches

- More concentrated solutions may cause more serious injury

Oral Effects

- Gastrointestinal irritation, with nausea, vomiting and diarrhoea

- Haematemesis, local irritation and oedema may occur

- Household bleaches are unlikely to cause severe irritation unless contact is prolonged or the amount ingested is large

- Severe oesophageal damage[3] occurs only rarely

- Corrosive injury of the stomach and hypernatraemia with hyperchloraemic acidosis have all been reported following large intentional ingestion

- In severe cases there may be hypotension, coma and gastrointestinal perforation with subsequent shock

- Pulmonary complications, usually resulting from aspiration, often contribute to death

Chronic Clinical Effects

Chronic gastroduodenitis has been reported in a 7-year-old child who chronically sucked her heavily bleached socks. When the socks were boiled rather than bleached the effects resolved.[4]

Management

Inhalation Management

- Maintain a clear airway, give humidified oxygen and ventilate if necessary

- If respiratory irritation occurs assess respiratory function and if necessary perform chest X-rays to check for chemical pneumonitis

- Patients should be kept at rest

- Treat pulmonary oedema with PEEP or CPAP ventilation

- Symptomatic and supportive care

Dermal Management

- Remove any remaining contaminated clothing, place in double, sealed, clear bags and label; store in a secure area away from patients and staff

- Irrigate with copious amounts of water

- Treat irritation symptomatically

Eye Management

- Irrigate thoroughly with running water or saline for 15 minutes

- Stain with fluorescein and refer to an ophthalmologist if there is any uptake of the stain

Oral Management

- NO GASTRIC LAVAGE OR EMETIC

- Neutralising chemicals should never be given because heat is produced during neutralisation and this could exacerbate any injury

- Patients must be admitted for observation until the extent of the injury (if any) can be determined

- Encourage oral fluids unless there is evidence of severe injury

- Oral feeding should be maintained if the patient is able to tolerate it, otherwise tube feeding or parenteral nutrition should be provided

- Aggressive intervention is essential for severely affected patients; urgent assessment of the airway is required

- Early gastro-oesophagoscopy should be undertaken within 12 hours of the event to assess the extent and severity of the injury

- Give plasma expanders/IV fluids for shock and check and correct the electrolytes and acid/base balance. Analgesia will almost certainly be needed

- Intubation and ventilation may be necessary for patients with respiratory distress

- On discharge all patients must be advised of the possibility of late onset sequelae and advised to return if necessary

- Strictures that prevent adequate nutritional intake and do not respond to dilatation require oesophagectomy and colonic interposition

- Oesophageal strictures which result in a lumen >10 mm do not impede normal life and should not require intervention[5]

- Surgical intervention may also be required for gastrointestinal perforation or haemorrhage

Summary of Environmental Hazards

- Sodium hypochlorite is highly soluble and relatively persistent in water, gradually dissociating to sodium and hypochlorite ions; hypochlorite is a powerful oxidising agent and can convert to chlorine at low pH

- Sodium hypochlorite is very toxic to fish in low concentrations

- It is unlikely to bioaccumulate in the food chain

- Drinking Water Standards:
  sodium: $150 \, mg \cdot l^{-1}$ (UK max)
  chlorite: $200 \, \mu g \cdot l^{-1}$ (WHO provisional guideline)

- Soil Guidelines: no data available

- Air Quality Standards: no data available

REFERENCES

1. Racioppi F, Daskaleros PA, Besbelli N, Borges A, Deraemaeker C, Magalini SI, Martinez Arrieta R, Pulce C, Ruggerone ML & Vlachos P, 1994. Household bleaches based on sodium hypochlorite: review of toxicology and poison control centre experience. *Fd Chem Toxicol*; 32(9): 845–61

2. Hostynek JJ, Younger PB & Maibach HI, 1989. Hypochlorite sensitivity in man. *Contact Dermatitis*; 20: 32–7

3. French RJ, Tabb HG & Rutledge LJ, 1970. Esophageal stenosis produced by ingestion of bleach: a report of two cases. *South Med J*; 63(10): 1140–4

4. Loeb FX & King TL, 1974. Chronic gastroduodenitis due to laundry bleach. Sock sucker syndrome. *Am J Dis Child*; 128: 256–7

5. Sarfat: E, Assens P & Celerier M, 1987. Management of caustic ingestion in adults. *Br J Surg*; 74: 146–8

Nicola Bates

STRYCHNINE

Key Points

- Strychnine exists as hard white crystals or powder; it is odourless with a bitter taste
- It is a potent convulsant
- Strychnine is rapidly absorbed by ingestion and through the mucous membranes
- Onset of symptoms is expected 10–30 minutes post ingestion
- *In the event of a large spill, stay upwind and out of low areas. Ventilate closed spaces. Protective clothing, eye protection and breathing apparatus should be worn*

FIRST AID

- Terminate exposure and support vital functions
- The casualty should be moved to an uncontaminated area
- Rescuers should, ideally, be trained personnel and must be careful *not to put themselves at risk* and *so wear appropriate protective clothing and, if available, breathing apparatus*
- If the casualty is unconscious a clear airway should be established and maintained; give 100% oxygen if available
- **Inhalation Exposure:** If the patient stops breathing, expired air resuscitation should be started immediately using a pocket mask with a one way valve, if available. It is important where the face is contaminated that expired air resuscitation is NOT attempted unless an airway with rescuer protection is used
- **Dermal Exposure:** Remove contaminated clothing, if possible under a shower and place in double, sealed, clear bags and label; store the bags in a secure area and away from patients and staff
- Wash the skin thoroughly with copious amounts of water
- **Eye Exposure:** Irrigate thoroughly with water or saline for 15 minutes
- **Oral Exposure:** Encourage small quantities of oral fluids (no more than 50–100 ml in total)

Detailed Information

- Strychnine exists as hard white crystals or powder; it is odourless with a bitter taste
- *Common synonyms* strychnin, strychnine alkaloid, strychnine sulphate, strychnos
- CAS 57-24-9
- UN 1692
- NIOSH/RTECS WL 2275000
- Molecular formula $C_{12}H_{22}N_2O_2$
- Molecular weight 334.45
- Strychnine is used as a rodenticide, commercial preparations contain 0.3–3% strychnine; it is usually refused by rats, but useful to control mice, squirrels, rabbits, moles and other predatory animals and birds

- Strychnine is a basic alkaloid derived from the ripe seeds of the plant *Strynos nux-vomica*, a tree native to India
- It is soluble in chloroform, slightly soluble in alcohol and benzene, slightly soluble in water and ether

Summary of Human Toxicity

- Strychnine is rapidly absorbed following ingestion and via the mucous membranes
- Poisoning is now relatively infrequent
- Poisoning has occurred when strychnine has been mistaken for cocaine[1]
- It is absorbed rapidly with the onset of symptoms within 10–30 minutes
- Elimination is mainly by hepatic metabolism with 1–20% excreted unchanged in the urine, depending upon amount ingested[2,3]
- Strychnine prevents the uptake of glycine at inhibitory synapses resulting in net excitatory effect, with minimal sensory stimulation, resulting in diffuse muscle contractions; this causes convulsive activity in an awake patient with no post-ictal phase[4]
- Sensory stimulus may lead to violent motor response which, in the early stages of intoxication tends to be co-ordinated with extensor thrusts and spasms, and later with tetanic convulsions
- Onset of symptoms following ingestion is between 10–30 minutes
- Convulsions generally subside within 12–24 hours post ingestion[5]
- Doses of 5–7 mg may lead to muscle tightness, especially in the neck and jaws with twitching of individual muscles, especially the little fingers[6]
- Reported fatal serum concentrations range between 3.8–11 mg.l^{-1}, although concentrations of 2.17 and 2.45 mg.l^{-1} have been survived[2]
- The mean lethal oral dose is 100–120 mg (1.5–2 mg.kg^{-1}); doses of 16 mg have been reported fatal although doses of 2000 mg have been survived[4,6]
- Death commonly occurs 1–3 hours post ingestion of a fatal dose; there is good patient prognosis if there is survival beyond 5–6 hours post ingestion[7]
- Occupational Exposure Standards:
  Long-term exposure limit: 0.15 mg.m^{-3}
  Short-term exposure limit: 0.45 mg.m^{-3}

Acute Clinical Effects

Inhalation, Oral and Systemic Effects

- Initial symptoms may include agitation, headache, heightened awareness and responsiveness to sensory stimulation

- Stiffness, tightness and cramps of muscles especially face, neck and back; chest pain and increased reflex excitability may occur

- Muscle spasms and convulsions with opisthotonos may occur without the initial symptoms; convulsions generally occur without a post-ictal phase[2]

- Consciousness is retained throughout the convulsions, which are usually painful and may be accompanied by dilated pupils, proptosis and conjugate or dissociated deviations of the eyes[5]

- Convulsions may result in sweating, hyperthermia, metabolic, lactic and respiratory acidosis, rhabdomyolysis and myoglobinuric renal failure[5]

- Sensory and visual stimulation may initiate muscle twitching, extensor spasms and severe pain[8]

- With severe or prolonged convulsions, sustained contractions of the chest wall muscles and diaphragm may lead to respiratory paralysis

- Between convulsions, muscular relaxation is complete, breathing resumes, pupils may contract and cyanosis lessens

- Tachycardia, hypertension and tachypnoea have been reported

- Apnoea with bradycardia, hypotension and cardiac arrest may occur

- Early deaths result from respiratory failure and subsequent cardiac arrest (resuscitation is rare); late deaths are due to anoxic brain damage or multi-organ failure due to hyperthermia, resulting from prolonged fits[5]

- Acute pancreatitis has also been reported, potentially due to hypercalcaemia[2]

Chronic Clinical Effects

No data available.

Management

Laboratory concentrations do not correlate with severity of toxic effects, but can confirm exposure when this is in doubt.

Inhalation, Oral and Systemic Management

- Maintain a clear airway, give humidified oxygen and ventilate if necessary

- Aggressive control of convulsions is essential; initially diazepam should be used, followed by barbiturates or neuromuscular blockage agents if necessary[4]

- Ventilate if convulsions are severe or uncontrollable

- Monitor electrolytes, blood gases, renal and liver function

- Symptomatic and supportive care, avoiding stimulation

Summary of Environmental Hazards

- In the atmosphere strychnine has the potential to be removed by direct photolysis or dry deposition

- In the soil this compound has the potential to photolyse on soil surfaces; volatilisation and chemical hydrolysis are not expected to be important fate processes in soil

- In water strychnine has the potential to photolyse; chemical hydrolysis, volatilisation, and bioaccumulation in fish are not expected to be important fate processes[5]

- Drinking Water Standards: no data available

- Soil Guidelines: no data available

- Air Quality Standards: no data available

REFERENCES

1. O'Callaghan WG, Joyce N, Counihan HE, Ward M, Lavelle P & O'Brien E, 1982. Unusual strychnine poisoning and its treatment: report of eight cases. *Brit Med J*; 285(6340): 478

2. Hernandez AF, Pomares J, Schiaffino S, Pla A & Villanueva E, 1998. Acute chemical pancreatitis associated with nonfatal strychnine poisoning. *Clin Toxicol*; 36(1–2): 67–71

3. Palatnick W, Meatherall R, Sitar D & Tenenbein M, 1997. Toxicokinetics of acute strychnine poisoning. *J Toxicol*; 35(6): 617–20

4. Ellenhorn MJ, Schonwalds S, Ordog G & Wasserberger J. *Ellenhorn's Medical Toxicology – Diagnosis and Treatment of Human Poisoning*, 2nd edn. Williams & Wilkins, London, 1997

5. Hall AH & Rumack BH (eds.) *TOMES System* ® Micromedex, Englewood, Colorado. CD ROM. vol. 41 (exp. 31 July 1999)

6. Hathaway GJ, Proctor NH & Hughes JP. *Proctor and Hughes' Chemical Hazards of the Workplace*, 4th edn. Van Nostrand Reinhold, New York, 1996

7. Gosselin RE, Smith RP & Hodge HC. *Clinical Toxicology of Commercial Products*, 5th edn. Williams & Wilkins, Baltimore, 1984

8. Nishiyama T & Nagase M, 1995. Strychnine poisoning: natural course of a nonfatal case. *Am J Emerg Med*; 13(2): 172–3

Henrietta Wheeler

STYRENE

Key Points

- Styrene is a colourless to yellowish, oily liquid with a penetrating aromatic odour
- It is irritating to eyes, the respiratory and gastrointestinal tracts, and the skin
- Central nervous system depression may occur following exposure
- 'Styrene sickness' with nausea, vomiting, appetite loss, and a sensation of drunkenness is common in chronic inhalation
- Styrene is thought to be carcinogenic
- *In the event of a large spill, stay upwind and out of low areas. Ventilate closed spaces. Protective clothing, eye protection and breathing apparatus should be worn*

First Aid

- Terminate exposure and support vital functions
- The casualty should be moved to an uncontaminated area
- Rescuers should, ideally, be trained personnel and must be careful *not to put themselves at risk* and *so wear appropriate protective clothing and, if available, breathing apparatus*
- If the casualty is unconscious a clear airway should be established and maintained; give 100% oxygen if available
- **Inhalation Exposure:** If the patient stops breathing, expired air resuscitation should be started immediately using a pocket mask with a one way valve, if available. It is important where the face is contaminated that expired air resuscitation is NOT attempted unless an airway with rescuer protection is used
- **Dermal Exposure:** Remove contaminated clothing, if possible under a shower and place in double, sealed, clear bags and label; store the bags in a secure area and away from patients and staff
- Wash the skin thoroughly with copious amounts of water
- **Eye Exposure:** Irrigate thoroughly with water or saline for 15 minutes
- **Oral Exposure:** Encourage small quantities of oral fluids (no more than 50–100 ml in total)

Detailed Information

- Styrene is a colourless to yellowish, highly refractive, oily liquid with a penetrating aromatic odour
- *Common synonyms* ethenyl benzene, pheneylethene, phenylethylene, vinylbenzene
- CAS 100-42-5
- UN 2055
- NIOSH/RTECS WL 3675000
- Molecular formula C_8H_8
- Molecular weight 104.14
- Styrene is used in the manufacture of plastics, paints, other protective coatings, resins, and synthetic rubbers; chemical intermediate, dental fillings, agricultural products and as a stabilising agent; widely used in products used for boat–building and boat repair
- It is an alkenylbenzene compound and it occurs naturally in the sap of styracaceous trees[1]
- Styrene is able to form peroxides when exposed to light or air or on contact with acids; reacts violently with strong oxidants
- Attacks copper slowly

Summary of Human Toxicity

- The majority of styrene exposure and absorption occurs by inhalation
- Styrene is well absorbed through the gastrointestinal tract and intact skin[2]
- In animal experiments, 1,300 ppm was the highest value which caused only irritant effects but no systemic toxicity in an 8 hour period of inhalation[3]
- To reach a state of unconsciousness, experimental animals required a 10 hour inhalation exposure to a concentration of 2,500 ppm[3]
- In animal experiments with oral administration, feeding of $2 \, g.kg^{-1}.d^{-1}$, death occurred after only a few days and severe irritation of the oesophagus and stomach were noted[1]
- In humans, there is a proportional relationship between air concentrations, duration of exposure, and initial blood styrene concentration[1]

Toxicity of Styrene[1]

Concentration	Clinical Effects
10 ppm	Odour threshold
100 ppm	Strong odour, no discomfort
376 ppm for 1 h	Transient neurological effects: impairment of reaction time
200–400 ppm	Objectionable odour, reversible eye irritation
600 ppm	Eye and nasal irritation
800 ppm	Eye and mucous membrane irritation, metallic taste, drowsiness and weakness

- Maximum Exposure Limits:
    Long-term exposure limit: 100 ppm (430 mg.m^{-3})
    Short-term exposure limit: 250 ppm (1080 mg.m^{-3})

Acute Clinical Effects

Inhalation Effects

- Exposure may lead to irritation of the nose and throat, coughing and wheezing
- CNS depression and coma have been reported in cases of inhalational poisoning[4]

- Drowsiness, weakness, depression, inertia, and unsteadiness may occur[2]

- Pulmonary oedema, has been observed in animals, but has not been reported in humans[5]

- Cardiac arrhythmias have been reported in animal inhalational studies, this has not been reported in humans[1]

Dermal Exposure

- Itching, dermatitis and erythematous papular dermatitis have been reported[2]

Eye Exposure

- Eye irritation, retrobulbar optic neuritis, central scotomas and loss of colour vision have been reported[6]

Oral Exposure

- No human case data available; animals fed 2 g.kg[-1].d[-1] died after a few days and had severe irritation of the oesophagus and stomach[1]

Chronic Clinical Effects

Prolonged skin contact has lead to cracked skin, itching and dermatitis. Workers may suffer 'styrene sickness': nausea, vomiting, appetite loss, fatigue, dizziness, ataxia and general weakness, and occupational asthma from prolonged respiratory exposure.[7]

Short-term memory impairment, disturbances in psychomotor performance, visual–motor accuracy and peripheral neuropathy have been reported in chronic exposures. Abnormalities in EEG tracings among styrene–exposed workers have been reported.[2] Mild liver function changes have also been reported.[1]

Case reports and some epidemiological studies have implied an increased risk for development of cancer in workers exposed to styrene although presently available data do not allow a direct causal relationship to be established in humans.[8]

Management

Measurement of blood styrene or urine metabolite concentrations have no clinical relevance other than confirming exposure when this is in doubt. Measurement of urine metabolite concentration is the preferred method.

Inhalation Management

- Monitor for respiratory distress

- If cough or difficulty in breathing develops, evaluate for respiratory tract irritation, bronchitis, or pneumonitis

- Carefully observe patients with inhalation exposure for the development of any systemic signs or symptoms and administer symptomatic treatment as necessary

- Give 100% humidified supplemental oxygen with assisted ventilation as required

- Respiratory tract irritation, if severe, can progress to pulmonary oedema which may be delayed in onset up to 24 to 72 hours after exposure in some cases

- Monitor arterial blood gases and chest X-ray if significant respiratory tract irritation occurs

Dermal Management

- Remove any remaining contaminated clothing, place in double, sealed, clear bags and label; store in a secure area away from patients and staff

- Irrigate with copious amounts of water for at least 30 minutes

- Treat irritation and erythema symptomatically

- Treat systemic symptoms

Eye Management

- Irrigate thoroughly with running water or saline for 15 minutes

- Stain with fluorescein and refer to an ophthalmologist if there is any uptake of the stain

Oral Management

- Encourage oral fluid for small ingestion

- DO NOT INDUCE EMESIS due to risk of oesophageal irritation

- Consider gastric lavage within 1 hour of a substantial ingestion, ensuring the airway is protected

- Symptomatic and supportive care

Summary of Environmental Hazards

- Styrene vapour in the atmosphere reacts rapidly with hydroxyl radicals and ozone, with a reaction half-life estimated at 3.5 and 9 hours respectively[9]

- Evaporation half-life of styrene from a well-mixed pool of water 1 m deep is estimated to be 5.9 hours[9]

- Styrene released into the soil is subject to biodegradation; it has been shown to persist in soil for up to 2 years[9]

- It is not expected to bioaccumulate or bioconcentrate in organisms and the food chain

- Drinking Water Standards:
  hydrocarbon total: 10 μg.l[-1] (UK max)
  styrene: 20 μg.l[-1] (WHO guideline)
  4–2600 μg.l[-1] (WHO level where customers may complain)

- Soil Guidelines: Dutch Criteria:
  0.1 mg.kg[-1] (target)
  100 mg.kg[-1] (intervention)

- Air Quality Standards:
  800 μg.m[-3] averaging time 24 hours (WHO guideline)

REFERENCES

1. Clayton GD & Clayton FE (eds.) *Patty's Industrial Hygeine & Toxicology* 4th edn. John Wiley & Sons Inc., New York, 1994

2. Snyder R (ed.) *Ethel Browning's Toxicity and Metabolism of Industrial Solvents* 2nd edn. Elsevier, Amsterdam, 1987

3. Sax NI. *Dangerous Properties of Industrial Materials*, 6th edn. Van Nostrand Reinhold, New York, 1984

4. Bakinson MA & Jones RD, 1985. Gassing due to methylene chloride, xylene, toluene, & styrene reported to Her Majesty's Factory Inspectorate 1961–1980. *Br J Ind Med*; 42: 184–90

5. Gosselin RE, Smith RP & Hodge HC. *Clinical Toxicology of Commercial Products*, 5th edn. Williams & Wilkins, Baltimore, 1984

6. Grant MW & Schuman JS. *Toxicology of the Eye*, 4th edn. Charles C Thomas, Springfield, 1993

7. Hayes JP, Lambourn L, Hopkirk JA, Durnham SR & Taylor AJ. 1991. Occupational asthma due to styrene. *Thorax*; 46(5): 396–7

8. Hodgson JT & Jones RD, 1985. Mortality of styrene production, polymerisation & processing workers at a site in Northwest England. *Scand J Work Environ*; 11: 347–52

9. Hall AH & Rumack BH (eds.) *TOMES System* ® Micromedex, Englewood, Colorado. CD ROM. vol. 41 (exp. 31 July 1999)

Henrietta Wheeler

SULPHUR DIOXIDE

Key Points

- Sulphur dioxide has a pungent, irritating, suffocating sulphur odour
- It is toxic by inhalation and is irritant to the respiratory tract, eyes and skin
- Sulphur dioxide may be fatal on inhalation or can cause acute pulmonary oedema or chronic obstructive pulmonary disease
- Skin contact with gas or liquefied gas may cause burns, and possible frostbite
- *In the event of a large spill, stay upwind and out of low areas. Ventilate closed spaces. Protective clothing, eye protection and breathing apparatus should be worn*

FIRST AID

- Terminate exposure and support vital functions
- The casualty should be moved to an uncontaminated area
- Rescuers should, ideally, be trained personnel and must be careful *not to put themselves at risk* and *so wear appropriate protective clothing and, if available, breathing apparatus*
- If the casualty is unconscious a clear airway should be established and maintained; give 100% oxygen if available
- **Inhalation Exposure:** If the patient stops breathing, expired air resuscitation should be started immediately using a pocket mask with a one way valve, if available. It is important where the face is contaminated that expired air resuscitation is NOT attempted unless an airway with rescuer protection is used
- **Dermal Exposure:** If frostbite occurs *do not* remove clothing, flush skin with water
- If frostbite is not present remove contaminated clothing, if possible under a shower and place in double, sealed, clear bags and label; store the bags in a secure area away from patients and staff
- Wash the skin thoroughly with copious amounts of water
- **Eye Exposure:** If the eye tissue is frozen seek medical advice as soon as possible
- If eye tissue is not frozen, irrigate thoroughly with water or saline for 15 minutes

Detailed Information

- Sulphur dioxide (an oxide of sulphur) is a colourless gas, with a pungent, irritating, suffocating sulphur odour
- *Common synonyms* sulphurous anhydride, sulphurous oxide
- CAS 7446-09-5
- UN 1079
- NIOSH/RTECS WS 4550000
- Molecular formula SO_2
- Molecular weight 64.06
- Sulphur dioxide is used in the manufacture of sulphuric acid, in casting non-ferrous metal, in wood pulp treatment in paper manufacture, as a bleaching agent, fungicide and food additive or preservative
- Sulphur dioxide is soluble in water, alcohol and ether, forming sulphurous acid
- It reacts violently with anhydrous ammonia, bases, amines and chlorine and is heavier than air
- It is easily compressed into a liquid and condenses at −100°C under normal pressure
- Sulphur dioxide is a major contributor to acid rain and is a major air pollutant near smelters and plants burning soft coal or oils high in sulphur content
- It is a non-flammable gas formed when sulphur-containing materials are burned

Summary of Human Toxicity

- About 90% of all sulphur dioxide inhaled is absorbed in the upper respiratory passages
- The highly irritating effects of sulphur dioxide are due to the rapid formation of sulphurous acid on contact with moist membranes (eyes, skin and respiratory tract)[1]
- Death occurs from suffocation due to reflex spasm of the larynx, respiratory paralysis or shock
- *Asthmatic patients may have bronchoconstriction with exposure to 0.5 to 1.0 ppm*[2]
- Asthmatics may have significant bronchoconstriction at exposure for 2 minutes or more to 1 ppm sulphur dioxide or less, especially following exercise[2]
- Exposure to a combination of sulphur dioxide and nitrogen dioxide has been shown to enhance the airway response to inhaled allergens in asthmatics[3]

Toxicity of Sulphur Dioxide[4]

Concentration	Clinical Effects
0.3–5 ppm	Odour threshold, taste threshold
8–12 ppm	Nose, throat and conjuctival irritation, lacrimation and choking
50 ppm	Severe throat and conjunctival irritation and lacrimation
400 ppm for one min	Minimum lethal exposure
30–100 ppm, frequent exposure	Reports of an alteration of the sense of smell and chronic respiratory tract irritation, as well as lacrimation and an uncomfortable conjunctivitis

- Occupational Exposure Standards:
  Long-term exposure limit: 2 ppm (5.3 mg . m⁻³)
  Short-term exposure limit: 5 ppm (13 mg . m⁻³)

Acute Clinical Effects

Inhalation Effects

- Sulphur dioxide is extremely irritating to the mucosa of the nasopharynx and respiratory tract

- Coughing is the most common symptom following exposures to low concentrations

- At higher concentrations sneezing, nausea, vomiting, sore throat and a sensation of choking are common

- Dyspnoea, chest discomfort, cyanosis, bronchoconstriction and wheezing may also occur

- Pulmonary oedema may be delayed for 24 hours

- Anoxia from obstruction was considered to be the cause of death in workers and experimental animals exposed to high concentrations

- Anosmia has been reported[4]

- A rash similar to a drug hypersensitivity reaction has been reported after inhalation

- Asthmatics may have significant bronchoconstriction at low concentrations, especially following exercise

- Reactive airway disease, obstructive and restrictive lung disease, or chronic bronchitis may develop in casualties who survive exposure to high concentrations[4]

- Bronchial hyperactivity occurs immediately after exposure and may persist for several years; damage to the pulmonary parenchyma also occurs, but appears to be transient in nature

Dermal Effects

- Burning sensation of the skin may occur and is sometimes followed by a vesicular eruption

- Dermal contact with liquid or gas can cause frostbite

Eye Effects

- Conjunctival irritation, inflammation of the lids and burns to the cornea may occur

Oral Effects

- Ingestion is not a likely route of exposure

Chronic Clinical Effects

Workers chronically exposed to sulphur dioxide may develop chronic bronchitis and erosion of dental enamel. Mortality from asthma and chronic bronchitis has been noted to increase in response to worsening air pollution with sulphur dioxide and to decrease with decreased levels of sulphur dioxide air pollution.[5]

Management

Sulphur dioxide blood concentrations are not a useful indicator of exposure or toxicity.

Inhalation Management

- Maintain a clear airway, give humidified oxygen and ventilate if necessary

- If respiratory irritation occurs assess respiratory function and if necessary perform chest X-rays to check for chemical pneumonitis

- Consider the use of steroids to reduce the inflammatory response

- Treat pulmonary oedema with PEEP or CPAP ventilation

- Symptomatic and supportive care

Dermal Management

- *If frostbite has occurred*: remove clothing carefully; these may need to be soaked off with tepid water; irrigate the area

- Surgical referral may be necessary

- *If frostbite has not occurred*: remove any remaining contaminated clothing

- Irrigate with copious amounts of water

- Treat burns symptomatically

- Place any contaminated clothes in double, sealed, clear bags and label; store in a secure area away from patients and staff

Eye Management

- Irrigate thoroughly with running water or saline for 15 minutes

- Stain with fluorescein and refer to an ophthalmologist if there is any uptake of the stain

Oral Management

- Not applicable

Summary of Environmental Hazards

- Sulphur dioxide as an air pollutant is oxidised in the environment to sulphuric acid which is a major component of acid rain

- Sulphur dioxide is a major air pollutant, noted especially in the vicinity of installations burning soft coal or oil which is high in sulphur content[4]

- A computer simulation model supported the hypothesis that acid-induced leaching of nutrients from the soil is a cause of nutritional imbalance in tree leaves[6]

- In a laboratory study the fate of sulphur dioxide in soils was determined after adsorption. Most of the sulphur dioxide retained in the soil was in the form of extractable sulphur or in the form of hydrogen sulphide; the study concluded that sorption was influenced by organic matter content in dry soils and by calcium carbonate in moist soils[7]

- Drinking Water Standards: no data available

- Soil Guidelines: no data available

- Air Quality Standards:
     100 ppb averaging time 15 min (UK)
     $125 \, \mu g \cdot m^{-3}$ averaging time 24 hours (WHO guideline)

REFERENCES

1. Hathaway GJ, Proctor NH & Hughes JP. *Proctor and Hughes' Chemical Hazards of the Workplace*, 4th edn. Van Nostrand Reinhold, New York, 1996

2. Horstman DH, Seal E, Folinsbee LJ, Ives P & Roger LJ, 1988. The relationship between exposure duration and sulfur dioxide-induced bronchoconstriction in asthmatic subjects. *Am Ind Hyg Assoc J*; 49(1): 38–47

3. Devalia JL, Rusznak C, Herdman MJ, Trigg CJ, Tarraf H & Davies RJ, 1994. Effect of nitrogen dioxide and sulphur dioxide on airway response of mild asthmatic patients to allergen inhalation. *Lancet*; 344: 1668–71

4. Hall AH & Rumack BH (eds.) *TOMES System®* Micromedex , Englewood, Colorado. vol 41 CD ROM (exp. 31 July 1999)

5. Imai M, Yoshida K & Kitabatake M, 1986. Mortality from asthma and chronic bronchitis associated with changes in sulfur oxides air pollution. *Arch Environ Health*; 41(1): 29–35

6. Vanoene H, 1992. Acid deposition and forest nutrient imbalances: a modeling approach. *Water Air Soil Pollut*; 63: 33–50

7. Cihacek LJ & Bremner JM, 1992. Characterization of sulfur retained by soils exposed to sulfur dioxide. *Commun Soil Sci Plant Anal*; 23(7&8): 805–16

Catherine Farrow

SULPHURIC ACID

Key Points

- Sulphuric acid is a colourless, oily, liquid
- It is corrosive to the respiratory tract, eyes, skin, mucous membranes and gastrointestinal tract, or any tissue with which it comes into contact
- Exposure to sulphuric acid on the skin, by inhalation or ingestion may be fatal
- Ingestion may cause haemorrhage and perforation of the gastrointestinal tract
- *In the event of a large spill, stay upwind and out of low areas. Ventilate closed spaces. Protective clothing, eye protection and breathing apparatus should be worn*

FIRST AID

- Terminate exposure and support vital functions
- The casualty should be moved to an uncontaminated area
- Rescuers should, ideally, be trained personnel and must be careful *not to put themselves at risk* and *so wear appropriate protective clothing and, if available, breathing apparatus*
- If the casualty is unconscious a clear airway should be established and maintained; give 100% oxygen if available
- **Inhalation Exposure:** If the patient stops breathing, expired air resuscitation should be started immediately using a pocket mask with a one way valve, if available. It is important where the face is contaminated that expired air resuscitation is NOT attempted unless an airway with rescuer protection is used
- **Dermal Exposure:** Remove contaminated clothing, if possible under a shower and place in double, sealed, clear bags and label; store the bags in a secure area away from patients and staff
- Wash the skin thoroughly with copious amounts of water
- **Eye Exposure:** Irrigate thoroughly with water or saline for 15 minutes
- **Oral Exposure:** Encourage small quantities of oral fluids (no more than 50–100 ml in total) unless perforation is suspected

Detailed Information

- Sulphuric acid is a colourless, non-flammable oily liquid when in the pure state and brownish when impure
- *Common synonyms* oil of vitriol, hydrogen sulphate, oleum
- CAS 7664-93-9
- UN 1830 (>51%)
- UN 1831 (fuming)
- UN 1832 (spent)
- UN 2796 (≤51%)
- NIOSH/RTECS WS 5600000
- Molecular formula H_2SO_4
- Molecular weight 98.08

- It is used in the manufacture of dyestuffs, fertilisers, parchment paper, food additives, electroplating, industrial explosives and battery acid
- Sulphuric acid is heavier than air, extremely corrosive and reacts violently with alkalis
- When mixing with water caution must be taken due to the heat produced which can cause explosive splattering; acid must always be added to water and never the reverse[1]
- It reacts violently with organic substances, solvents, metals and many other substances, with a risk of fire and explosion; releases toxic and corrosive sulphur oxide when involved in fire
- It is not combustible on its own but in concentrated form it has a high affinity to water and will char most organic materials[2]

Summary of Human Toxicity

- Sulphuric acid may be fatal following inhalation, ingestion and dermal exposure
- It exerts its effects by virtue of its strong acidity
- Concentrated sulphuric acid chars the tissue by removing water
- Individuals living in heavily polluted areas may receive chronic inhalation exposures to sulphuric acid
- Individuals may become acclimatised to sulphuric acid and tolerate higher concentrations
- Its odour threshold is 1–3 mg.m^{-3}
- 5 mg.m^{-3} may produce choking and coughing
- Concentrations of 80 mg.m^{-3} are immediately dangerous to life
- The lowest published lethal dose for humans is 135 mg.kg^{-1}
- Occupational Exposure Standard:
  Long-term exposure limit: 1 mg.m^{-3}
  Short-term exposure limit: no data available

Acute Clinical Effects

Inhalation Effects

- Sulphuric acid is a severe respiratory tract irritant
- Initially there may be tickling of the nose and sneezing, with sore throat and eyes, a cough, tight chest, headache, ataxia, and confusion
- Dyspnoea may develop between 3 to 30 hours post exposure
- Hypoxia and cyanosis may potentially occur
- Inhalation of low concentrations (0.35 to 5 mg.m^{-3}) of sulphuric mist may lead to reflex changes resulting in shallow and more rapid breathing[3]
- Pneumonitis, pulmonary and laryngeal oedema may occur in acute exposure

- Higher concentrations may lead to delayed development of pulmonary fibrosis, residual bronchitis and pulmonary emphysema

Dermal Effects

- Mild exposures to a dilute form may cause irritation and erythema to the skin

- The liquid or concentrated vapours may cause immediate, severe and penetrating burns

- Concentrated solutions on the skin may lead to thermal burns and deep ulcers

- Severe and fatal skin burns can occur with necrosis and scarring

- Sudden circulatory collapse can occur with shock if large areas of the skin have been burnt

Eye Effects

- Sulphuric acid vapours and fumes may cause irritation and conjunctivitis and even necrosis of the conjunctiva at low concentrations

- Liquid sulphuric acid causes severe pain, corneal ulcers, corneal clouding, or severe burns of the corneal epithelium if splashed in the eye

- Reduced visual field or blindness may occur from direct contact with the eye

- Perforation of the globe and loss of ocular contents may occur

- Permanent damage and visual impairment can occur from sulphuric acid splashes in the eye[4]

Oral Effects

- Ingestion of small quantities may lead to local mucosal irritation, epigastric pain, nausea and vomiting, which may be mucoid and coffee-ground in nature

- Ingestion of a larger quantity may lead to oesophageal corrosion, stricture, necrosis and perforation of the gastrointestinal tract, typically more severe in the stomach and intestinal tract rather than the oesophagus

- Severe metabolic acidosis and shock may occur

- Pyloric stenosis often follows several weeks to years later in patients who survive an acute episode of ingestion[5]

Chronic Clinical Effects

Prolonged low concentrations of sulphuric acid may lead to changes in pulmonary function, chronic bronchitis, pulmonary fibrosis, emphysema and chemical pneumonitis. Conjunctivitis, rhinorrhoea, gastritis and overt symptoms may occur and can resemble acute viral respiratory tract infections. Discoloration and erosion of dental enamel has been observed. There is some evidence that chronic low exposures may cause upper respiratory cancer.[6]

Management

Inhalation Management

- Maintain a clear airway, give humidified oxygen and ventilate if necessary

- If respiratory irritation occurs assess respiratory function and if necessary perform chest X-rays to check for chemical pneumonitis

- Consider the use of steroids to reduce the inflammatory response

- Treat pulmonary oedema with PEEP or CPAP ventilation

- Symptomatic and supportive care

Dermal Management

- Remove any remaining contaminated clothing, place in double, sealed, clear bags and label; store in a secure area away from patients and staff

- Irrigate with copious amounts of water

- Treat burns symptomatically

Eye Management

- Irrigate thoroughly with running water or saline for 15 minutes

- Stain with fluorescein and refer to an ophthalmologist if there is any uptake of the stain

Oral Management

- NO GASTRIC LAVAGE OR EMETIC

- Encourage oral fluids, unless perforation is suspected

- Give plasma expanders/blood or IV fluids for shock and analgesics for pain

- Consider the use of steroids to reduce the inflammatory response

- Take abdominal X-ray to check for perforation

- Symptomatic and supportive care

- If facilities are available, early gastro-oesophagoscopy should be undertaken within 12 hours of the event to assess the extent and severity of the injury

Summary of Environmental Hazards

- Sulphuric acid is formed in photochemical smog from the oxidation of sulphur dioxide to sulphur trioxide and a subsequent reaction with water

- Sulphuric acid is highly soluble in water

- It is persistent in water and dissociates to give hydrogen and hydrogen sulphate ions

- It has moderate acute toxicity to aquatic life

- Drinking Water Standards:
      pesticide: $0.1\,\mu g.l^{-1}$ (UK max)

- Soil Guidelines: no data available

- Air Quality Standards: no data available

REFERENCES

1. Sax NI & Lewis RJ. *Hawley's Condensed Chemical Dictionary*, 11th edn. Van Nostrand Reinhold Company, New York, 1987

2. Hall AH & Rumack BH (eds.) *TOMES Plus® System*, Micromedex Inc, Englewood, Colorado, CD ROM. vol 41, (exp. 31 July 1999)

3. Harbison RD (ed.) *Hamilton and Hardy's Industrial Toxicology*, 5th edn. Mosby-Year Book Inc., St Louis, 1998

4. Grant MW & Schuman JS. *Toxicology of the Eye*, 4th edn. Charles C Thomas, Springfield, 1993

5. Gosselin RE, Smith RP & Hodge HC. *Clinical Toxicology of Commercial Products*, 5th edn. Williams & Wilkins, Baltimore, 1984

6. Hathaway GJ, Proctor NH & Hughes JP. *Proctor and Hughes' Chemical Hazards of the Workplace*, 4th edn. Van Nostrand Reinhold, New York, 1996

Catherine Farrow

SULPHUR TRIOXIDE

Key Points

- Sulphur trioxide can exist as a fuming, colourless to white crystalline solid or as a colourless liquid, with a pungent odour
- Sulphur trioxide is a severe irritant of the respiratory tract, eyes, and skin
- It may cause death or permanent injury after even short exposures to small quantities
- Sulphur trioxide toxicity resembles that of sulphuric acid
- *In the event of a large spill, stay upwind and out of low areas. Ventilate closed spaces. Protective clothing and breathing apparatus should be worn*

FIRST AID

- Terminate exposure and support vital functions
- The casualty should be moved to an uncontaminated area
- Rescuers should, ideally, be trained personnel and must be careful *not to put themselves at risk* and *so wear appropriate protective clothing and, if available, breathing apparatus*
- If the casualty is unconscious a clear airway should be established and maintained; give 100% oxygen if available
- **Inhalation Exposure:** If the patient stops breathing, expired air resuscitation should be started immediately using a pocket mask with a one way valve, if available. It is important where the face is contaminated that expired air resuscitation is NOT attempted unless an airway with rescuer protection is used
- **Dermal Exposure:** If frostbite occurs *do not* remove clothing, flush skin with water
- If frostbite is not present remove contaminated clothing, if possible under a shower and place in double, sealed, clear bags and label; store the bags in a secure area away from patients and staff
- Wash the skin thoroughly with copious amounts of water
  Eye Exposure: If the eye tissue is frozen seek medical advice as soon as possible
- If eye tissue is not frozen, irrigate thoroughly with water or saline for 15 minutes
- **Oral Exposure:** Encourage small quantities of oral fluids (no more than 50–100 ml in total)

Detailed Information

- Sulphur trioxide can exist as a fuming, colourless to white crystalline solid or as a colourless liquid, with a pungent odour
- *Common synonyms* sulphuric anhydride, sulphur oxide
- CAS 7446–11–9
- UN 1829
- NIOSH/RTECS WT 4830000
- Molecular formula SO_3
- Molecular weight 80.07

- Sulphur trioxide is used in the sulphonation of organic compounds, especially nonionic detergents, solar energy collectors and as an oxidising agent
- Sulphur trioxide vapour is heavier than air and reacts with moist air to form sulphuric acid, giving off corrosive fumes (white mist)
- The anhydride combines with water forming sulphuric acid and evolving heat
- When dissolved in water it is a strong acid, reacts strongly with alkalis, and is corrosive
- It is a strong oxidant which reacts violently with combustible substances and reducing agents with the risk of fire and explosion
- Attacks metals in the presence of moisture, giving off hydrogen[1]

Summary of Human Toxicity

- Sulphur trioxide can be absorbed by inhalation and ingestion
- It is corrosive to the eyes, the skin and the respiratory tract
- The majority of toxic effects of sulphur trioxide mimic those of sulphuric acid
- Lowest acute toxic inhalation dose of 30 mg.kg^{-1} has been reported[2]
- In the eyes sulphur trioxide is a severe irritant, even at concentrations as low as 1 ppm[3]
- Occupational Exposure Standards: no data available

Acute Clinical Effects

Inhalation Effects

- Sulphur trioxide is a severe respiratory tract irritant
- Initially there may be a dry mouth with sore throat and eyes, a cough, tight chest, headache, ataxia, and confusion
- Dyspnoea may develop between 3 to 30 hours post exposure
- Hypoxia and cyanosis may potentially occur
- Increased concentrations may cause pneumonitis, laryngeal spasm and pulmonary oedema

Dermal Effects

- Mild exposures to a dilute form may cause irritation and erythema to the skin
- The liquid or concentrated vapours may cause immediate, severe and penetrating burns
- Concentrated solutions on the skin may lead to thermal burns and deep ulcers
- Frostbite injury may occur with dermal exposure to liquefied sulphur trioxide

Eye Effects

- Sulphur trioxide vapours and fumes may cause irritation and conjunctivitis and even necrosis of the conjunctiva at low concentrations

- Liquid sulphur trioxide causes severe pain, corneal ulcers, corneal clouding, or severe burns of the corneal epithelium if splashed in the eye

- Reduced visual fields or blindness may occur from direct contact with the eye

- Perforation of the globe and loss of ocular contents may occur

Oral Effects

- There is no data or information on sulphur trioxide ingestion, however on contact with moisture sulphur trioxide forms sulphuric acid

- Ingestion of small quantities of sulphur trioxide may lead to local mucosal irritation, epigastric pain, nausea and vomiting, which may be mucoid and coffee-ground in nature

Chronic Clinical Effects

Prolonged low concentrations of sulphur trioxide may lead to chronic bronchitis, pulmonary fibrosis and chemical pneumonitis. Conjunctivitis and overt symptoms may resemble acute viral respiratory tract infections.

Management

Inhalation Management

- Maintain a clear airway, give humidified oxygen and ventilate if necessary

- If respiratory irritation occurs assess respiratory function and if necessary perform chest X-rays to check for chemical pneumonitis

- Consider the use of steroids to reduce the inflammatory response

- Treat pulmonary oedema with PEEP or CPAP ventilation

- Symptomatic and supportive care

Dermal Management

- *If frostbite has occurred*: remove clothes carefully, these may need to be soaked off with tepid water; irrigate the area

- Surgical referral may be necessary

- *If frostbite has not occurred*: remove any remaining contaminated clothing, place in double, sealed, clear bags and label; store in a secure area away from patients and staff

- Irrigate with copious amounts of water

- Treat burns symptomatically

Eye Management

- Irrigate thoroughly with running water or saline for 15 minutes

- Stain with fluorescein and refer to an ophthalmologist if there is any uptake of the stain

Oral Management

- NO GASTRIC LAVAGE OR EMETIC

- Encourage oral fluids, unless perforation is suspected

- Give plasma expanders/blood or IV fluids for shock and analgesics for pain

- Consider the use of steroids to reduce the inflammatory response

- Take abdominal X-ray to check for perforation

- Symptomatic and supportive care

- If facilities are available, early gastro-oesophagoscopy should be undertaken within 12 hours of the event to assess the extent and severity of the injury

Summary of Environmental Hazards

- Sulphur trioxide is a highly reactive gas, in the presence of moisture in the air it is hydrated to sulphuric acid[4]

- Drinking Water Standards: no data available

- Soil Guidelines: no data available

- Air Quality Standards: no data available

REFERENCES

1. Dutch Institute for the Working Environment and the Dutch Chemical Industry Association (eds.) *Chemical Safety Sheets – Working Safely with Hazardous Chemicals*. Samson Chemical Publishers, The Netherlands, 1991

2. Sax NI. *Dangerous Properties of Industrial Materials*, 6th edn. Van Nostrand Reinhold, New York, 1984

3. Grant MW & Schuman JS. *Toxicology of the Eye*, 4th edn. Charles C Thomas, Springfield, 1993

4. International Programme on Chemical Safety. *Environmental Health Criteria 8: Sulfur oxides and Suspended Particulate Matter*. WHO, Geneva 1979

Henrietta Wheeler

TETRACHLOROETHYLENE

Key Points

- Tetrachloroethylene is a clear, colourless liquid that has an ether-like odour
- It is irritating to the respiratory and GI tract, as well as the skin and mucous membranes
- Tetrachloroethylene causes CNS depression and can sensitise the myocardium to endogenous catecholamines
- It may cause liver and renal damage
- *In the event of a large spill, stay upwind and out of low areas. Ventilate closed spaces. Protective clothing, eye protection and breathing apparatus should be worn*

FIRST AID

- Terminate exposure and support vital functions
- The casualty should be moved to an uncontaminated area
- Rescuers should, ideally, be trained personnel and must be careful *not to put themselves at risk* and *so wear appropriate protective clothing and, if available, breathing apparatus*
- If the casualty is unconscious a clear airway should be established and maintained; give 100% oxygen if available
- **Inhalation Exposure:** If the patient stops breathing, expired air resuscitation should be started immediately using a pocket mask with a one way valve, if available. It is important where the face is contaminated that expired air resuscitation is NOT attempted unless an airway with rescuer protection is used
- **Dermal Exposure:** Remove contaminated clothing, if possible under a shower and place in double, sealed, clear bags and label; store the bags in a secure area away from patients and staff
- Wash the skin thoroughly with copious amounts of water
- **Eye Exposure:** Irrigate thoroughly with water or saline for 15 minutes
- **Oral Exposure:** Encourage small quantities of oral fluids (no more than 50–100 ml in total)

Detailed Information

- Tetrachloroethylene is a clear, colourless non-flammable liquid that has an ether-like odour
- *Common synonyms* carbon dichloride, ethylene tetrachloride, PCE, perc, perchloroethylene, perk, tetrachloroethene
- CAS 127-18-4
- UN 1897
- NIOSH/RTECS KX 3850000
- Molecular formula C_2Cl_4
- Molecular weight 165.83
- Tetrachloroethylene is used as a dry cleaning agent, in textile processing, as a scouring solvent, in manufacturing fluorocarbons, as a drying agent for metals, as a fumigant for insects and rodents, and as a degreasing solvent

- It is active against hookworms (*Ancylostoma* and *Necator*) and is still used in endemic areas although it has generally been superseded by drugs that are less toxic and easier to administer
- Tetrachloroethylene sinks in water and is not soluble; it is soluble in ethanol, ethyl ether, chloroform and benzene; it is miscible with alcohol, benzene, chloroform, ether, and most solvents and oils
- Tetrachloroethylene decomposes when heated, or when exposed to UV light, to give off phosgene and hydrochloric acid
- Mixtures of tetrachloroethylene and dinitrogen tetroxide are explosive when subject to shock; tetrachloroethylene reacts violently or explosively with certain alkali or alkaline earth metals

Summary of Human Toxicity

- Tetrachloroethylene is absorbed by inhalation, ingestion and to a lesser extent dermally
- 80 to 100% of inhaled tetrachloroethylene is excreted unchanged by the lungs; once absorbed, the highest concentrations of tetrachloroethylene are found in the adipose tissue, liver, brain, kidney and lungs, reflecting its high lipid solubility[1]
- The metabolism takes place mainly in the liver and is excreted in the urine
- The biological half-life of tetrachloroethylene on inhalation is between 3 to 72 hours and on ingestion 144 hours[2]
- It causes myocardial depression and sensitises the myocardium to endogenous catecholamines via inhalation and ingestion, this cardiac sensitisation is unpredictable in terms of exposure, duration and quantity
- The minimum lethal human exposure for tetrachloroethylene has not been established; an oral dose of 500 mg.kg^{-1} was not fatal[3]
- Alcohol may increase tetrachloroethylene toxicity
- Tetrachloroethylene has been ingested and inhaled in the quest for pleasure and elation; habitual inhalation may lead to dependence
- Odour tolerance is exhibited in chronically exposed humans

Toxicity of Tetrachloroethylene [3,4,5]

Concentration	Clinical effects
0.17 mg.l^{-1}	Taste threshold
1.0 ppm, in air	Odour threshold
0.3 ppm, in water	Odour threshold
50 ppm for 8 h	No effect, faint odour
400 ppm for 2 h	Eye and nasal irritation, ataxia and strong tolerable odour
600 ppm for 10 min	Numb mouth, dizzy, ataxia and very strong exposure
1500 ppm for 30 min	Gagging, eye and respiratory irritation, loss of consciousness and almost intolerable odour
5,000 ppm for 6 min	Vertigo, nausea and confusion
60–450 ppm for 2–20 years	Abnormal ECG recordings

- Occupational Exposure Standards:
  Long-term exposure limit: 50 ppm (345 mg.m^{-3})
  Short-term exposure limit: 100 ppm (689 mg.m^{-3})

Acute Clinical Effects

Inhalation Effects

- Depending on the dose and duration of exposure, inhalation leads to respiratory irritation, coughing, wheezing, pulmonary oedema, increasing cyanosis and respiratory failure

- Tetrachloroethylene vapours are irritant to nasal, ocular and respiratory mucosa

- Headache, fatigue, ataxia, dizziness, nausea, vomiting, hypotension, mental confusion and temporary blurred vision have been reported after inhalation

- Inhalation may lead to systemic effects; see below

Dermal Effects

- Tetrachloroethylene is irritant to the skin; produces dry, scaly skin, blisters and dermal burns

- Erythema and a severe burning sensation may occur if tetrachloroethylene is left on the skin for 40 minutes or longer

- Dermatitis is caused by defatting of the skin

- Significant exposures may cause systemic symptoms; see below

Eye Effects

- Tetrachloroethylene vapours are irritating to eyes at high concentrations

- An ocular splash exposure is expected to cause lacrimation and burning but no permanent damage

Oral Effects

- Ingestion of tetrachloroethylene may cause gastric irritation with nausea and vomiting

- Aspiration may lead to respiratory irritation, coughing, wheezing, pulmonary oedema and increasing cyanosis

- Ingestion of tetrachloroethylene has been associated with the development of toxic epidermal necrolysis[6]

- Ingestion may cause systemic effects; see below

Systemic Effects

- CNS depression, dizziness, inebriation, light-headedness, mental dullness, ataxia, drowsiness and coma, respiratory depression or death may occur

- Hypotension, cardiac arrhythmias, hepatic and renal injury, proteinuria and haematuria

- Tetrachloroethylene may sensitise the myocardium to adrenaline and other catecholamines which may cause arrhythmias and sudden death after massive acute exposures

Chronic Clinical Effects

Chronic occupational exposure has been known to produce multiple ventricular premature beats, peripheral neuropathy, hepatitis, confusion, disorientation, muscle cramps, fatigue, agitation and damage to liver, kidney and spleen.[4]

Dependence may follow habitual inhalation of small quantities of tetrachloroethylene vapour.

Chronic skin exposure may cause reddening and chapping of the skin. Dry, scaly and fissured dermatitis may also occur from repeated skin contact.

Management

Blood tetrachloroethylene concentrations have no clinical relevance other than to confirm exposure when this is in doubt.

Inhalation Management

- Maintain a clear airway, give humidified oxygen and ventilate if necessary

- If respiratory irritation occurs assess respiratory function and if necessary perform a chest X-rays to check for chemical pneumonitis

- See systemic management below

Dermal Management

- Remove any remaining contaminated clothing, place in double, sealed, clear bags and label; store in a secure area away from patients and staff

- Irrigate with copious amounts of water

- An emollient may be required

- Treat irritation symptomatically

- For significant exposure, see systemic management below

Eye Management

- Irrigate thoroughly with running water or saline for 15 minutes

- Stain with fluorescein and refer to an ophthalmologist if there is any uptake of the stain

Oral Management

- Encourage oral fluids for ingestion of small quantities

- DO NOT INDUCE EMESIS due to risk of aspiration

- Consider gastric lavage within 1 hour of a substantial ingestion, ensuring airway is protected

- An abdominal X-ray may be of use in confirming ingestion as tetrachloroethylene is radiopaque

- Controlled hyperventilation to enhance pulmonary elimination may be considered although this has been demonstrated to be effective in only one case; this has not been demonstrated elsewhere[7]

- Symptomatic and supportive care; see systemic management below

Systemic Management

- Patients should be kept at rest

- With severe exposures the patient should be kept on a cardiac monitor for 12 hours, avoiding the use of all stimulants, except for resuscitation

- Monitor level of consciousness and respiration, ECG, renal and liver function

- Daily urinalysis for proteinuria and haematuria may be useful after massive exposures

Summary of Environmental Hazards

- Tetrachloroethylene is persistent in the atmosphere

- Tetrachloroethylene reacts with photochemically produced hydroxyl radicals or chlorine ions with a half-life of 1 hour to 2 months

- Volatilisation also appears to be the major pathway by which tetrachloroethylene is lost from water. The half-life of tetrachloroethylene is estimated to be 3–30 days for river water and 30–300 days for lake- and groundwater[8]

- Spills on the soil may be expected to leach slowly into the groundwater

- Tetrachloroethylene does not concentrate in the food chain

- Drinking Water Standards:
  hydrocarbon total: $10\,\mu g.l^{-1}$ (UK max)
  pesticide: $0.1\,\mu g.l^{-1}$ (UK max)
  tetrachloroethylene: $10\,\mu g.l^{-1}$ (UK max)
  $40\,\mu g.l^{-1}$ (WHO guideline)

- Soil Guidelines: Dutch Criteria:
  $0.01\,mg.kg^{-1}$ (target)
  $4\,mg.kg^{-1}$ (intervention)

- Air Quality Standards:
  $5\,mg.m^{-3}$ averaging time 24 hours (WHO guideline)

REFERENCES

1. Ware GW (ed.) In *Reviews of Environmental Contamination and Toxicology*, vol 106. Springer–Verlag, 1988

2. Gosselin RE, Smith RP & Hodge HC. *Clinical Toxicology of Commercial Products*, 5th edn. Williams & Wilkins, Baltimore, 1984

3. Clayton GD & Clayton FE (eds.) *Patty's Industrial Hygiene and Toxicology*, 4th edn. John Wiley & Sons, Inc., New York, 1994

4. Baselt RC & Cravey RH. *Disposition of Toxic Drugs and Chemicals in Man*, 4th edn. Chemical Toxicology Institute, 1995

5. Ellenhorn MJ, Schonwalds S, Ordog G & Wasserberger J. *Eilenhorn's Medical Toxicology – Diagnosis and Treatment of Human Poisoning*, 2nd edn. Williams & Wilkins, London, 1997

6. Potter B, Auerback R & Lorincz AL, 1960. Toxic epidermal necrolysis. *Arch Dermatol*; 82: 903–7

7. Koppel C, Arendt U & Koeppe P, 1958. Acute tetrachloroethylene poisoning – blood elimination kinetics during hyperventilation therapy. *Clin Toxicol*; 23(2-3): 103–15

8. Hall AH & Rumack BH (eds.) *TOMES System®*, Micromedex, Englewood, Colorado. CD ROM. vol.41 (exp. 31 July 1999)

Henrietta Wheeler

THALLIUM

Key Points

- Thallium is a bluish–white, very soft, inelastic, easily fusible, heavy metal
- It is highly toxic by ingestion, inhalation, and dermal exposure
- The main toxic effects include gastrointestinal disturbances, alopecia, psychosis, neurological and circulatory effects
- It is mostly used in the form of soluble salts such as thallium acetate and thallium sulphate

FIRST AID

- Terminate exposure and support vital functions
- The casualty should be moved to an uncontaminated area
- Rescuers should, ideally, be trained personnel and must be careful *not to put themselves at risk* and *so wear appropriate protective clothing and, if available, breathing apparatus*
- If the casualty is unconscious a clear airway should be established and maintained; give 100% oxygen if available
- **Inhalation Exposure:** If the patient stops breathing, expired air resuscitation should be started immediately using a pocket mask with a one way valve, if available. It is important where the face is contaminated that expired air resuscitation is NOT attempted unless an airway with rescuer protection is used
- **Dermal Exposure:** Remove contaminated clothing , if possible under a shower and place in double, sealed, clear bags and label; store the bags in a secure area away from patients and staff
- Wash the skin thoroughly with copious amounts of water
- **Eye Exposure:** Irrigate thoroughly with water or saline for 15 minutes
- **Oral Exposure:** Encourage small quantities of oral fluids (no more than 50–100 ml in total) unless perforation is suspected

Detailed Information

- Thallium is a bluish–white, very soft, inelastic, easily fusible, heavy metal; it leaves a streak on paper
- *Common synonyms* Tl
- CAS 7440-28-0
- UN 1707
- NIOSH/RTECS XG 3425000
- Atomic symbol Tl
- Atomic weight 204.383
- Thallium salts were formerly used as active ingredients in rodenticides and insecticides, sales are now regulated due to toxicity; it is still used in the manufacture of optical lenses and imitation jewellery; isotopes are used in cardiac scanning
- It is incompatible with oxidisers, especially fluorine, chlorine, and bromine
- Thallium reacts on contact with strong acids such as hydrofluoric, sulphuric, and nitric
- Thallium burns but does not easily ignite, it produces toxic fumes when burning[1]

Summary of Human Toxicity

- Thallium is rapidly absorbed, by inhalation, by ingestion, from dermal exposure and through the mucous membranes
- The exact mechanism of thallium toxicity is unclear; it has been shown to have a high affinity for sulphydryl groups in the mitochondrial membrane; therefore it inhibits many enzymes
- It exchanges with potassium (this may be due to the similarity in size of their ionic radii) and interferes with oxidative phosphorylation
- Tissues with high potassium concentrations accumulate thallium (nerve, kidney and muscle tissue)
- Water-soluble thallium salts (malonate, sulphate, acetate, carbonate) are more toxic than the less soluble salts (sulphide, iodide)[2]
- Most fatalities occur after ingestion of 8–20 mg . kg^{-1} [2,3]
- There are three main phases in thallium toxicity: [4,5,6]

    Phase 1: lasts about four hours and represents intravascular distribution

    Phase 2: lasts 4–48 hours during which time thallium is distributed to the central nervous system; generally the distribution phase is completed by 24 hours post ingestion

    Phase 3: the elimination phase, starts at about 24 hours post ingestion, with its duration depending on the therapeutic intervention used; it can be prolonged due to the enteral re-circulation

- Prussian blue (potassium ferrihexacyanoferrate II) is the chelation treatment of choice for thallium toxicity
- Thallium ions are exchanged for potassium ions in the lattice of the Prussian blue molecule and are subsequently excreted in the faeces
- Repeat doses of Prussian blue are administered to interrupt the cycle of intestinal secretion and re-absorption seen in thallium poisoning
- Occupational Exposure Standards (soluble compounds):
    Long-term exposure limit: 0.1 mg . m^{-3}
    Short-term exposure limit: no data available

Acute Clinical Effects

Inhalation Effects

- Thallium is a respiratory irritant
- Exposure may lead to nasal discharge, chest tightness, coughing, shortness of breath and fever
- Systemic effects may occur; see below

Dermal Effects

- Dermal contact leads to irritation at site and may cause dry scaly skin; severe exposure may lead to acne

- Systemic effects may occur with prolonged contact; see below

Eye Effects

- Severe irritation, itching and burning sensation may occur

- Systemic effects may occur with significant exposure; see below

Oral Effects

- Ingestion may initially lead to nausea, vomiting, diarrhoea, abdominal pain, and gastrointestinal haemorrhage

- Symptoms usually occur within 1–3 days

- Systemic effects may occur; see below

Systemic Effects

- Symptoms are usually non-specific due to multi-organ involvement

- Initial symptoms of poisoning are abdominal pain, gastro-enteritis, tachycardia and headache, which usually occur within the first 12 hours[3]

- Blood sugars may be raised temporarily[5]

- Characteristic dark, pigmented band appears in the scalp hair within 4 days

- Confusion, hallucinations, and convulsions usually occur at 2 to 5 days post exposure

- In severe cases coma, respiratory paralysis, and death occurs within 1 week

- Smaller doses cause ataxia and painful sensorimotor neuropathy

- Other neurological features include cranial nerve palsies, optic neuropathy, choreoathetosis, tremor, and encephalopathy[7]

- Severe and abrupt alopecia begins 10 days after exposure, and is completed within 1 month

- Skin may be involved, including acneiform eruptions, a papulomacular rash, and dystrophy of the nails (Mee's lines)[8]

- Automotor dysfunction, hypertension, cardiomyopathy, ECG changes, testicular toxicity, hypokalaemia, renal failure, abnormal liver function, leukocytosis, and thrombocytopenia have been reported[4]

- Ophthalmic signs include loss of the lateral half of the eyebrows, eyelid skin lesions, blepharoptosis, facial nerve palsy, internal and external ophthalmoplegia, and nystagmus[8]

- Non-inflammatory keratitis and lens opacities have also been described

- Optic atrophy as a sequel to thallium-induced toxic optic neuropathy is well reported, functional changes include impairment of contrast sensitivity, abnormal colour vision, impaired visual acuity, and central or cecocentral scotomas[9]

- Thallium crosses the placenta barrier, and produces alopecia and nail abnormalities in the foetus exposed in the last trimester

Chronic Clinical Effects

Following repeated exposures, the patient insidiously develops a distal neuropathy initially with sensory and motor loss, which spreads proximally. Respiratory paralysis may develop if cranial nerves are affected. Ataxia, tremor, and marked tenderness of the soles, together with personality changes and severe loss of intellectual function may result.[2]

Management

Normal serum, plasma or urine thallium concentrations are $< 1 \mu g.l^{-1}$

Exposure is usually monitored by estimation of urine concentrations (which can exceed 20 mg.l^{-1} in severe cases)

Inhalation Management

- Maintain a clear airway, give humidified oxygen and ventilate if necessary

- If respiratory irritation occurs assess respiratory function and if necessary perform chest X-rays to check for chemical pneumonitis

- Treat pulmonary oedema with PEEP or CPAP ventilation

- Consider the use of steroids to reduce inflammatory response

- Systemic toxicity is possible; see systemic management below

Dermal Management

- Wash affected area with copious amounts of water or saline

- Symptomatic and supportive care

- Systemic effects may occur; see systemic management below

Eye Management

- Irrigate thoroughly with running water or saline for 15 minutes

- Stain with fluoroscein and refer to an ophthalmologist if there is any uptake of the stain

- Systemic effects may occur; see systemic management below

Oral Management

- Thallium salts are radio-opaque, ingestion may be confirmed by X-ray

- If radiopacities are seen in the stomach, consider gastric lavage followed by instillation of Prussian blue; see chelation therapy below

- Contact a poisons information service for further guidance on gut decontamination

- The absence of radiopacities does not eliminate the possibility of ingestion

- Systemic effects may occur; see systemic management below

Systemic Management

- The main aim of treatment is to support the patient's vital functions and facilitate the removal of thallium from the tissues

- Prussian blue should be administered; see below

- Monitor blood sugars, urea, creatinine and electrolytes, especially potassium (as the Prussian blue therapy may result in a hypokalaemia)

- Cardiotoxicity should be assessed with ECG, echocardiography and chest X-rays

- Serum, urine and faecal samples should be taken for thallium estimation 3 times a week

- If there is evidence of ascending polyneuropathy (of the Guillain-Barré syndrome type), monitor the vital capacity daily and ventilate the patient electively if it is equal to or less than 1 litre

- Physiotherapy is essential to prevent muscle contractures and speed recovery of power

- Haemodialysis has been shown to reduce the thallium concentration in some patients; its use is controversial, and only indicated in cases of thallium poisoning in the presence of renal failure

- Hair re-growth from the alopecia occurs and may be more luxuriant and a different shade than before

Chelation Therapy:

Contact a poisons information service for further guidance and paediatric doses.

Soluble Prussian blue is the treatment of choice. Prussian blue is given orally at a rate of 250 mg.kg^{-1} per day in divided doses via a fine bore nasogastric tube. Prussian blue and the Prussian blue-thallium complex are poorly absorbed via the gastrointestinal tract, so no adverse effects would be expected. Therapy should be continued until urine thallium concentration is < 0.5 mg/24 hours

Summary of Environmental Hazards

- Thallium (I) compounds tend to have high solubility

- Significant compartments likely to be soil and water

- Drinking Water Standards: no data available

- Soil Guidelines: no data available

- Air Quality Standards: no data available

REFERENCES

1. Sax NI. *Dangerous Properties of Industrial Materials*, 6th edn. Van Nostrand Reinhold, New York, 1984

2. Aw TC & Vale JA. Poisoning from Metals in *Oxford Textbook of Medicine*. DJ Wetherall, JGG Ledingham & DA Warrell, 3rd edn., vol 1, Ch. 836: 1113–4

3. Ellenhorn MJ, Schonwalds S, Ordog G & Wasserberger J. *Ellenhorn's Medical Toxicology – Diagnosis and Treatment of Human Poisoning*, 2nd edn. Williams & Wilkins, London, 1997

4. Cavanagh JB, 1991. What have we learnt from Graham Fredrick Young? Reflections on the mechanisms of thallium toxicity. *Neuropath Appl Neurobio*; 17: 3-9

5. Vergauwe PL, Knockaert DC & Van Tittlebloom TJ, 1990. Near fatal subacute thallium poisoning necessitating prolonged mechanical ventilation. *Am J Emerg Med*; 8: 548–50

6. Moore D, House I & Dixon A, 1993. Thallium poisoning. *Br Med J*; 306: 1527–9

7. Roby DS, Fein AM, Bennett RH, Morgan LS, Zatuchni J & Lippman ML, 1984. Cardiopulmonary effects of acute thallium poisoning. *Chest*; 84: 236–40

8. Heyl T & Barlow RJ, 1989. Thallium poisoning: a dermatological perspective. *Brit J Dermatol;* 121: 787–92

9. Grant MW & Schuman JS. *Toxicology of the Eye*, 4th edn. Charles C Thomas, Springfield, 1993

Jennifer Butler

TITANIUM TETRACHLORIDE

Key Points

- Titanium tetrachloride is a colourless to light yellow liquid which fumes strongly when exposed to air, forming a dense and persistent white cloud
- On contact with moisture it releases hydrochloric acid and is corrosive to all tissues
- Fatalities have occurred in workers exposed by both inhalation and dermal contact
- Titanium tetrachloride causes irritation to the lungs, eyes, skin, mucous membranes or tissues with which it is in direct contact
- *In the event of a large spill, stay upwind and out of low areas. Ventilate closed spaces. Protective clothing, eye protection and breathing apparatus should be worn*

First Aid

- Terminate exposure and support vital functions
- The casualty should be moved to an uncontaminated area
- Rescuers should, ideally, be trained personnel and must be careful *not to put themselves at risk* and *so wear appropriate protective clothing and, if available, breathing apparatus*
- If the casualty is unconscious a clear airway should be established and maintained; give 100% oxygen if available
- **Inhalation Exposure:** If the patient stops breathing, expired air resuscitation should be started immediately using a pocket mask with a one way valve, if available. It is important where the face is contaminated that expired air resuscitation is NOT attempted unless an airway with rescuer protection is used
- *Neutralising solutions (e.g. NaHCO₃) should never be used due to the heat generated*
- **Dermal Exposure:** Remove contaminated clothing, if possible under a shower and place in double, sealed, clear bags and label; store the bags in a secure area away from patients and staff
- Due to the fact that titanium tetrachloride reacts with water to release highly corrosive hydrochloric acid, exposed areas should first be *wiped with a dry cloth* and then washed thoroughly with copious amounts of water
- **Eye Exposure:** Eyelids and face should be *wiped dry* to prevent burns; irrigate thoroughly with water or saline for 15 minutes
- **Oral Exposure:** Encourage small quantities of oral fluids (no more than 50–100 ml in total) unless perforation is suspected

Detailed Information

- Titanium tetrachloride is a colourless to light yellow liquid which fumes strongly when exposed to air, forming a dense and persistent white cloud
- *Common synonyms* titanic chloride, titanium chloride
- CAS 7550-45-0
- UN 1838

- NIOSH/RTECS XR 1925000
- Molecular formula Cl_4Ti
- Molecular weight 189.73
- Titanium tetrachloride is used as an intermediate in the manufacture of the white, inert pigment, titanium dioxide; used in the manufacture of paints, lacquers, leathers, inks, rubber, textiles and plastic
- Vapours mix readily with air and form hydrochloric acid
- It is decomposed by moisture to form heat, hydrochloric acid and titanium dioxide
- Soluble in dilute hydrochloric acid, soluble in water with evolution of heat, concentrated aqueous solutions are stable and corrosive, dilute solutions precipitate insoluble basic chlorides[1]
- At high temperatures with pure oxygen it oxidises to titanium dioxide

Summary of Human Toxicity

- The toxicity of titanium tetrachloride is due to the release of hydrochloric acid on contact with moisture; it is thus corrosive to tissues
- Fatalities have occurred in workers exposed by both inhalation and dermal contact, but the concentration of material was not specified[2]
- Even low airborne concentrations are intolerable; it causes corrosive damage to the lungs, skin, mucous membranes or tissues with which it is in direct contact
- Direct eye and skin splashes can cause severe burns; the severity of the burns are similar to strong alkali rather than acid burns[3]
- A two minute inhalation of fumes has resulted in ARDS, chemical pneumonitis, diffuse endobronchial polyposis and endobronchial erythema, and tracheal stenosis requiring stenting[4]
- Long-term exposure, even at low concentrations may cause respiratory irritation and acute or chronic bronchitis
- Occupational Exposure Standards:
    Long-term exposure limit: no data available
    Short-term exposure limit: no data available

Acute Clinical Effects

Inhalation Effects

- Exposure to hydrochloric acid fumes from the decomposition of titanium tetrachloride can cause respiratory tract irritation
- The fumes can cause irritation to the mucous membranes, cough, bronchoconstriction with wheezing and chemical pneumonitis and/or noncardiogenic pulmonary oedema can occur[1,5]
- Pulmonary oedema may be delayed to up to 24 to 72 hours
- Pyrexia may occur from acute inhalation[4]

- Inhalation may cause oedema to the pharynx, vocal cords and trachea; sequelae may include stenosis of the larynx, trachea and upper bronchi[1,4]

Dermal Effects

- Dry skin has been reported following direct contact with pure titanium tetrachloride[2]

- Direct skin contact can cause irritation or severe corrosive dermal burns, especially if the material is in contact with water

Eye Effects

- Contact has been reported to cause corneal damage, conjunctivitis and keratitis

- Liquid titanium tetrachloride in the eye can cause severe injuries including corneal scarring and vascularisation, raised intraocular pressure, conjunctival ischaemia, anterior uveitis, opacification of the lens, corneal perforation and permanent blindness[6]

- Splashes have resulted in entropion and trichiasis of the upper and lower eyelid[6]

- Exposure to dense fumes may also cause corneal burns[3]

Oral Effects

- Ingestion can cause irritation or burns in the mouth, throat, larynx, oesophagus and gastrointestinal tract

- Symptoms may include nausea, vomiting, diarrhoea and abdominal pain

- Severe early complications are haemorrhage and perforation and late development of oesophageal or gastric strictures

Chronic Clinical Effects

Long-term inhalation of low concentrations can cause cough, bronchoconstriction with wheezing and bronchitis, and chemical pneumonitis. Long-term inhalation may cause pulmonary injury similar to silicosis.[2]

Management

Inhalation Management

- Maintain a clear airway, give humidified oxygen and ventilate if necessary

- If respiratory irritation occurs assess respiratory function and if necessary perform chest X-rays to check for chemical pneumonitis

- Consider the use of steroids to reduce the inflammatory response

- Treat pulmonary oedema with PEEP or CPAP ventilation

- Symptomatic and supportive care

Dermal Management

- Dry wipe any exposed area

- Wash the exposed area thoroughly with soap and water

- Treat as a thermal burn; analgesia may be required

Eye Management

- Dry wipe the exposed area then immediately irrigate with copious amounts of normal saline or water for at least 15 minutes

- Stain with fluorescein and refer to an ophthalmologist if there is any uptake of the stain

Oral Management

- NO GASTRIC LAVAGE OR EMETIC

- Encourage oral fluids, unless perforation is suspected

- Give plasma expanders/blood or IV fluids for shock and analgesics for pain

- Consider the use of steroids to reduce the inflammatory response

- Take abdominal X-ray to check for perforation

- Symptomatic and supportive care

- If facilities are available, early gastro-oesophagoscopy should be undertaken within 12 hours of the event to assess the extent and severity of the injury

Summary of Environmental Hazards

- In water titanium tetrachloride converts to a lower valence form of titanium and hydrochloric acid

- When added to water a violent reaction occurs producing a fuming cloud and heat

- Dangerous to aquatic life in high concentrations

- Titanium bioaccumulates in fish from surrounding water

- Drinking Water Standards:
    chloride: 400 mg.l^{-1} (UK max)
    250 mg.l^{-1} (WHO guideline)

- Soil Guidelines: no data available

- Air Quality Standards: no data available

REFERENCES

1. Harbison RD (ed.) *Hamilton and Hardy's Industrial Toxicology*, 5th edn. Mosby-Year Book Inc., St Louis, 1998

2. Hall AH & Rumack BH (eds.) *TOMES System®*, Micromedex, Englewood, Colorado. CD ROM. vol.41 (exp. 31 July 1999)

3. Grant MW & Schuman JS. *Toxicology of the Eye*, 4th edn. Charles C Thomas, Springfield, 1993

4. Park T, DiBenedetto R, Morgan K, Colmers R & Sherman E, 1984. Diffuse endobronchial polyposis following a titanium tetrachloride inhalation injury. *Am Rev Resp Dis*; 130(2): 315–17

5. Sax NI. *Dangerous Properties of Industrial Materials*, 6th edn. Van Nostrand Reinhold, New York, 1984

6. Chitkara DK & McNeela BJ, 1992. Titanium tetrachloride burns to the eye. *Br J Ophthalmol*; 76(6): 380–2

Henrietta Wheeler

TOLUENE

Key Points

- Toluene is a clear, colourless, volatile liquid with a benzene-like odour
- It is irritating to the eyes, skin, respiratory system and gastrointestinal tract
- Toluene causes CNS depression and can sensitise the myocardium to endogenous catecholamines
- Chronic toluene exposure may cause liver and renal damage
- Toluene vapour mixes readily with air
- *In the event of a large spill, stay upwind and out of low areas. Ventilate closed spaces. Protective clothing, eye protection and breathing apparatus should be worn*

FIRST AID

- Terminate exposure and support vital functions
- The casualty should be moved to an uncontaminated area
- Rescuers should, ideally, be trained personnel and must be careful *not to put themselves at risk* and *so wear appropriate protective clothing and, if available, breathing apparatus*
- If the casualty is unconscious a clear airway should be established and maintained; give 100% oxygen if available
- **Inhalation Exposure:** If the patient stops breathing, expired air resuscitation should be started immediately using a pocket mask with a one way valve, if available. It is important where the face is contaminated that expired air resuscitation is NOT attempted unless an airway with rescuer protection is used
- **Dermal Exposure:** Remove contaminated clothing, if possible under a shower and place in double, sealed, clear bags and label; store the bags in a secure area away from patients and staff
- Wash the skin thoroughly with copious amounts of water
- **Eye Exposure:** Irrigate thoroughly with water or saline for 15 minutes
- **Oral Exposure:** Encourage small quantities of oral fluids (no more than 50–100 ml in total)

Detailed Information

- Toluene is a clear, colourless, volatile liquid with a benzene-like odour
- *Common synonyms* methylbenzene, phenylmethane
- CAS 108-88-3
- UN 1294
- NIOSH/RTECS XS 5250000
- Molecular formula $C_6H_5CH_3$
- Molecular weight 92.14
- Toluene is used as a solvent in paints and coatings, as a glue solvent and as a constituent of petrol
- Vapours mix readily with air with a risk of fire

- Toluene is soluble in alcohol, benzene and ether; very slightly soluble in water
- It reacts violently with strong oxidants with a risk of fire and explosion
- Toluene is flammable and explosive in air

Summary of Human Toxicity

- Toluene is absorbed readily through the lungs and gastrointestinal tract and, rarely, from dermal exposure
- Inhalation is the most common route of exposure
- Ethanol acutely inhibits toluene metabolism although, after repeated administration, ethanol enhances the metabolism of toluene in rats by inducing hepatic drug metabolising enzymes[1]
- Over 70% of oral toluene is excreted in the urine as hippuric acid[2]

Toxicity of Toluene[3]

Concentration	Clinical effects
2.5 ppm	Odour threshold
200 ppm	Mild throat and eye irritation; detectable impairment of motor and cognitive function
300 ppm	Inco-ordination
400 ppm	Lacrimation; skin paraesthesiae, gross inco-ordination, confusion
500–600 ppm	Anorexia, ataxia, anxiety, momentary amnesia, impaired reaction time, nausea
800–10,000 ppm	Confusion, extreme fatigue and insomnia, narcosis, death

- Occupational Exposure Standards
  Long-term exposure limit: 50 ppm (191 mg.m⁻³)
  Short-term exposure limit: 150 ppm (574 mg.m⁻³)

Acute Clinical Effects

Inhalation Effects

- Toluene vapour is a respiratory irritant and is well absorbed via the lungs
- It may cause irritation to the eyes, nose, throat and respiratory tract
- At higher concentrations, respiratory failure from bronchospasm or pulmonary oedema may occur
- Systemic effects occur from large or prolonged exposure; see below

Dermal Effects

- Toluene on the skin may cause irritation and has a defatting action, causing erythema and dryness

- Prolonged contact may cause oedema and blistering to the skin and may potentially cause systemic effects; see below

Eye Effects

- Toluene can cause corneal injury which is usually reversible

- Symptoms may include burning pain, blepharospasm, conjunctival hyperaemia and corneal oedema

Oral Effects

- An uncommon route of exposure

- Symptoms may include oropharyngeal and gastric irritation, vomiting and systemic effects; see below

Systemic Effects

- Exposure to toluene may sensitise the myocardium to endogenous catecholamines which may cause cardiac arrythmias and sudden death

- Myocardial infarction has been reported following abuse of toluene[4]

- Nausea, vomiting, dizziness, headache, narcosis, CNS depression and convulsions can result from large exposure

- Hypotension, ventricular fibrillation, cardiac arrest and myocardial infarction have been reported

Chronic Clinical Effects

Chronic exposure may lead to fatigue, headache, dizziness, nausea and loss of appetite. Chronic exposure may possibly lead to bone marrow depression, occasionally leading to aplastic anaemia or leukaemia. Possible liver and brain damage as well as autoimmune renal damage has been reported.[3]

Management

Measurement of urine hippurate, methyl–hippurates and trichloroacetate concentrations are not of clinical relevance other than to confirm exposure when this is in doubt.

Inhalation Management

- Maintain a clear airway, give humidified oxygen and ventilate if necessary

- If respiratory irritation occurs assess respiratory function and if necessary perform chest X-rays to check for chemical pneumonitis

- For systemic management, see below

Dermal Management

- Remove any contaminated clothing, place in double, sealed, clear bags and label; store in a secure area away from patients and staff

- Irrigate the skin with copious amounts of water

- Use emollients as necessary to counter defatting action of toluene

- Treat chemical burns symptomatically

- For systemic management, see below

Eye Management

- Irrigate thoroughly with running water or saline for 15 minutes

- Stain with fluorescein and refer to an ophthalmologist if there is any uptake of the stain

Oral Management

- Encourage oral fluids

- DO NOT INDUCE EMESIS due to risk of aspiration

- Consider gastric lavage within 1 hour of a substantial ingestion, ensuring airway is protected

- Contact a poisons information service for further guidance on gut decontamination

- See systemic management below

Systemic Management

- Patients should be kept at rest

- With severe exposures the patient should be kept on a cardiac monitor for 24 hours, avoiding the use of all stimulants except for resuscitation

- Treat pulmonary oedema with PEEP or CPAP ventilation

- Symptomatic and supportive care

Summary of Environmental Hazards

- Toluene released in the atmosphere will degrade by reaction with photochemically produced hydroxyl radicals (half-life 3 hours to slightly over 1 day) or be washed out in rain[5]

- It is not subject to direct photolysis

- Toluene released into water will evaporate and biodegrade; this removal can be rapid or take several weeks, depending on temperature, mixing conditions, and acclimatisation of micro-organisms

- Toluene released to soil will be lost by evaporation from near-surface soil and by leaching to the groundwater; biodegradation occurs both in soil and groundwater, but is apt to be slow especially at high concentrations, which may be toxic to micro-organisms

- It is not significantly adsorbed to sediment or bioconcentrate in aquatic organisms

- Drinking Water Standards:
  hydrocarbon total: $10\,\mu g.l^{-1}$ (UK max)
  toluene: $700\,\mu g.l^{-1}$ (WHO guideline)
  $24–170\,\mu g.l^{-1}$ (WHO level where customers may complain)

- Soil Guidelines: Dutch Criteria:
  $0.05\,mg.kg^{-1}$ (detection limit) (target)
  $130\,mg.kg^{-1}$ (intervention)

REFERENCES

1. Synder R (ed.) *Ethel Browning's Toxicity and metabolism of industrial solvents*, 2nd edn. Elsevier, Amsterdam, 1987

2. Baelum J, Molhaw L, Hansen SH & Dossing M, 1993. Hepatic metabolism of toluene after gastrointestinal uptake in humans. *Scand J Work Environ Health*; 19: 55–62

3. Hathaway GJ, Proctor NH & Hughes JP. *Proctor and Hughes' Chemical Hazards of the Workplace*, 4th edn. Van Nostrand Reinhold, New York, 1996

4. Cunningham SR, Daizell GW, McGirr P & Khan MM, 1987. Myocardial infarction and primary ventricular fibrillation after glue sniffing. *Brit Med J*; 294: 739–80

5. Hall AH & Rumack BH (eds.) *TOMES System®*, Micromedex, Englewood, Colorado. CD ROM. vol.41 (exp. 31 July 1999)

Peter Barber

TOLUENE DIISOCYANATES

Key Points

- Toluene diisocyanates (TDIs) are colourless liquids or crystals that turn pale yellow on exposure to air
- TDI vapour is irritating to the respiratory tract and a potential respiratory sensitiser
- TDI is a strong irritant to the eyes, mucous membranes, and skin
- *In the event of a large spill, stay upwind and out of low areas. Ventilate closed spaces. Protective clothing, eye protection and breathing apparatus should be worn*

FIRST AID

- Terminate exposure and support vital functions
- The casualty should be moved to an uncontaminated area
- Rescuers should, ideally, be trained personnel and must be careful *not to put themselves at risk* and *so wear appropriate protective clothing and, if available, breathing apparatus*
- If the casualty is unconscious a clear airway should be established and maintained; give 100% oxygen if available
- **Inhalation Exposure:** If the patient stops breathing, expired air resuscitation should be started immediately using a pocket mask with a one way valve, if available. It is important where the face is contaminated that expired air resuscitation is NOT attempted unless an airway with rescuer protection is used
- **Dermal Exposure:** Remove contaminated clothing, if possible under a shower and place in double, sealed, clear bags and label; store the bags in a secure area away from patients and staff
- Wash the skin thoroughly with copious amounts of water
- **Eye Exposure:** Irrigate thoroughly with water or saline for 15 minutes
- **Oral Exposure:** Encourage small quantities of oral fluids (no more than 50–100 ml in total)

Detailed Information

- TDIs are colourless liquids or crystals that turn pale yellow on exposure to air, with a sweet, pungent, fruity odour
- *Common synonyms* diisocyanates, isocyanic acid, methyl phenylene ester, TDI

Identification of Toluene Diisocyanates

Isomer	CAS	UN	NIOSH/RTECS
2,4-TDI	584-84-9	2078	CZ 6300000
2,6-TDI	91-08-7	2078	CZ 6310000
Commercial mixture (80:20)	26471-62-5	2078	CZ 6200000

- Molecular formula $C_9H_6N_2O_2$
- Molecular weight 174.15

- Toluene diisocyanates are produced as two isomers: 2,4 toluene diisocyanate (2,4-TDI) and 2,6-toluene (2,6-TDI); TDI is generally used as a mixture of 80% 2,4-TDI and 20% 2,6-TDI
- TDI is used for polyurethane foams, plastics, adhesives, sealant, boat building and paints
- Vapours react readily with air
- TDI is insoluble in water but is soluble in ether, acetone and other organic solvents
- Reacts with water, producing carbon dioxide; reacts with compounds containing active hydrogen
- TDI reacts violently with strong oxidants, amines, alcohols, acids and warm water with a risk of fire and explosion
- Products of combustion include hydrogen cyanide, carbon monoxide and oxides of nitrogen

Summary of Human Toxicity

- TDI is absorbed by inhalation and tends to have a cumulative effect; there are two classes of reactions to TDI:
    - *irritation* or pharmacodynamic action to which all those exposed are susceptible to some degree
    - *sensitisation* reaction or allergic response in those persons who have become sensitised to TDIs during earlier exposure[1]
- TDI is a severe irritant to the mucous membranes, eyes and gastrointestinal and respiratory tracts; it also has a marked inflammatory reaction on direct skin contact[2]
- Respiratory sensitisation occurs in susceptible individuals after repeated exposure to concentrations of 0.002 ppm[3]

Toxicity of Toluene Diisocyanates[4,5,6]

Concentration	Clinical effects
0.05–0.1 ppm	Threshold for eye irritation
0.5 ppm	Respiratory response imminent
>0.5 ppm for 4–8 h	Respiratory symptoms may be delayed
0.05–0.13 ppm	Odour threshold
2.5 ppm	Immediately dangerous to life

- Maximum Exposure Limits for isocyanates:
    Long-term exposure limit: 0.02 mg.m^{-3}
    Short-term exposure limit: 0.07 mg.m^{-3}

Acute Clinical Effects

Inhalation Effects

- Symptoms are often delayed, commonly by 4 to 8 hours and may persist for several days, months or, rarely, years post exposure

- Exposure may lead to irritation or burning sensation in the nose and throat, tight chest, shortness of breath, coughing, choking and headache

- Severe gastrointestinal disturbances including nausea, vomiting, and abdominal pain

- Prolonged exposure may lead to dry painful cough, chest pain and occasionally production of bloody sputum

- High concentrations may result in bronchitis, severe asthmatic symptoms, including bronchospasm, and pulmonary oedema

- Neurological signs may include euphoria, ataxia, drowsiness and loss of consciousness

- Lingering problems may include poor memory, personality changes, irritability and depression

- Sensitisation is more common after repeated exposure; re-exposure after sensitisation may cause a delayed severe asthma-like attack, which can persist for days

Dermal Effects

- TDI is a severe irritant to the skin with erythema, inflammation, itching and blisters

Eye Effects

- Low vapour concentrations result in a pricking sensation in the eyes[7]

- Eye contact with TDI (vapour, aerosols or liquid) may cause mild irritation, itching, lacrimation, conjunctivitis and keratoconjunctivitis

- Oculorhinitis may occur and be delayed for several hours[8]

- Iridocyclitis and secondary glaucoma have been reported following TDI splashes to the eye[7]

Oral Effects

- There have been no reports of human ingestion

- Ingestion is likely to result in gastrointestinal irritation with nausea, vomiting and abdominal pain

Chronic Clinical Effects

A chronic syndrome consisting of coughing, wheezing, tightness or congestion in the chest and shortness of breath has been characterised with repeated exposures of low concentrations of TDI.

Onset of symptoms of sensitisation may be insidious, becoming progressively more pronounced with continued exposure over a period of days to months. Initial symptoms are often nocturnal dyspnoea and/or nocturnal cough with progression to asthmatic bronchitis. Time from initial employment to the development of symptoms suggestive of asthma has been reported to vary from 6 months to 20 years.[4]

Skin sensitisation on repeated exposure to TDIs may occur; urticaria, dermatitis, skin lesions of an eczematous and erythematous nature, and allergic contact dermatitis have been reported.[9]

Long-lasting personality changes, irritability, depression, loss of memory and impotency may occur. TDI is a suspected carcinogen.

Management

Inhalation Management

- Maintain a clear airway, give humidified oxygen, bronchodilators and ventilate if necessary

- If respiratory irritation occurs assess respiratory function, monitor arterial blood gases and if necessary perform chest X-rays to check for chemical pneumonitis

- Treat pulmonary oedema with PEEP or CPAP ventilation

- In sensitised individuals administration of steroids up to 4 hours post exposure may be of use to inhibit delayed reaction

- Symptomatic and supportive care

Dermal Management

- Remove any remaining contaminated clothing, place in double, sealed, clear bags and label; store in a secure area away from patients and staff

- Irrigate the skin with water

- Emollients may ease irritation

- Symptomatic and supportive care

Eye Management

- Irrigate thoroughly with running water or saline for 15 minutes

- Stain with fluorescein and refer to an ophthalmologist if there is any uptake of the stain

Oral Management

- Encourage oral fluids

- DO NOT INDUCE EMESIS due to risk of aspiration

- Symptomatic and supportive care

Summary of Environmental Hazards

- In the atmosphere TDI reacts with photochemically produced hydroxyl radicals (half-life 3.3 hours) and is also removed by dry deposition[4]

- In water a crust forms around the liquid with < 0.5% of the original material remaining after 35 days; low concentrations of TDIs hydrolyse in the aqueous environment in approximately a day[4]

- If spilled on wet land TDI is rapidly degraded

- Drinking Water Standards:
        hydrocarbon total: $10 \, \mu g.l^{-1}$ (UK max)

- Soil Guidelines: Dutch Criteria: no data available

- Air Quality Standards: no data available

REFERENCES

1. Butcher BT, Jones RN, O'Neil CE, Glindmeyer HW, Diem JE, Dharmarajan V, Weill H & Salvaggio JE, 1977. Longitudinal studies of workers employed in the manufacture of toluene diisocyanate. *Am Rev Respir Dis*; 116(3): 411–21

2. Hathaway GJ, Proctor NH & Hughes JP. *Proctor and Hughes' Chemical Hazards of the Workplace*, 4th edn. Van Nostrand Reinhold, New York, 1996

3. Elkins HB, McCarl GW, Brugsch HG & Fahy JP, 1962. Massachusetts experience with toluene diisocyanate. *Am Ind Hyg Assoc J*; 23: 265–72

4.. Hall AH & Rumack BH (eds.) *TOMES System®*, Micromedex, Englewood, Colorado. CD ROM. vol.41 (exp. 31 July 1999)

5. Haddad LM & Winchester JF (eds.) *Clinical Management of Poisoning and Drug Overdose*, 2nd edn. WB Saunders Co., Philadelphia, 1990

6. Clayton GD & Clayton FE (eds.) *Patty's Industrial Hygiene and Toxicology*, 4th edn. John Wiley & Sons, Inc., New York, 1994

7. Grant MW & Schuman JS. *Toxicology of the Eye*, 4th edn. Charles C Thomas, Springfield, 1993

8. Paggiaro PL, Rossi O, Lastricci L, Pardi F, Pezzini A & Buschieri L, 1985. TDI-induced oculorhinitis and bronchial asthma. *J Occup Med*; 27(1): 51–2

9. Calas E, Casteain PY, Lapointe HR, Ducos P, Cavelier C, Duprat P & Poitou P, 1977. Allergic contact dermatitis to a photopolymerizable resin used in printing. *Contact Dermat*; 3(4): 186–94

Henrietta Wheeler

TRICHLOROETHYLENE

Key Points

- Trichloroethylene is a colourless liquid with a slightly sweet, chloroform-like odour
- The most significant absorption occurs through the inhalation of the vapour, but dermal and gastrointestinal absorption can also occur
- Trichloroethylene causes CNS depression and can sensitise the myocardium to endogenous catecholamines
- Trichloroethylene exposure can cause liver and renal damage
- *In the event of a large spill, stay upwind and out of low areas. Ventilate closed spaces. Protective clothing, eye protection and breathing apparatus should be worn*

FIRST AID

- Terminate exposure and support vital functions
- The casualty should be moved to an uncontaminated area
- Rescuers should, ideally, be trained personnel and must be careful *not to put themselves at risk* and *so wear appropriate protective clothing and, if available, breathing apparatus*
- If the casualty is unconscious a clear airway should be established and maintained; give 100% oxygen if available
- **Inhalation Exposure:** If the patient stops breathing, expired air resuscitation should be started immediately using a pocket mask with a one way valve, if available. It is important where the face is contaminated that expired air resuscitation is NOT attempted unless an airway with rescuer protection is used
- **Dermal Exposure:** Remove contaminated clothing, if possible under a shower and place in double, sealed, clear bags and label; store the bags in a secure area away from patients and staff
- Wash the skin thoroughly with copious amounts of water
- **Eye Exposure:** Irrigate thoroughly with water or saline for 15 minutes
- **Oral Exposure:** Encourage small quantities of oral fluids (no more than 50–100 ml in total)

Detailed Information

- Trichloroethylene is a clear, colourless, non corrosive, volatile liquid with a chloroform-like odour
- *Common synonyms* TCE, ethylene trichloride, 1,1,3-trichloroethylene, trichloroethene
- CAS 79-01-6
- UN 1710
- NIOSH/RTECS KX 4550000
- Molecular formula C_2HCl_3
- Molecular weight 131.40
- Trichloroethylene is used as a degreasing agent, an industrial solvent and as a dry cleaning fluid; an extracting solvent for oils and waxes; fumigant and in paints and adhesives

- Miscible with common organic solvents, slightly soluble in water
- Trichloroethylene is heavier than air and reacts violently with alkali metals and metal powders; reacts violently with sodium hydroxide producing dichloroacetylene with risk of fire and explosion
- Non-flammable liquid, but decomposes in flame or on hot surface, producing phosgene and hydrochloric acid[1]

Summary of Human Toxicity

- Trichloroethylene is absorbed rapidly by inhalation and ingestion; dermal absorption is slow and unlikely to cause toxicity
- It is lipophilic and accumulates in the liver, brain, lungs and adipose tissue
- The elimination of trichloroethylene is in two phases with half-lives of 3 and 30 hours respectively[2]
- Following inhalation, about 13% is exhaled unchanged[3]
- Following oral exposure, 90% is metabolised in the liver; the metabolites trichloroethanol and trichloroacetic acid are excreted in the urine
- Alcohol is believed to potentiate the effects of trichloroethylene intoxication
- The lethal oral dose is estimated at 7 g.kg^{-1}[4]

Toxicity of Trichloroethylene[2,5]

Concentration	Clinical effects
100 ppm	Odour threshold, barely perceptible to the unacclimatised
200 ppm	Odour apparent, not unpleasant, mild eye irritation
400 ppm for 3 h	Odour very definite, not unpleasant, slight eye irritation and minimal light headedness
1000–1200 ppm for 6 min	Very strong odour, unpleasant; definite eye and nasal irritation with light headedness and dizziness
2000 ppm for 5 min	Odour very strong, not tolerable; marked eye and respiratory irritation with drowsiness, dizziness and nausea

- Maximum Exposure Limits:
  Long-term exposure limit: 100 ppm (550 mg.m^{-3})
  Short-term exposure limit: 150 ppm (820 mg.m^{-3})

Acute Clinical Effects

Inhalation Effects

- Respiratory irritation may include cough, shortness of breath, cyanosis, respiratory depression and pulmonary oedema
- See systemic effects below

Dermal Effects

- Irritation to the skin may occur if trichloroethylene is allowed to evaporate

- Extensive exposures can cause a burning sensation, pain, fissured lesions and burns[6]

Eye Effects

- Contact with the eye produces irritation and injury to the corneal epithelium, but permanent injury is unlikely[5]

- First and second degree burns have been attributed to the vapour[7]

- Double vision, blurred vision, optic neuritis and blindness have been reported following inhalation

Oral Effects

- Ingestion may cause nausea and vomiting with a risk of aspiration

- Large exposures can cause oedema of the larynx and pharynx

- Ingestion can cause systemic toxicity, see below

Systemic Effects

- CNS effects include headache, light-headedness, dizziness, ataxia, euphoria, confusion, drowsiness and coma

- Nausea, vomiting, hypotension, bradycardia or tachycardia and hepatitis may occur

- Death may occur from ventricular arrhythmias as at high concentrations the solvent sensitises the myocardium to adrenaline and other catecholamines

- Recovery from the narcotic effects of trichloroethylene is usually rapid following cessation of the exposure

- Trigeminal nerve impairment, peripheral neuropathy have been reported

- Trichloroethylene can lead to hepatic and renal dysfunction

Chronic Clinical Effects

Peripheral paraesthesia, headache, vertigo, tremor, anorexia, fatigue, difficulty in concentrating, memory impairment and liver damage have all been reported from chronic occupational exposures. Chronic skin contact may cause rash, eczema, dermatitis and rough skin. Intolerance to alcohol may develop following repeat exposure to trichloroethylene.

Management

Exposure to trichloroethylene can be confirmed by measurement of blood concentrations of trichloroethylene and/or trichloethanol or measurement of urine concentrations of trichloroacetic acid. Measurement of urine metabolite concentrations is the preferred method.

Inhalation Management

- Maintain a clear airway, give humidified oxygen and ventilate if necessary

- If respiratory irritation occurs assess respiratory function and if necessary perform chest X-rays to check for chemical pneumonitis

- Treat pulmonary oedema with PEEP or CPAP ventilation

- Symptomatic and supportive care

- See systemic management below

Dermal Management

- Remove any remaining contaminated clothing, place in double, sealed, clear bags and label; store in a secure area away from patients and staff

- Irrigate with copious amounts of water

- Treat erythema symptomatically; it should subside within 24 hours[6]

- Treat lesions and burns symptomatically, they normally recover within 2 weeks[6]

Eye Management

- Irrigate thoroughly with running water or saline for 15 minutes

- Stain with fluorescein and refer to an ophthalmologist if there is any uptake of the stain

Oral Management

- Encourage oral fluids

- DO NOT INDUCE EMESIS due to risk of aspiration

- Consider gastric lavage within 1 hour of a substantial ingestion, ensuring airway is protected

- Contact a poisons information service for further guidance on gut decontamination

- Symptomatic and supportive care

- See systemic management below

Systemic Management

- Patients should be kept at rest

- With severe exposures the patient should be kept on a cardiac monitor and all stimulants should be avoided except for resuscitation

Summary of Environmental Hazards

- Trichloroethylene in air will react quickly, especially under smog conditions; atmospheric residence time has been reported at 5 days, with the formation of phosgene, dichloroacetyl chloride, and formyl chloride

- It is degraded by photo-oxidation with a half-life of 7 days[2]

- In water biodegeneration of trichloroethylene occurs

- Spills of trichloroethylene on to soil will evaporate quickly, or is highly mobile and may be leached into groundwater; it has a low adsorbtion potential

- Trichloroethylene may undergo moderate bioconcentration

- Drinking Water Standards:
  trichloroethylene: 30 $\mu g.l^{-1}$ (UK max)
  70 $\mu g.l^{-1}$ (WHO provisional guideline)

  hydrocarbon total: 10 $\mu g.l^{-1}$ (UK max)

- Soil Guidelines: Dutch Criteria:
  0.001 mg.kg^{-1} (target)
  60 mg.kg^{-1} (intervention)

- Air Quality Standards:
  1 mg.m^{-3} averaging time 24 hours (WHO guideline)

REFERENCES

1. Dutch Institute for the Working Environment and the Dutch Chemical Industry Association. *Chemical Safety Sheets – Working Safely with hazardous Chemicals*. Samson Chemical Publishers, The Netherlands, 1991

2. Hall AH & Rumack BH (eds.) *TOMES System*®, Micromedex, Englewood, Colorado. CD ROM. vol.41 (exp. 31 July 1999)

3. Baselt RC & Cravey RH. *Disposition of Toxic Drugs and Chemicals in Man*, 4th edn. Chemical Toxicology Institute, California, 1995

4. Yoshida M, Fukabori S, Hera K, Yuasa H, Nakacki K, Yamamura Y & Yoshida K, 1996. Concentrations of trichloroethylene and its metabolites in blood and urine after acute poisoning by ingestion. *Hum Exper Toxicol*; 15(3): 254–8

5. Hathaway GJ, Proctor NH & Hughes JP. *Proctor and Hughes' Chemical Hazards of the Workplace*, 4th edn. Van Nostrand Reinhold, New York, 1996

6. Goh CL & Ng SK, 1988. A cutaneous manifestation of trichloroethylene toxicity. *Contact Derm*; 18(1): 59–61

7. Grant MW & Schuman JS. *Toxicology of the Eye*, 4th edn. Charles C Thomas, Springfield, 1993

Henrietta Wheeler

VINYL CHLORIDE

Key Points

- Vinyl chloride is a colourless gas with a characteristic, pungent, but pleasant odour
- It is irritating via inhalation, to the skin, eyes and mucous membranes
- Vinyl chloride causes tissue burns by rapid evaporation and consequent freezing
- It may cause CNS and respiratory depression at high concentrations
- There may be a latent period between exposure and symptom onset
- Vinyl chloride is carcinogenic to humans and animals
- *Vinyl chloride vapours are heavier than air and may accumulate in low or confined areas; in the event of a large spill, stay upwind and out of low areas. Ventilate closed spaces. Protective clothing, eye protection and breathing apparatus should be worn*

FIRST AID

- Terminate exposure and support vital functions
- The casualty should be moved to an uncontaminated area
- Rescuers should, ideally, be trained personnel and must be careful *not to put themselves at risk* and *so wear appropriate protective clothing and, if available, breathing apparatus*
- If the casualty is unconscious a clear airway should be established and maintained; give 100% oxygen if available
- **Inhalation Exposure:** If the patient stops breathing, expired air resuscitation should be started immediately using a pocket mask with a one way valve, if available. It is important where the face is contaminated that expired air resuscitation is NOT attempted unless an airway with rescuer protection is used
- **Dermal Exposure:** If frostbite occurs *do not* remove clothing, flush skin with water
- If frostbite is not present remove contaminated clothing, if possible under a shower and place in double, sealed, clear bags and label; store the bags in a secure area away from patients and staff
- Wash the skin thoroughly with copious amounts of water
- **Eye Exposure:** If the eye tissue is frozen seek medical advice as soon as possible
- If eye tissue is not frozen, irrigate thoroughly with water or saline for 15 minutes
- **Oral Exposure:** Encourage small quantities of oral fluids (no more than 50–100 ml in total)

Detailed Information

- Vinyl chloride is a colourless, liquefied gas with characteristic sweet odour
- *Common synonyms* chlorethene, chloroethylene, ethylene monochloride, monochlorethene, VC
- CAS 75-01-4

- UN 1086
- NIOSH/RTECS KU 9625000
- Molecular formula C_2H_3Cl
- Molecular weight 62.50
- Vinyl chloride is used as a refrigerant gas and as a chemical intermediate for PVC
- It is heavier than air and spreads at ground level, with a risk of ignition at a distance
- Slightly soluble in water, soluble in alcohol and ether
- Vinyl chloride is able to form peroxides and thus polymerise; able to polymerise when moderately heated and exposed to air and light; reacts violently with strong oxidants
- Vinyl chloride is able to explode in air; products of combustion are phosgene and hydrochloric acid
- Large fires are practically inextinguishable[1]

Summary of Human Toxicity

- Vinyl chloride can be absorbed via inhalation and through the skin
- The active metabolite (chloroacetaldehyde), not vinyl chloride itself, is responsible for its toxicity
- The majority of inhaled vinyl chloride is excreted unchanged; dose depending, a varying amount is metabolised in a saturated process[2]
- It is metabolised by epoxidation, with subsequent production of chloroacetaldehyde; further oxidation and conjugation with glutathione are responsible for the metabolites found in the urine[2]
- Although vinyl chloride has an odour at high concentrations it is of no value in preventing excessive exposure
- Detectable vapour concentrations have never been adequately determined and seem to vary on an individual basis

Toxicity of Vinyl Chloride[3]

Concentration	Clinical effects
500 ppm for 7.5 h	No changes or abnormal neurological response
8,000–10,000 ppm	CNS Depression
20,000 ppm for 5 min	Dizziness, light-headedness, nausea and dulling of vision and auditory cues

- Maximum Exposure Limits:
  Long-term exposure limit: 7 ppm
  Short-term exposure limit: no data available

Acute Clinical Effects

Inhalation Effects

- Vinyl chloride is a respiratory irritant causing coughing, wheezing and breathlessness

- Exposure may lead to CNS depression, including fatigue, dyspnoea, headache, ataxia, inebriation, drowsiness and coma

- Visual disturbances, numbness and tingling of the extremities may occur

- Ventricular fibrillation may cause sudden death following acute exposure

- Convulsions have occurred in deeply anaesthetised patients

Dermal Effects

- Dermal contact may cause irritation, pain and burns

- Frostbite can occur due to rapid evaporation

- Contact dermatitis has been reported

Eye Effects

- Vinyl chloride is irritating to the eye

- Pain and frostbite may occur

- One case of corneal injury has been reported, this healed completely within 48 hours[4]

Oral Effects

- Ingestion may lead to nausea, vomiting, diarrhoea, abdominal pain and haematemesis

- Perforation may occur following large exposure

Chronic Clinical Effects

Occupational exposure to vinyl chloride is associated with an increased incidence of angiosarcoma of the liver and other malignant tumours, acroosteolysis, Reynaud's syndrome, scleroderma, and impaired liver function.[3]

Pulmonary changes including asthma and pneumoconiosis have been reported from chronic exposures.

Management

Biological and haematological tests are of limited value in predicting angiosarcoma but may be of value in other hepatotoxic effects.

Inhalation Management

- Maintain a clear airway, give humidified oxygen and ventilate if necessary

- If respiratory irritation occurs assess respiratory function and if necessary perform chest X-rays to check for chemical pneumonitis

- Consider the use of steroids to reduce the inflammatory response

- Treat pulmonary oedema with PEEP or CPAP ventilation

- Symptomatic and supportive care

Dermal Management

- *If frostbite has occurred*: remove clothing carefully, these may need to be soaked off with tepid water; irrigate the area

- Surgical referral may be necessary

- *If frostbite has not occurred*: remove any remaining contaminated clothing

- Irrigate with copious amounts of water

- Treat burns symptomatically

- Place any contaminated clothes in double, sealed, clear bags and label; store in a secure area away from patients and staff

Eye Management

- Irrigate thoroughly with running water or saline for 15 minutes

- Stain with fluorescein and refer to an ophthalmologist if there is any uptake of the stain

Oral Management

- NO GASTRIC LAVAGE OR EMETIC

- Encourage oral fluids, unless perforation is suspected

- Give plasma expanders/blood or IV fluids for shock and analgesics for pain

- Consider the use of steroids to reduce the inflammatory response

- Take abdominal X-ray to check for perforation

- Symptomatic and supportive care

Summary of Environmental Hazards

- In the atmosphere vinyl chloride is expected to exist mainly in the vapour-phase and to degrade rapidly in air by gas-phase reaction with photochemically produced hydroxyl radicals with an estimated half-life of 1.5 days[5]

- Products of reaction in the atmosphere are hydrochloric acid, chloroacetaldehyde and chloroethylene oxide

- In water it will be subject to rapid volatilisation with an estimated half-life of 0.805 h for evaporation from a river 1 m deep with a current of 3 m.sec^{-1} and a wind velocity of 3 m.sec^{-1} [5]

- If vinyl chloride is released to soil, it will be subject to rapid volatilisation with half-lives of 0.2 and 0.5 days for evaporation from soil at 1 and 10 cm incorporation, respectively

- Vinyl chloride which does not evaporate will be expected to be mobile in soil and may leach to the groundwater

- It is not expected to bioaccumulate in the food chain

- Drinking Water Standards:
    hydrocarbon total: 10 µg.l^{-1} (UK max)
    vinyl chloride: 5 µg.l^{-1} (WHO guideline)

- Soil Guidelines: Dutch Criteria:
    0.1 mg.kg^{-1} (intervention)

- Air Quality Standards: no safe level recommended due to carcinogenic properties

REFERENCES

1. Sax NI. *Dangerous Properties of Industrial Materials*, 6th edn. Van Nostrand Reinhold, New York, 1984

2. Clayton GD and Clayton FE (eds.) *Patty's Industrial Hygiene and Toxicology*, 4th edn. John Wiley & Sons, Inc., New York, 1994

3. Hathaway GJ, Proctor NH & Hughes JP. *Proctor and Hughes' Chemical Hazards of the Workplace*, 4th edn. Van Nostrand Reinhold, New York, 1996

4. Grant MW & Schuman JS. *Toxicology of the Eye*, 4th edn. Charles C Thomas, Springfield, 1993

5. Hall AH & Rumack BH (eds.) *TOMES System®*, Micromedex, Englewood, Colorado. CD ROM. vol.41 (exp. 31 July 1999)

Henrietta Wheeler

XYLENE

Key Points

- Xylene is a colourless volatile liquid at room temperature with an aromatic odour
- It is irritating to the eyes, skin, respiratory system and gastrointestinal tract
- Xylene causes CNS depression and can sensitise the myocardium to endogenous catecholamines
- Severe acute over exposure and chronic xylene exposure may cause liver and renal damage
- *In the event of a large spill, stay upwind and out of low areas. Ventilate closed spaces. Protective clothing, eye protection and breathing apparatus should be worn*

FIRST AID

- Terminate exposure and support vital functions
- The casualty should be moved to an uncontaminated area
- Rescuers should, ideally, be trained personnel and must be careful *not to put themselves at risk* and *so wear appropriate protective clothing and, if available, breathing apparatus*
- If the casualty is unconscious a clear airway should be established and maintained; give 100% oxygen if available
- **Inhalation Exposure:** If the patient stops breathing, expired air resuscitation should be started immediately using a pocket mask with a one way valve, if available. It is important where the face is contaminated that expired air resuscitation is NOT attempted unless an airway with rescuer protection is used
- **Dermal Exposure:** Remove contaminated clothing, if possible under a shower and place in double, sealed, clear bags and label; store the bags in a secure area away from patients and staff
- Wash the skin thoroughly with copious amounts of water
- **Eye Exposure:** Irrigate thoroughly with water or saline for 15 minutes

Detailed Information

- Xylene is a colourless volatile liquid at room temperature with an aromatic odour similar to benzene
- Xylene has three isomers:

Standard name	Synonyms	CAS	UN	NIOH/RTECS
xylene (mixed)		1330-20-7	n.a.	ZE 2100000
o-xylene	o-methyltoluene; 1,2-dimethylbenzene	95-47-6	1307	ZE 2450000
m-xylene	m-methyltoluene; 1,3-dimethylbenzene	108-38-3	1307	ZE 2275000
p-xylene	p-methyltoluene; 1,4-dimethylbenzene	106-42-3	1307	ZE 2625000

n.a. = number not available

- Molecular formula C_8H_{10}
- Molecular weight 106.17
- Used as a solvent, degreaser, cleaning agent and aviation fuel; as a starter for production of synthetic fibres, plastics and enamel
- Xylenes are usually supplied for use as a mixture of the three isomers and a proportion of ethylbenzene (which has the same empirical chemical formula)
- Vapours readily mix with air
- Xylene is soluble in alcohol and ether; insoluble in water
- Reacts violently with strong oxidants

Summary of Human Toxicity

- Xylene is absorbed readily through the lungs, skin and gastrointestinal tract
- Xylene is a CNS depressant; it is metabolised by microsomal enzymes and alcohol dehydrogenase
- Co-ingestion of alcohol decreases metabolic clearance of xylene by about 50%
- Over 95% is excreted via the kidneys[1]

Toxicity of Xylene[2]

Concentration	Clinical effects
1 ppm	Odour threshold
207 ppm	Impaired vestibular and visual function and reaction time
298 ppm	Impaired performance
689 ppm	Dizziness and irritation
10,000 ppm	Loss of consciousness expected

- Occupational Exposure Standards:
  Long-term exposure limit: 100 ppm (441 mg.m^{-3})
  Short-term exposure limit: 150 ppm (662 mg.m^{-3})

Acute Clinical Effects

Inhalation Effects

- Xylene vapour is a respiratory irritant and is well absorbed via the lungs
- It may cause irritation to the eyes, nose and throat
- Pulmonary oedema from high concentrations may be delayed up to 48 hours
- CNS depression and systemic effects may occur; see below

Dermal Effects

- Liquid xylene is a skin irritant and has a defatting action, causing erythema and dryness
- Prolonged contact may cause formation of vesicles and possibly necrosis

- Prolonged exposure may potentially cause systemic effects; see below

Eye Effects

- Xylene vapour causes irritation to the eye
- Liquid splashes can cause blepharospasm and hyperaemia[3]
- Transient superficial corneal damage may occur, but no lasting damage is reported

Oral Effects

- Not a common route of poisoning
- Local irritation to the mucous membranes would be expected; burning sensation of the oropharynx and stomach with larger exposures
- CNS depression and systemic effects may occur; see below

Systemic Effects

- CNS symptoms, including dizziness, excitement or drowsiness, inco-ordination, ataxia, short-term memory loss, anorexia, respiratory depression
- Metabolic acidosis, hypokalaemia, hypobicarbonataemia and hypophosphataemia may be observed
- Acute renal failure and liver damage have been reported
- Death may occur from ventricular arrhythmias as at high concentrations the solvent sensitises the myocardium to adrenaline and other catecholamines

Chronic Clinical Effects

Leucopaenia and lymphocytosis has been observed. Scaling, blistering and necrosis of skin following long-term exposure. Personality changes and short-term memory loss may occur from chronic low level exposure. Chronic bronchitis and gastrointestinal disturbances may occur. There is evidence of mutagenicity due to xylene.

Management

Measurement of one of the major urinary metabolites, methyl hippuric acid is not of clinical relevance other than to confirm exposure when this is in doubt.

Inhalation Management

- Maintain a clear airway, give humidified oxygen and ventilate if necessary
- If respiratory irritation occurs assess respiratory function and if necessary perform chest X-rays to check for chemical pneumonitis
- See systemic management below

Dermal Management

- Remove any contaminated clothing, place in double, sealed, clear bags and label; store in a secure area away from patients and staff
- Irrigate the skin with copious amounts of water
- Use emollients as necessary to counter defatting action of xylene
- Treat chemical burns symptomatically

- See systemic management below

Eye Management

- Irrigate thoroughly with running water or saline for 15 minutes
- Stain with fluorescein and refer to an ophthalmologist if there is any uptake of the stain

Oral Management

- Encourage oral fluids for small ingestion
- DO NOT INDUCE EMESIS due to risk of aspiration
- Consider gastric lavage within 1 hour of a substantial ingestion, ensuring airway is protected
- Symptomatic and supportive care
- See systemic management below

Systemic Management

- Patients should be kept at rest
- With severe exposures the patient should be kept on a cardiac monitor for 12 hours, avoiding the use of all stimulants except for resuscitation
- Monitor potassium, bicarbonate, pH and phosphate; check renal and liver function
- Haemodialysis or peritoneal dialysis is of use in acute renal failure
- Treat pulmonary oedema with PEEP or CPAP ventilation
- Symptomatic and supportive care
- Contact a poisons information service for further guidance on gut decontamination

Summary of Environmental Hazards

- Xylene is expected to exist entirely in the vapour phase in the atmosphere where it is degraded by reaction with photochemically produced hydroxyl radicals with an estimated atmospheric lifetime of about 1–2 days[2]
- Xylene rapidly separates from water due to low solubility and high volatility
- Estimated half-lives for a model river and model lake are 3 and 99 hours, respectively[2]
- Biodegradation in groundwater depends on the dissolved oxygen concentration
- Xylene has a low potential for bioaccumulation
- Drinking Water Standards:
    hydrocarbon total: $10 \, \mu g.l^{-1}$ (UK max)
    xylene: $500 \, \mu g.l^{-1}$ (WHO guideline)
        $20–1800 \, \mu g.l^{-1}$ (WHO level where customers may complain)
- Soil Guidelines: Dutch Criteria:
    $0.05 \, mg.kg^{-1}$ (detection limit) (target)
    $25 \, mg.kg^{-1}$ (intervention)
- Air Quality Standards: no data available

REFERENCES

1. Ellenhorn MJ, Schonwalds S, Ordog G & Wasserberger J. *Ellenhorn's Medical Toxicology – Diagnosis and Treatment of Human Poisoning*, 2nd edn. Williams & Wilkins, London, 1997

2. International Programme on Chemical Safety. *Environmental Health Criteria 190: Xylenes.* WHO, Geneva, 1997

3. Grant MW & Schuman JS. *Toxicology of the Eye*, 4th edn. Charles C Thomas, Springfield, 1993

Henrietta Wheeler

ZINC AND ZINC OXIDE

Key Points

- Zinc is a bluish-white lustrous metal; it is stable in dry air and becomes coated with the basic carbonate in moist air; zinc oxide is a white powder
- Metal fume fever due to inhalation of fresh zinc oxide fumes is the most common manifestation of zinc exposure; symptoms resemble influenza
- Zinc oxide is absorbed via the respiratory and gastrointestinal tract
- Ingestion may cause gastrointestinal upset; chronic ingestion may lead to copper deficiency
- Zinc and zinc oxide are of low toxicity in eye and dermal exposure
- *In the event of a large spill, stay upwind and out of low areas. Ventilate closed spaces. Protective clothing, eye protection and breathing apparatus should be worn*

FIRST AID

- Terminate exposure and support vital functions
- The casualty should be moved to an uncontaminated area
- Rescuers should, ideally, be trained personnel and must be careful *not to put themselves at risk* and *so wear appropriate protective clothing and, if available, breathing apparatus*
- If the casualty is unconscious a clear airway should be established and maintained; give 100% oxygen if available
- **Inhalation Exposure:** If the casualty is suffering from respiratory difficulties give 100% oxygen, if available
- **Dermal Exposure:** Remove contaminated clothing, if possible under a shower and place in double, sealed, clear bags and label; store the bags in a secure area away from patients and staff
- **Eye Exposure:** Irrigate thoroughly with water or saline for 15 minutes
- **Oral Exposure:** Encourage small quantities of oral fluids (no more than 50–100 ml in total)

Detailed Information

Zinc

- Zinc is found as a bluish-white, lustrous metal or as a grey powder
- *Common synonyms* granular zinc, blue powder
- CAS 7440-66-6
- UN 1436
- NIOSH/RTECS ZG 8600000
- Atomic symbol Zn
- Atomic weight 65.38
- Uses for zinc include galvanising and other anti-corrosive coating for other metals, alloys including bronze and brass, batteries, kitchen utensils, castings, printing plates, automotive equipment, as a reducing agent in organic

chemistry, producing insulin zinc salts, and as an analytical reagent

- Zinc is stable in dry air; it spontaneously forms a coating of basic carbonate in moist air
- It reacts slowly with ethanoic acid, and vigorously with nitric acid
- It reacts with alkaline hydroxides to form 'zincates' or hydroxo compounds

Zinc oxide

- Zinc oxide is a white powder
- *Common synonyms* calamine, flowers of zinc
- CAS 1314-13-2
- UN number not available
- NIOSH/RTECS ZH 4810000
- Molecular formula ZnO
- Molecular weight 81.39
- Uses of zinc oxide include accelerator, pigment and reinforcer in rubber tyres; as a white pigment in paint and ceramic glaze; as a semiconductor; as a photoconductor in photocopying equipment; and in cosmetics

Summary of Human Toxicity

- Zinc and zinc oxide are of low toxicity in eye and dermal exposure
- Zinc is an essential trace element, needed for nucleic acid synthesis and for metalloenzyme function
- Zinc toxicity from chronic ingestion usually takes the form of copper deficiency
- Metal fume fever occurs when ultrafine (<1 μm) particles of zinc oxide are produced by heating zinc to boiling point; influenza-like symptoms of metal fume fever may develop several hours after exposure

Toxicity of Zinc Oxide[1]

Concentration	Clinical effects
5 mg.m^{-3} for 2 h	Metal fume fever (potentially delayed for 6 to 10h)
23 to 171 mg.m^{-3} for 30 min	Increased white cell counts

- Occupational Exposure Standards :
  Long-term exposure limit: 5 mg.m^{-3} (zinc oxide, fumes)
  Short-term exposure limit: 10 mg.m^{-3} (zinc oxide, fumes)

Acute Clinical Effects

Inhalation Effects

- Symptoms during exposure to zinc oxide fumes include a sweet taste, throat dryness and cough

- Influenza-like symptoms of metal fume fever may develop several hours after exposure and include weakness, generalised aching, chills, nausea, vomiting and headache[1]

- Signs of metal fume fever include pyrexia, sweating, dyspnoea, tachycardia and a leucocytosis

Dermal Effects

- Zinc is not considered to be a skin irritant

Eye Effects

- Zinc is of low toxicity in the eye; reversible changes in the electroretinogram were observed when metallic zinc was introduced into the vitreous humour[2]

Oral Effects

- Gastrointestinal upset is the most likely consequence of ingestion of zinc

Chronic Clinical Effects

Excessive (especially chronic) oral intake of zinc reduces absorption of copper and immune function[3]

Management

Inhalation Management

- If respiratory irritation occurs assess respiratory function

- Symptomatic and supportive care

- Recovery is usually complete within 24–72 hours

Dermal Management

- Remove any remaining contaminated clothing, place in double, sealed, clear bags and label; store in a secure area away from patients and staff

- Symptomatic and supportive care

Eye Management

- Irrigate thoroughly with running water or saline for 15 minutes

- Stain with fluorescein and refer to an ophthalmologist if there is any uptake of the stain

Oral Management

- Encourage oral fluids

- Symptomatic and supportive care

Summary of Environmental Hazards

- Zinc is present in air at 10–100 ng.m^{-3} (rural) and 100–500 ng.m^{-3} (urban)[4]

- No health-based limits for zinc content of drinking water have been proposed; however drinking water containing more than 3 mg.l^{-1} of zinc (present as the sulphate) has an opalescent appearance and an unpleasant, astringent taste

- Zinc content of tapwater may be greatly increased by leaching of zinc from piping and fittings

- Zinc is very toxic to fish and is expected to bioaccumulate

- Drinking Water Standards:
    zinc: 5000 µg.l^{-1} (UK max)
        3000 µg.l^{-1} (WHO level where customers may complain)

- Soil Guidelines: Dutch Criteria:
    zinc: 140 mg.kg^{-1} (target)
        720 mg.kg^{-1} (intervention)

- Air Quality Standards: no data available

REFERENCES

1. Hathaway GJ, Proctor NH & Hughes JP. *Proctor and Hughes' Chemical Hazards of the Workplace*, 4th edn. Van Nostrand Reinhold, New York, 1996

2. Grant MW & Schuman JS. *Toxicology of the Eye*, 4th edn. Charles C Thomas, Springfield, 1993

3. Clayton GD & Clayton FE (eds.) *Patty's Industrial Hygiene and Toxicology*, 4th edn. John Wiley & Sons, Inc, New York, 1994

4. WHO. *Guidelines for drinking-water quality, Vol 2: Health criteria and other supporting information.* World Health Organisation, Geneva, 1993

Peter Barber

ZINC CHLORIDE

Key Points

- Zinc chloride is found as white, odourless and deliquescent crystals; solutions are acidic
- It is potentially toxic by all routes of exposure although the common route of exposure is by inhalation of zinc chloride fumes
- Symptoms of inhalation range from mild irritation, coughing and chest pain to respiratory distress, which may be fatal
- Corrosive damage may be sustained in eye, dermal or oral exposure
- *In the event of a large spill, stay upwind and out of low areas. Ventilate closed spaces. Protective clothing, eye protection and breathing apparatus should be worn*

FIRST AID

- Terminate exposure and support vital functions
- The casualty should be moved to an uncontaminated area
- Rescuers should, ideally, be trained personnel and must be careful *not to put themselves at risk* and *so wear appropriate protective clothing and, if available, breathing apparatus*
- If the casualty is unconscious a clear airway should be established and maintained; give 100% oxygen if available
- **Inhalation Exposure:** If the patient stops breathing, expired air resuscitation should be started immediately using a pocket mask with a one way valve, if available. It is important where the face is contaminated that expired air resuscitation is NOT attempted unless an airway with rescuer protection is used
- **Dermal Exposure:** Remove contaminated clothing, if possible under a shower and place in double, sealed, clear bags and label; store the bags in a secure area away from patients and staff
- Wash the skin thoroughly with copious amounts of water
- **Eye Exposure:** Irrigate thoroughly with water or saline for 15 minutes
- **Oral Exposure:** Encourage small quantities of oral fluids (no more than 50–100 ml in total) unless perforation is suspected

Detailed Information

- Zinc chloride is found as white, odourless deliquescent granules or fused pieces or rods
- *Common synonyms* butter of zinc, zinc butter, zinc dichloride
- CAS 7646-85-7
- UN 1840 zinc chloride, solution
- UN 2331 zinc chloride, anhydrous
- NIOSH/RTECS ZH 1400000
- Molecular formula $ZnCl_2$
- Molecular weight 136.30

- Uses of zinc chloride include soldering flux, smoke bombs (hexite), deodorant, disinfectant, fireproofing, etching, vulcanising, galvanising, mordant in printing and dyeing and the textile industry and as a dehydrating agent in chemical synthesis
- Zinc chloride is readily soluble in water
- The aqueous solution of $ZnCl_2$ is acidic, due to formation of hydrochloric acid and zinc oxychloride[1]
- Zinc salts react with alkaline hydroxides including ammonium hydroxide to give zinc hydroxide precipitate

Summary of Human Toxicity

- Exposure to zinc chloride is usually by inhalation of fumes or hexite smoke
- Symptoms of inhalation range from mild irritation of respiratory tract to severe respiratory distress
- Zinc chloride is irritant and potentially corrosive by oral or dermal exposure; however its emetic effect is reported to reduce the potential for oral toxicity
- The minimum lethal human exposure to this chemical has not been established
- Exposure to zinc chloride (hexite) smoke leads to ARDS and increased plasma zinc concentrations; can be fatal[2]
- Workplace exposure to zinc chloride fume concentrations of $50 \ mg.m^{-3}$ or greater are believed to be immediately dangerous to life or health
- Occupational Exposure Standards :
  Long-term exposure limit: $1 \ mg.m^{-3}$
  Short-term exposure limit: $2 \ mg.m^{-3}$

Acute Clinical Effects

Inhalation Effects

- Symptoms include irritation of nose and throat, coughing, copious sputum, dyspnoea, retrosternal and chest pain
- Pulmonary oedema and bronchopneumonia, stridor, fever, cyanosis, tachypnoea and ARDS may occur
- Severe symptoms may be associated with few physical lung signs; deterioration may occur 2 to 4 days post exposure and after an initial period of improvement[1]

Dermal Effects

- Zinc chloride causes irritation to the skin
- Contact may lead to ulceration and burns

Eye Effects

- Exposure may lead to immediate pain, conjunctivitis, corneal ulceration and oedema
- Greyish stromal opacities with decreased visual acuity, increased intraocular pressure and sometimes persistent severe pain may occur

- Eye changes similar to acute closed-angle glaucoma have been reported[3]

- In one case pain and loss of vision was reported, the eye was enucleated at 30 days post exposure[3]

Oral Effects

- The most common symptom of oral exposure is gastrointestinal irritation including nausea, vomiting, epigastric pain and diarrhoea

- Mild superficial mucosal damage has occurred

- Oesophagitis, erosive pharyngitis and oesophageal burns result from ingestion

- Gastric necrosis, posterior perforation, pyrexia, hypotension or hypertension may be delayed for up to 7 days

- Stricture formation may occur[4]

- CNS depression including lethargy and confusion and renal damage has been reported

Chronic Clinical Effects

Effects of chronic inhalation exposure to zinc chloride include pulmonary fibrosis and cor pulmonale. Focal alveolitis, consolidation, emphysema, infiltration with macrophages and fibrosis were observed in guinea pigs exposed to zinc chloride for 1 hour daily for 5 weeks.

Chronic excessive intake of zinc may cause hypocupraemia; zinc accumulates in the pancreas, and chronic pancreatic exocrine deficiency should be considered.[5]

Occupational dermatitis has been reported, with ulceration of fingers, hands and forearms observed.

Management

Measurement of serum and urine zinc concentrations are not of clinical relevance other than to confirm exposure when this is in doubt.

Normal serum zinc concentration: $11-24\,\mu mol.l^{-1}$
$(0.7-1.6\,mg.l^{-1})$

Normal urine zinc concentration: $4.5-9\,\mu mol\,/\,24\,hours$
$(0.3-0.6\,mg\,/\,24\,hours)$

Inhalation Management

- If respiratory irritation occurs assess respiratory function and if necessary perform chest X-rays to check for chemical pneumonitis

- Consider the use of steroids to reduce the inflammatory response

- Treat pulmonary oedema with PEEP or CPAP ventilation

- Symptomatic and supportive care

Dermal Management

- Remove any remaining contaminated clothing, place in double, sealed, clear bags and label; store in a secure area away from patients and staff

- Irrigate with copious amounts of water

- Treat symptomatically

Eye Management

- Irrigate thoroughly with running water or saline for 15 minutes

- Stain with fluorescein and refer to an ophthalmologist if there is any uptake of the stain

- Corneal damage may take several months to resolve, and visual acuity may not return to previous levels

Oral Management

- NO GASTRIC LAVAGE OR EMETIC

- Neutralising chemicals should never be given because heat is produced during neutralisation and this could exacerbate any injury

- Patients must be admitted for observation until the extent of the injury (if any) can be determined; consider discussion with a poisons information service

- Encourage oral fluids unless there is evidence of severe injury

- Aggressive intervention is essential for severely affected patients. Urgent assessment of the airway is required

- Early gastro-oesophagoscopy should be undertaken within 12 hours of the event to assess the extent and severity of the injury

- Give plasma expanders/IV fluids for shock and check and correct the acid/base balance; analgesia will almost certainly be needed

- A supraglottic-epiglottic burn with erythema and oedema is usually a sign that further oedema will occur which will lead to airway obstruction and is an indication for early intubation or tracheostomy[6]

- Intubation and ventilation may be necessary for patients with respiratory distress

- On discharge all patients must be advised of the possibility of late onset sequelae and advised to return if necessary

- Oesophageal strictures which result in a lumen >10 mm do not impede normal life and should not require intervention.[7] Surgical intervention may be required for gastrointestinal perforation or haemorrhage

- Consider the use of steroids to reduce the inflammatory response

- Treat pulmonary oedema with PEEP or CPAP ventilation

- Symptomatic and supportive care

Summary of Environmental Hazards

- No health-based limits for zinc content of drinking water have been proposed; however drinking water containing more than $3\,mg.l^{-1}$ of zinc (present as the sulphate) may have an unpleasant taste

- Zinc content of tap water may be greatly increased by leaching of zinc from piping and fittings

- Zinc is very toxic to fish and is expected to bioaccumulate

- Drinking Water Standards:
  zinc: $5000\,\mu g.l^{-1}$ (UK max)
  $3000\,\mu g.l^{-1}$ (WHO level where customers may complain)

chloride: 400 mg.l⁻¹ (UK max)
250 mg.l⁻¹ (WHO guideline)

- Soil Guidelines: Dutch Criteria:
zinc: 140 mg.kg⁻¹ (target)
720 mg.kg⁻¹ (intervention)

- Air Quality Standards: no data available

REFERENCES

1. Hathaway GJ, Proctor NH & Hughes JP. *Proctor and Hughes' Chemical Hazards of the Workplace*, 4th edn. Van Nostrand Reinhold, New York, 1996

2. Hjortso E, Qvist J, Bud MI, Thomsen JL, Andersen JB, Wiberg-Jorgenson F, Jensen NK, Jones R, Reid LM & Zapol WM, 1988. ARDS after accidental inhalation of zinc chloride smoke. *Intens Care Med*; 14: 17–24

3. Grant MW & Schuman JS. *Toxicology of the Eye*, 4th edn. Charles C Thomas, Springfield, 1993

4. Gillis DA, Higgins G & Kennedy R, 1985. Gastric damage from ingested acid in children. *J Pediatr Surg*; 20(5): 494–6

5. McKinney PE, Brent J & Kulig K, 1995. Zinc chloride ingestion in a child: exocrine pancreatic insufficiency. *Ann Emerg Med*; 25(4): 562

6. Meredith JW, Kon ND & Thompson JN, 1988. Management of injuries from liquid lye ingestion. *J Trauma*; 28(8): 1173–80

7. Sarfati E, Assens P & Celerier M, 1987. Management of caustic ingestion in adults. *Br J Surg* 74: 146–8

Peter Barber

Indexes

CHEMICALS, SYNONYMS AND TRADE NAMES INDEX

★ For antidotes and supportive therapy in general see: Fisher J, Morgan-Jones D, Murray V & Davies G. *Chemical Incident Management for Accident and Emergency Clinicians. Appendix 2.* London: The Stationary Office, 1999

CAS NUMBERS INDEX

UN NUMBERS INDEX

NIOSH/RTECS NUMBERS INDEX

Printed in the United Kingdom by The Stationery Office

TJ64   C6   9/00   19585   545668